AF658374

Plant Protection in Tropical Root and Tuber Crops

P. Parvatha Reddy

Plant Protection in Tropical Root and Tuber Crops

P. Parvatha Reddy
Indian Institute of Horticultural
Research
Bangalore, Karnataka, India

ISBN 978-81-322-2388-7 ISBN 978-81-322-2389-4 (eBook)
DOI 10.1007/978-81-322-2389-4

Library of Congress Control Number: 2015942675

Springer New Delhi Heidelberg New York Dordrecht London

Printed on acid-free paper

Springer (India) Pvt. Ltd. is part of Springer Science+Business Media (www.springer.com)

Preface

In the year 2012, the major root and tuber (R&T) crops occupied about 53 million hectares, producing 797 million metric tons worldwide, 42 % of which was from Asia and 6.45 % from India. Tropical R&T crops consist of both dicots like cassava/tapioca (*Manihot esculenta*) and sweet potato (*Ipomoea batatas*) and monocots like yams (*Dioscorea* spp.) and edible aroids like taro (*Colocasia esculenta*), tannia (*Xanthosoma sagittifolium*), and elephant foot yam (*Amorphophallus* spp.). This group of crops provides a substantial part of the world's food supply (contributes about 6 % of the world's dietary calories). On a global basis, approximately 45 % of root and tuber crop production is consumed as food, with the remainder used as animal feed or for industrial processing for products such as starch, distilled spirits, and a range of minor products.

India produced a record of 252 million tons of food grains during 2011–2012. Nevertheless, about 217 million of our people (17.50 % of total population) remained chronically undernourished in 2012. India pledged to reduce the undernourished population from 26.90 % in 1990–1992 by half in 2015 under the Millennium Development Goals (MDGs), which is unlikely to be realized in spite of satisfactory overall economic growth. Many of the poorest farmers and most undernourished households in India depend on R&T crops as a contributing if not principal source of food and nutrition. These farm households value R&T crops because they produce large quantities of dietary energy and have stable yields under conditions in which other crops may fail. According to FAO, "nutrition-sensitive" growth in a country can be ensured by supporting increased dietary diversity. In that context, tropical R&T crops (source of vitamins, minerals, antioxidants, and essential amino acids such as lysine) become important for ensuring food and nutritional security of the country.

The pre- and postharvest losses of R&T crops due to pests and diseases are very high. Some of the important pests and diseases in R&T crops include cassava mosaic virus, sweet potato weevil, yam anthracnose and dry rot nematode, taro leaf blight, and elephant foot yam collar rot. These pests are carried through infected planting material from one season to the next season and lead to loss in yield and quality. Hence, crop protection in R&T crops has a major role to play in fulfilling the MDGs identified by the United Nations.

The information on integrated pest management in tropical R&T crops is very much scattered, and there is no book at present which comprehensively deals with the above aspects. The present book deals with insect/mite/

vertebrate pests and fungal/bacterial/viral/mycoplasma/nematode diseases of R&T crops. The book is divided into 15 chapters such as Tuber Crops: An Overview, Cassava, Sweet Potato, Yams, Taro, Elephant Foot Yam, Chinese Potato, Yam Bean, Achira, Winged Bean, Enset, Tannia, Arrowroot, Crepe Ginger, and The Way Forward. For each crop pest, information on the distribution, symptoms/damage, host range, biology, survival and spread, and management methods (regulatory, physical, cultural, chemical, biological, host resistance, and integrated) is provided. The book is extensively illustrated with excellent quality photographs enhancing the quality of publication. The book is written in lucid-style, easy-to-understand language along with adoptable recommendations involving eco-friendly control measures.

This book can serve as a useful reference to policy makers, research and extension workers, practicing farmers, and students. The material can also be used for teaching postgraduate courses. Suggestions to improve the contents of the book are most welcome (e-mail: reddy_parvatha@yahoo.com). The publishers, Springer, New Delhi, India, deserve commendation for their professional contribution.

Bangalore, India
January 30, 2015

P. Parvatha Reddy

Contents

About the Author

Dr. P. Parvatha Reddy obtained his Ph.D. degree jointly from the University of Florida, USA, and the University of Agricultural Sciences, Bangalore.

Dr. Reddy served as the Director of the prestigious Indian Institute of Horticultural Research (IIHR) at Bangalore from 1999 to 2002 during which period the Institute was honored with the "ICAR Best Institution Award." He also served as the Head, Division of Entomology and Nematology, at IIHR and gave tremendous impetus and direction to research, extension, and education in developing biointensive integrated pest management strategies in horticultural crops. These technologies are being practiced widely by farmers across the country since they are effective, economical, eco-friendly, and residue free. Dr. Reddy has about 34 years of experience working with horticultural crops and has been involved in developing an F1 tomato hybrid "Arka Varadan" resistant to root-knot nematodes.

Dr. Reddy has over 250 scientific publications to his credit, which also include 30 books. He has also guided two Ph.D. students at the University of Agricultural Sciences, Bangalore.

Dr. Reddy is serving as Chairman, Research Advisory Committee (RAC), Indian Institute of Vegetable Research, Varanasi; Member, RAC of National Centre for Integrated Pest Management, New Delhi; Member of the Expert Panel for monitoring the research program of National Initiative on Climate Resilient Agriculture (NICRA) in the theme of Horticulture including Pest Dynamics and Pollinators; and Senior Scientific Advisor, Dr. Prem Nath Agricultural Science Foundation, Bangalore. He had also served as a Member of the Research Advisory Committee of the National Research Centre for Citrus, Nagpur, and the Project Directorate of Biological Control, Bangalore. He served as a Member, Quinquennial Review Team (QRT), to review the progress of the Central Tuber Crops Research Institute, Trivandrum; All India Co-ordinated Research Project (AICRP) on Tuber Crops; AICRP on Nematodes; and All India Network Research Project (AINRP) on Betel vine. He is the Honorary Fellow of the Society for Plant Protection Sciences, New Delhi; Fellow of the Indian Phytopathological Society, New Delhi; and Founder President of the Association for Advancement of Pest Management in Horticultural Ecosystems (AAPMHE), Bangalore.

Dr. Reddy has been awarded with the prestigious "Association for Advancement Pest Management in Horticultural Ecosystems Award," "Dr. G.I. D'Souza Memorial Lecture Award," "Prof. H.M. Shah Memorial Award," and "Hexamar Agricultural Research and Development Foundation Award" for his unstinted efforts in developing sustainable, biointensive, and eco-friendly integrated pest management strategies in horticultural crops.

Dr. Reddy has organized "Fourth International Workshop on Biological Control and Management of *Chromolaena odorata*," "National Seminar on Hi-tech Horticulture," "First National Symposium on Pest Management in Horticultural Crops: Environmental Implications and Thrusts," and "Second National Symposium on Pest Management in Horticultural Crops: New Molecules and Biopesticides."

1 Tropical Root and Tuber Crops: An Overview

Abstract

Root and tuber crops provide a substantial part of the world's food supply and are also an important source of animal feed and industrial products. The population of India is projected to grow to 1.62 billion by 2050, and the demand for food grains, which is projected to increase to 345 million tons in 2030, can even go up to 360 million tons by 2050. The likely gap in demand-supply could be bridged by crops like cassava and sweet potato, which are concentrated sources of energy. Most of the tuber crops are bestowed with resilience to global warming and climate change, potential for better return under adverse soil and weather conditions, diverse industrial applications and nutritional attributes, and most of all exceptionally high CO_2 fixation potential. In terms of annual volume of production, cassava and sweet potato rank among the top 10 food crops produced in developing countries. The information on World and Indian scenario, area, production, and productivity, SWOT analysis, gap areas identified, and thrust areas for tropical root and tuber crops, is provided.

Keywords

Tropical root and tuber crops • SWOT analysis • Cassava • Sweet potato

1.1 Introduction

1.1.1 World Scenario

Tropical root and tuber (R&T) crops play a critical role in the global food system, particularly in the developing world, where they rank among the top 10 food crops (Table 1.1). Among the tuber crops, cassava is the most important one in the tropics, and it ranks fourth, after rice, sugarcane, and maize, as a source of calories for human consumption. It is a major carbohydrate food for about 500 million people in the world, and in Africa, it is the most important source of calories in the human diet. By 1997, the production of roots and tubers in developing countries had an estimated annual value of more than 41 billion US dollars or nearly one-fourth the values of the major cereals (Scott et al. 2000). Roots and tubers contribute to the energy and nutrition require-

P.P. Reddy, *Plant Protection in Tropical Root and Tuber Crops*,
DOI 10.1007/978-81-322-2389-4_1

Table 1.1 Area, production, and productivity of the ten most important food crops in the world in 2011

Crop	Area (million ha)	Production (million tons)	Productivity (t/ha)
Wheat	220.4	704.1	3.19
Maize	170.4	883.5	5.18
Rice	164.1	722.8	4.40
Barley	48.6	134.3	2.76
Sorghum	35.3	54.2	1.53
Millet	31.9	27.7	0.87
Cassava	**19.6**	**252.2**	**12.84**
Potatoes	19.2	374.4	19.45
Oats	9.7	22.5	2.33
Sweet potato	**8.0**	**104.3**	**13.11**

FAOSTAT (2013)

ments of more than two billion people; constitute an important source of income in rural and marginal areas; have multiple uses, most notably as food security crops, regular food crops, and cash crops; and are increasingly used as livestock feed and raw material for industrial purposes. They have long served as the principal source of food and nutrition for many of the world's poorest and undernourished households and are generally valued for their stable yields under conditions in which other crops may fail.

Tropical root and tuber crops consist of both dicots like sweet potato (*Ipomoea batatas*) and cassava/tapioca (*Manihot esculenta*) and monocots like yams (*Dioscorea* spp.) and edible aroids like taro (*Colocasia esculenta*), tania (*Xanthosoma sagittifolium*), and elephant foot yam (*Amorphophallus* spp.). The important edible products harvested from yams are tubers and from cassava and sweet potatoes are roots, while aroids give corms or cormels. Individually, cassava, sweet potato, and yam rank among the most important food crops worldwide. This group of crops provide a substantial part of the world's food supply (contributes about 6 % of the world's dietary calories) and are also an important source of animal feed and raw materials for industrial products. On a global basis, approximately 45 % of root and tuber crop production is consumed as food, with the remainder used as animal feed or for industrial processing for products such as starch, distilled spirits, and a range of minor products. These species belong to different botanical families but, for the purpose of scientific research, are grouped together because they are vegetatively propagated, produce underground food, and are bulky and perishable (Table 1.2).

The pattern of root and tuber crop utilization varies considerably among countries. In the developing countries (with the exception of China and Brazil), relatively small amounts (less than 20 %) are fed to livestock. Most of the remainder is used locally as food. The relatively high cost of transportation, processing, and storage, as well as the considerable time needed in food preparation, frequently makes unprocessed root and tuber crops less attractive to urban consumers.

The consumption of root and tuber crops as food in developed countries is considerably smaller than it is in developing countries, but their use as animal feeds is relatively higher. A very small proportion of root and tuber crop production (approximately 5 %) is traded internationally. More than two-thirds of those exports come from developing countries, with Thailand's cassava exports accounting for more than half of the total.

There are considerable differences in the agroclimatic conditions suitable for the production for the different root and tuber crops. Cassava is grown across a broad range of agroclimatic conditions from sea level to 1,800 m and from areas with as little as 500 mm of rainfall to tropical rain forest areas with more than 2,000 mm per year.

Sweet potato is understood to have originated in the Americas (as did cassava), and it too is grown over a considerable range of latitude and elevation (up to 2,500 m). Conversely, yams have a relatively narrower range of production, being mainly confined to the tropical region throughout the world from sea level to 1,400 m. The main production of yam is in the savanna region of West Africa, where more than 90 % of the crop is grown. Unlike cassava and sweet potato, the white and yellow yams are thought to be indigenous to West Africa, whereas the water yam is thought to have originated in Southeast Asia.

Most of the cultivated edible aroids are well adapted to high rainfall (and occasional flooding)

Table 1.2 Characteristics of the tropical root and tuber crops

	Cassava	Sweet potato	Yams	Aroids
	Manihot esculenta	*Ipomoea batatas*	*Dioscorea* spp.	*Colocasia esculenta* and *Xanthosoma sagittifolium*
Characteristics	Euphorbiaceae	Convolvulaceae	Dioscoreaceae	Araceae
World production in 2007 (million tons)[a]	226	124	51	13
World cultivated area (million ha)	18.6	9.0	4.6	1.9
World average yield (fresh t/ha)	12.2	13.7	11.2	6.7
Yield potential (fresh t/ha)	90	120	110	110
Planting material (propagule)	Stems	Vine cuttings	Tubers	Corms, suckers
Growth period (months)	8–36	3–6	8–36	6–16
Optimal rainfall (mm)	1,000–1,500	750–1,000	1,200–1,500	2,500–3,500
Optimal temperature (°C)	25–30	20–25	30	20–35
Drought resistant	Yes	Yes	Yes	No
Waterlogged tolerant	No	No	No	Yes
Shade tolerant	No	No	No	Yes
Soil fertility requirements	Low	Low	High	High
Seasonality of crop cycle	No	Yes	Yes	No
In-ground storage life	Long	Moderate	Moderate	Long
Postharvest storage life	Very short	Short	Long	Moderate
Leaves used for human consumption	Yes	Yes, fairly common in West Africa and East Asia	No	Yes
Leaves used for animal feed	Yes	Yes	No	Yes
Dry matter (% fresh weight, FW)	30–40	20–35	20–40	20–30
Starch (% FW)	27–37	18–28	20–25	15–25
Starch grain (in microns)	5–50	2–40	1–70	1–6
Amylose (% starch)	15–30	8–32	10–30	3–45
Gelatinization temperature (°C)	49–73	58–65	69–88	68–75
Total sugars (% FW)	0.5–2.5	1.5–5.0	0.5–2.0	2.0–3.0
Proteins (% FW)	0.5–2.0	1.0–3.0	2.0–4.0	1.5–3.0
Fibers (% FW)	1.0	1.0	0.6	0.5–3.0
Vitamin A (µg/100 g/FW)	17	900	117	0–42
Vitamin C (mg/100 g/FW)	50	35	25	10
Minerals (% FW)	0.5–1.5	1.0	0.5–1.0	0.5–1.5
Energy (kj/100 g/FW)	600	500	440	400
Anti-nutritional compounds	Cyanogens	Trypsin inhibitors	Alkaloids, tannins	Oxalate crystals

Lebot (2009)
[a]Bradbury and Holloway (1988)

and can be cultivated in temperatures ranging between 16° and 30 °C, at elevations up to 1,600 m.

The lesser-known Andean root and tuber crops, of which there are more than a dozen, vary considerably in their ranges of adaptation and tolerance to environmental conditions. They are primarily considered to be medium-to-higher elevation crops of moderate temperature regimes and water requirements and have greater tolerance to frost than do other root and tuber crops.

1.1.2 Indian Scenario

India produced a record 252 million tons of food grains during 2011–2012. Nevertheless, about 217 million of our people (17.50 % of total population) remained chronically undernourished during 2012 (FAO et al. 2012). India pledged to reduce undernourished population from 26.90 % in 1990–1992 by half in 2015 under millennium development goal (MDG), which is unlikely to be realized in spite of satisfactory overall economic growth. According to FAO, "nutrition-sensitive" growth in a country can be ensured by supporting increased dietary diversity. In that context, tropical tuber crops like cassava, sweet potato, elephant foot yam, yams, taro, tania, yam bean, arrowroot, etc. become important for ensuring food and nutritional security of the country. These crops contributed significantly in food and nutritional security of rural India during pre-green revolution era. In fact, tuber crops constituted 1.58 and 3.62 % of cereal consumption in urban and rural areas, respectively, during the years 1973–1974 that have reduced to 0.27 and 0.33 %, respectively, during 2009–2010. It is necessary to reverse this trend in order to achieve MDG target. The population of India is projected to grow to 1.62 billion by 2050, and the demand for food grains, which is projected to increase to 345 million tons in 2030, can even go up to 360 million tons by 2050. The likely gap in demand-supply could be bridged by crops like cassava and sweet potato, which are concentrated sources of energy. Most of the tuber crops are bestowed with resilience to global warming and climate change, potential for better return under adverse soil and weather conditions, diverse industrial applications and nutritional attributes, and most of all exceptionally high CO_2 fixation potential. The exceptional soil carbon sequestration property of cassava makes it a potential crop for ameliorating green revolution fatigue. Besides, there is immense potential of cassava as raw material for bioethanol production. Cassava may contribute in a big way toward fulfilling targets for production of green energy without disturbing food security situation.

Cassava and sweet potatoes are the two important tuber crops in India with 0.221 and 0.104 million ha area under cultivation and 8.076 and 1.04 million tons annual production, respectively (FAOSTAT 2012, 2013). While cassava continues to be a vital subsistence crop for small-scale farmers and tribal communities, its utilization pattern has changed dramatically in recent years largely due to its use as a primary raw material for starch, sago, adhesive, and animal feed industries as well as its potential for biofuel production. The prospect of using cassava for biofuel production has yet remained largely unexplored in India. Sweet potato is primarily utilized as a vegetable and subsidiary food by the tribal communities and poor people. Other tubers are grown as vegetable crops in homestead or in semicommercial scale. Most of them are reservoirs of resistant starch, minerals, vitamins, antioxidants, and dietary fibers.

Tuber crops in general and cassava in particular are well adapted to conditions of drought. They can also be cultivated under wide range of soil conditions, and their ability to grow under a wide range of agroclimatic situations enhances the scope of extending the cultivation to the non-traditional regions of the country like Maharashtra, Gujarat, Bihar, and Karnataka. Nevertheless, tuber crops in general are labor-intensive and require large amount of labor for cultivation. Reducing the cost of production through less dependence on labor and inputs is necessary to tide over this situation, and the extent to which this can be achieved decides the future of these root crops in India. Being vegetatively propagated crops, quality seeds are limiting factors for rapid spread of new varieties and the crop. Minisett technology for rapid multiplication of planting material for tuber crops is yet to take off in many R&T crop-growing states, stressing the need to create awareness and dissemination about the technology through government machinery and NGOs.

Demand for cassava in the human consumption sector has been declining drastically especially in the traditional cassava-consuming state, Kerala, in India. This creates a situation where the future of tuber crops especially cassava lies in the industrial and livestock sectors. Although, cassava and sweet potato leaves have high protein content, their potential as animal feed has not been properly exploited. Sweet potato, despite being a health food, is seldom consumed by the urban and elite, and factory-level processing does not exist presently in India, unlike in China, Japan, Korea, and parts of America. The immense medicinal value of tuber crops due to their anthocyanin and carotenoid content has so far not been exploited. Cassava cultivation is expanding to nontraditional areas of India especially in view of its newfound importance as a biofuel crop. However, there is a need to develop suitable varieties as well as management practices for these regions to enhance the productivity of the crop.

Presently, there is a loosely knit linkage between the research, extension, and clientele system in the case of tuber crops. There is no well-defined government policy for root and tuber crop development, both at Central or State Government levels except for 2–3 states. In order to synchronize production of tuber crops with marketing, it is necessary to open new avenues for its better utilization under the present context where traditional uses have almost stabilized. Tuber crops as such provide a vast scope for diversification and value addition, and there lies a great opportunity for nontraditional uses of tuber crops in the form of convenience foods, functional foods, biofuels, starch-based innovative products like biofilms, thermofoams, etc. A greater thrust is needed in policies and programs which support research, extension, and value chain management, besides a concerted effort to bring down the cost of cultivation of tuber crops and expand their utilization spectrum which will help in the sustainable development of these crops in India.

Root and tuber crops are the most important food crops after cereals. They have the highest rate of dry matter production per day and are major calorie contributors. Tuber crops find an important place in the dietary habits of small and marginal farmers especially in the food security of tribal population. Tuber crops not only enrich the diet of the people but also possess medicinal properties to cure many ailments or check their incidence. Many tropical tuber crops are used in the preparation of stimulants, tonics, carminatives, and expectorants. The tuber crops are rich in dietary fiber and carotenoids, viz., β-carotene and anthocyanin. India holds a rich genetic diversity of tropical root and tuber crops, viz., cassava, sweet potato, aroids, yams, and several minor tuber crops. The Indo-Burma region is the center of origin of taro and Asiatic edible yams. The two hot spots of global biodiversity, viz., Northeastern Himalayas and Western Ghats, are particularly rich in wild relatives of tropical R&T crops. Safe conservation and sustainable use of plant biodiversity are essential for meeting the present and future needs of tuber crop improvement in India.

The economically and socially important tropical tuber crops are cassava; sweet potato; yams; aroids which include elephant foot yam, taro, and tania; and other minor tuber crops, namely, Chinese potato, arrowroot, yam bean, etc. There are five major areas of distribution of root and tuber crops in India. These are (1) Southwestern hilly and coastal region, (2) Southern peninsular region, (3) Eastern coastal region, (4) Northeastern region, and (5) Northwestern region. The important tuber crops grown in India and the regions of biodiversity are given in Table 1.3.

1.2 Area, Production, and Productivity of Tuber Crops

In the year 2012, the major R&T crops occupied about 53 million ha producing 797 million metric tons worldwide, 42 % of which was from Asia and 6.45 % from India (Table 1.4). Cassava is cultivated in 18.6 million ha and spread over the continents of South America, Africa, and Asia, producing 226 million tons of tubers. The average productivity in the world is 12.2 t ha^{-1}.

Table 1.3 The important tuber crops grown in India and the regions of biodiversity

Common name	Scientific name	Family	Places/areas grown
Cassava	*Manihot esculenta*	Euphorbiaceae	Southern region, occasionally in Northeastern and Western regions
Sweet potato	*Ipomoea batatas*	Convolvulaceae	Introduced and found all over but mostly concentrated in Eastern UP, Bihar, West Bengal, and Orissa
Greater yam	*Dioscorea alata*	Dioscoreaceae	Southern and Northeastern region
White yam	*D. rotundata*	Dioscoreaceae	Introduced to India and spread to Southern and Northeastern region
Lesser yam	*D. esculenta*	Dioscoreaceae	Southern, NE, and Eastern region
Potato yam	*D. bulbifera* var. sativa	Dioscoreaceae	Southern, Northeast, and Eastern region
Taro	*Colocasia esculenta*	Araceae	Throughout India with greater diversity in the Northeast, Eastern region, and Southern
Tania	*Xanthosoma sagittifolium*	Araceae	Southern and Northeastern region
Elephant foot yam	*Amorphophallus paeoniifolius*	Araceae	Southern, Northeast, and Eastern region
Chinese potato	*Plectranthus rotundifolius*	Labiatae	Southern parts of India
Yam bean	*Pachyrhizus erosus*	Leguminosae	Northeastern region
Winged bean	*Psophocarpus tetragonolobus*	Leguminosae	Southern and Northeast
Arrowroot	*Maranta arundinacea*	Marantaceae	Adapted to plain areas with high rain fall; shade loving

Table 1.4 Area, production, and productivity of root and tuber crops in the world

Tuber crop	Area (million ha)	Production (million tons)	Productivity (t/ha)	Potential yield (t/ha)
Cassava	18.6	226	12.2	90
Sweet potato	9.0	124	13.7	120
Yams	4.6	51	11.2	110
Aroids	1.9	13	6.7	110

Table 1.5 Crop-wise area, production and productivity of different tuber crops in India during 2012

Crops	Area (000 ha)	% of area under tuber crops	Production (000 t)	% of production under tuber crops	Productivity (t/ha)	B: C ratio
Cassava	226.70	52.96	8,746.56	80.60	38.582	3.176
Sweet potato	110.39	25.79	1,040.84	9.59	9.50	2.366
Yams	3.00	0.70	46	0.43	15.0	0.536
Elephant foot yam	36.00	8.41	364	3.36	10.0	0.525
Taro	52.00	12.14	654	6.02	13.0	2.350
Total	**428.09**		**10,851.40**			

Tuber crops are cultivated on a large scale throughout India, except in extreme north under varied production systems. The area, production, and productivity of different tuber crops in India during 2012 are presented in Table 1.5.

Many of the poorest farmers and most undernourished households in India depend on R&T crops as a contributing, if not principal, source of food and nutrition. These farm households value R&T crops because they produce large quantities of dietary energy and have stable yields under conditions in which other crops may fail. R&T crops produce remarkable quantities of energy per day in comparison to cereals. In addition, some R&T crops are important source of vitamins, minerals, antioxidants, and essential amino acids such as lysine. R&T crops, therefore, have a major role to play in fulfilling the millennium development goals (MDGs) identified by the United Nations (UN).

1.3 SWOT Analysis for Tuber Crops

1.3.1 Strengths

- Cassava plays an important role in starch and sago manufacture, which are produced from nearly 450 factories in Tamil Nadu and Andhra Pradesh. The starch and sago production requirement may be more than 0.3 million tons in the coming 2015–2016.
- Hitherto, cassava, considered to be concentrated in South Indian states like Kerala, Tamil Nadu, and Andhra Pradesh, has started spreading to nontraditional areas like Maharashtra, Gujarat, and Northeastern states like Assam. Realizing the industrial potential of cassava especially for biofuel purpose, many entrepreneurs started establishing cassava starch/sago factories in nontraditional states and also encouraging the farmers to cultivate cassava.
- A novel technology for the cost-effective production of ethanol from cassava requiring only 48–50 h processing time has been perfected recently at Central Tuber Crops Research Institute (CTCRI), Thiruvananthapuram, Kerala, which has to be scaled up for commercialization.
- Cassava is mainly used in industrial (60 %), human consumption (28 %), and animal feed sectors (12 %). It was estimated that cotton yarn sizing industry is currently consuming nearly 85,000 t of cassava starch. Projection of cassava starch demand in the sizing industry, based on population projections (Census Commissionerate, Govt. of India) and projections of per capita availability of cotton cloth, shows that by 2016, sizing industry would require 90,383 t of cassava starch; by 2021, the requirement would be 101,368 t; and by 2026, the sizing industry requires 115,793 t of cassava starch.
- Cassava starch-based adhesive finds important place for application in corrugation box industry, paper conversion industry, and liquid gum industry for domestic use. Projected cassava starch demand (based on a growth of 10 %) in corrugation box industry by 2016, 2021, and 2025 would be 0.185, 0.298, and 0.436 million tons, respectively.
- Several value-added products including health foods have been developed from cassava and sweet potato, and these are in various stages of commercialization.

1.3.2 Weaknesses

- Though cassava is used as a raw material for production of starch and sago in Tamil Nadu and Andhra Pradesh, majority of the cassava-based factories are operating much below their full capacity utilization mainly because of lack of adequate tuber supply, continued use of age-old machineries, and lack of appropriate industrial policies, especially in view of the two big factories that have come up. Besides, the conventional factories require a large quantity of water for starch extraction, and the effluent discharge from these factories pose environmental problems.
- Demand for cassava in the human consumption sector is declining drastically especially in the traditional cassava-consuming state, Kerala, in India. The drastic reduction in consumption is because of increased availability of cereal food grains, increased per capita income, and thereby increased standard of living. This creates a situation where the future of R&T crops especially cassava lies in the industrial and livestock sectors.
- The very poor shelf life of cassava poses problems to the cassava starch factories in Tamil Nadu and Andhra Pradesh and is a major

drawback in its long-distance transport to other states, where the industrialists wish to set up factories based on cassava. Although it is established that the oxidative changes stimulated in the tubers by polyphenol oxidases and peroxidases are primarily responsible for the onset of PPD, a lot more needs to be understood about the mechanism.

- The area under sweet potato has been declining continuously owing to lack of commercial exploitation of the crop. Till today, sweet potato is grown for consuming as vegetable only, and the same is the case with yams and aroids also. This has also led to an increase in the market price of these crops, which ultimately affects the production cost of all value-added products, including alcohol.
- There exists a loosely knit linkage between the research, extension, and clientele system at present in the case of tuber crops. There is no well-defined government development policy for R&T crops development, both at Central or State Government levels except for 2–3 states unlike in countries such as Thailand and Indonesia.

1.3.3 Opportunities

- Cost-effective production of starch and its diversification to new products having domestic and international demand are thus essential for the sustainability of processing industries. There is an increasing awareness about the vast pollution caused by plastic materials as well as the inefficient and unhealthy ways of its disposal. There is a need to develop 100 % biodegradable packages/allied materials like biofilms, containers, etc.
- There is immense potential of cassava as raw material for bioethanol production. In fact, China has taken a policy of substituting corn with cassava for bioethanol production. Considering the fact that access to modern affordable energy is essential for the achievement of the internationally agreed development goals, the United Nations General Assembly declared 2012 the "International Year of Sustainable Energy for All." Cassava may contribute in a big way toward fulfilling targets for production of green energy without disturbing food security situation.
- With the rapidly changing food habits and increased migration of people to urban areas coupled with the projected rise in per capita income, there is also a projected increase in lifestyle diseases. Demand for processed and ready-to-eat convenience foods may also simultaneously increase in the next 30–40 years. One of the major challenges for the country will be to provide nutritious, safe, and healthy food to people. There exists potential to develop prophylactic and therapeutic functional foods from tuber crops, as unlike fruits, the starch and flour open large avenues for processing, fortification, etc.
- Cassava starch has got a lot of potential as corrugated adhesives and textile adhesives, and the R&D activities have to be streamlined to develop technologies suitable for the industry, and this can be one of the ways to reduce the competition from maize starch in the industry.
- Application of nanotechnology for developing tuber starch-based novel products can go a long way in developing new lifesaving ventures like safe drug delivery systems. Nanotechnology can also be applied in encapsulation of bioactive compounds derived from tuber crops.
- Sweet potato, despite being a health food, is seldom consumed by the urban and elite, and factory-level processing does not exist presently in India, unlike in China, Japan, Korea, and parts of America. The immense nutraceutical value of tuber crops due to its anthocyanin and carotenoid contents has so far not been exploited. Vitamin A deficiency in children and tribal rural population of India is in alarming proportion. As an alternative to combat this problem, there is a big scope for orange-fleshed sweet potato to be popularized among children of rural and tribal population as snack food and to increase the area under orange-fleshed sweet potato.

1.3.4 Threats

- Tuber crops are rated as low-status food and there is no price support or guaranteed price. With increasing availability of cereals, the demand for tuber crops, especially cassava, has declined to a great extent, resulting in rapid shrinkage of area and at the same time phenomenal increase in the price due to low availability. As a consequence, utilization of tuber crops as animal feed and industrial processing is not picking up the expected momentum. Tuber crops have to compete with other starch sources whether for food, feed, or industrial use.
- A major challenge faced by the cassava growers and industrialists alike (in the context of WTO agreement) is the increasing import of starch from countries like Thailand, where the cost of production is less due to cheap labor. India started facing stiff competition from Thailand in the international trade for cassava starch after liberalization and opening up of our markets. Cheaper Thailand cassava starch started entering Indian markets thereby affecting domestic cassava starch industries. This is likely to lead to an alarming situation, where many of the traditional starch factories will have to be closed down, which will indirectly lead to a further decrease in the cultivated area under cassava and unemployment for many who depend on the industry.

The outcome of SWOT analysis of different tuber crops carried out is detailed in Table 1.6.

1.4 Gap Areas Identified

- Basic research on biology, genetics, and physiological aspects of tuber crops
- Biodiversity documentation and exploitation of traditional knowledge
- Application of biotechnological tools in tuber crop improvement
- High-yielding varieties with tolerance to biotic stresses (cassava mosaic disease, sweet potato weevil, anthracnose in greater yam, nematode in white yam, leaf blight in taro, collar root in elephant foot yam) and abiotic stresses (drought, salinity)
- Varieties for specific purpose, viz., good fried chip quality, high amylopectin content, and climate-resilient varieties with high input-use efficiency
- Integrated farming system approach involving tuber crops
- Quality planting material production
- Postharvest management and product diversification
- Exploitation of nutraceutical potential of tuber crops
- Scaling up of technologies developed to industrial/commercial levels.
- Pre- and postharvest mechanization
- Efficient technology transfer of good agricultural practices
- Development of economic models for different sizes of holdings
- Strengthening of both forward and backward linkages for sustainable development of tuber crops
- Skilled human resource and development
- A national-level policy on tuber crop production and utilization
- Mission mode R&D program on tuber crops

1.5 Thrust Areas

1.5.1 Crop Improvement

- Exploration, collection, characterization, and maintenance of the germplasm including wild species should be undertaken to cater to the breeding objectives for the future, especially for locating sources of resistance to some of the pests and diseases.
- Molecular characterization of the tuber crop germplasm, viz., DNA finger printing, needs to be done using only SSR markers. DNA barcoding may also be done for all the accessions wherever feasible on priority basis.
- Protocols for in vitro conservation of germplasm for major tuber crops should be standardized and adopted.

Table 1.6 SWOT analysis of different tuber crops

Strengths	Weaknesses	Opportunities	Threats
Cassava			
1. High productivity	1. Poor shelf life	1. Great potential to become a secondary staple food	1. Cassava mosaic disease
2. Adaptability to wide soil, climatic, and environmental conditions	2. Nonavailability of quality planting material	2. Scope for area expansion in traditional and nontraditional areas due to wider adaptability	2. Bulky and perishable nature of tubers limit the scope for export of raw tubers
3. Strong research base	3. Longer crop duration	3. Big scope for product diversification and value addition	3. Stiff competition from other sources of starch like maize which are cheaper whether for food, feed, or industrial use
4. Rich genetic resources	4. Inadequate processing facilities in nontraditional areas	4. Potential to develop environment-friendly biodegradable packages/ allied materials like biofilms, containers, etc.	4. Decline in area over the last two decades
5. Food security crop	5. Inadequate government policy and extension program support	5. Opportunity for industrial-level bioethanol production	5. Import of cassava starch from countries like Thailand, where the cost of production is less
6. Generates employment in rural sector	6. Absence of regulated market or market federation	6. Scope to develop prophylactic and therapeutic functional foods	6. Wide price fluctuation
7. Plays an important role in starch and sago manufacture	7. Poor linkages among stakeholders		7. High labor cost
8. Great demand for diverse industrial applications	8. Lack of reliable data on area, production, and industrial use		
9. Rich raw material for biofuel production	9. Poor socioeconomic base of farmers		
10. High potential for several value-added food products			
Sweet potato			
1. High caloric yield per unit area	1. Declining area under cultivation	1. Enormous scope as a nutritional food	1. Sweet potato weevil
2. Source of staple and nutrition food for poor and marginalized	2. Nonexploitation of industrial applications	2. Scope for area expansion	2. High labor cost
3. Strong research base	3. Susceptible to weevil infestation	3. Can be fitted well in different cropping systems	3. Lack of mechanization
4. Rich genetic resources	4. Poor socioeconomic base of farmers	4. Exploitation of nutraceutical value due to its anthocyanin and carotenoids content	4. Lack of technology for industrial applications

(continued)

Table 1.6 (continued)

Strengths	Weaknesses	Opportunities	Threats
5. Good source of livestock feed	5. Inadequate government policies and extension program support	5. Scope for popularization of orange-fleshed sweet potato among children of rural and tribal population to combat Vitamin A deficiency	5. Poor market linkage
6. Source of carotene and anthocyanins		6. Can be a source for ethanol and wine production	6. Poor storage facilities
7. Wider adoptability and short duration			
Yams and aroids			
1. Adaptability to wide soil, climatic and environmental conditions	1. Lack of reliable data on area and production	1. Area can be expanded due to wider adaptability of the crops	1. Taro leaf blight
2. Strong research base	2. Restricted area under cultivation	2. Can fit well in different cropping systems	2. Collar rot in elephant foot yam
3. Rich genetic diversity	3. Lack of industrial utilization	3. High medicinal value	3. Anthracnose leaf spot in yams
4. High dietary and nutraceutical value	4. Lack of public awareness on nutraceutical value	4. Scope for providing good income to tribals	4. Nematode infestation
5. Staple dietary source for tribal and poor people	5. Poor socioeconomic base of the farmers		5. Nonavailability of quality planting material
6. Rich traditional knowledge	6. Inadequate government policies and extension program support		

- Registration and protection of varieties and geographical indicators (GIs) in tuber crops have to be pursued vigorously as per norms of Protection of Plant Varieties and Farmers' Right Act (PPVFRA), and custodian farmers and tribal communities shall be given required recognition rewards.
- Emphasis may be given for developing hybrids/varieties with the following traits:
  - Resistance/tolerance to mosaic disease, high starch content for industrial purposes, as well as better keeping quality and drought tolerance/resistance in cassava
  - Tolerance to salinity in sweet potato
  - Enhanced levels of nutraceuticals, viz., β-carotene in sweet potato and anthocyanin in greater yam, high protein in cassava, etc.
  - Climate-resilient varieties with high input-use efficiency
- The two simple sequence repeat (SSR) markers identified in cassava should be tested further for their utility in the conventional breeding program. More molecular markers may be identified.
- Molecular characterization of minor tuber crops for pharmaceutical applications needs to be undertaken.
- Programs should be taken up to modify genes in starch metabolism. Feasibility of developing methods for early screening of transformants for starch quality should be explored.
- In vitro propagation techniques should be standardized in elephant foot yam and taro for seed multiplication, and the potential of aeroponics for seed multiplication in yam and taro needs to be explored.
- Work on genetic improvement of yam bean, sweet potato, and taro for valued traits needs to be pursued with required refinement.
- Breeding for high-yielding, short-duration, and climate-resilient tuber crop varieties needs emphasis.
- Enhancing the amylose/amylopectin content in tuber starches through genetic transformation needs to be undertaken.

1.5.2 Crop Production

- In view of high cost of cultivation (for planting, weeding, and harvesting) of tuber crops, efforts must be made to cut the cost through farm mechanization and through integrated nutrient management. Suitable agrotechniques such as modification of crop geometry that suits farm mechanization also need to be developed. The cassava harvester developed by Tamil Nadu Agricultural University, Coimbatore, requires improvement and refinement. Since Thailand has good harvester, ways shall be found to collaborate with them in developing cassava harvester.
- Research should be undertaken on cultivation of tuber crops under minimum cultivation, conservation agriculture system, use of live mulches, etc.
- The mulching technology in cassava which has given impressive impact shall be popularized through brochure and video emphasizing its multiple benefits.
- Studies should be taken up on the long-term effect of mulching on productivity, soil health, pest incidence, and weeds with special emphasis on weed seed bank in the soil.
- The research conducted on organic farming should be looked into critically, taking into account the demand for organic produce, the premium the product commands, the cost-benefit analysis, pest and disease incidence, certification, etc.
- Considerable research has been carried out in intercropping of tuber crops in orchards and planting of short-duration intercrops between rows of tuber crops. It is time to take stock of all the findings before conducting further experiments on this aspect, and consolidated package should be developed. An assessment should be made on adoption level by farmers and reasons thereof if not practiced by them.
- Considering that tuber crops are used to supplement feed for livestock, it is worthwhile to initiate farming system research involving tuber crops with animal component. In view of practicality, it is best taken up as an on-farm trial.
- Detailed studies should be conducted on the impact of climate change on cassava and sweet potato cultivation. Modeling work on mapping of cassava cultivation under varying scenarios of climate change should also be attempted. In-depth field investigations are needed to explore the uncertainties in tuberization. Unstable performance of sweet potato under different climate change regime should also be considered.
- The site-specific nutrient management (SSNM) technology developed should be taken forward by including it in adoptive trials.
- Quality planting material is a major constraint in tuber crops for large-scale spread of new varieties developed. A full-fledged seed production and distribution system should be established especially for cassava, elephant foot yam, and yams at the earliest. Rapid multiplication technologies developed for aroids and yams should be scaled up. A participatory seed production plan based on seed village concept should be adopted with required modifications. A seed production chain should be built up from breeders' seed to certified seed production and supply.
- There is a need for working out economics of cultivation of important tuber crops for different size holdings, and it should be updated from time to time. Besides, crop economic models should be developed and published regularly for the benefit of stakeholders.

1.5.3 Crop Protection

- Development of robust, sensitive, and cost-effective diagnostic kits for quick and reliable detection of viruses in tuber crops needs to be undertaken for indexing and production of quality planting material [for important viruses like Sri Lankan Cassava Mosaic Virus (SLCMV) and Dasheen Mosaic Virus (DMV) in elephant foot yam].
- Refinement of the integrated disease management strategies for *Amorphophallus* collar rot, *Colocasia* leaf blight, and yam anthracnose

including organic amendments, biofertilizers, and biological control agents needs to be taken up.

- Strict quarantine should be followed for monitoring the movement of cassava planting material in order to prevent the entry of cassava brown streak virus (CBSV) which is not reported in India till now but is widely prevalent in Africa and causes major havoc in tuber production.
- Disease-free cassava and elephant foot yam mother cultures shall be established for all the released cultivars and maintained for further mass multiplication.
- The biopesticides developed by CTCRI, viz., Nanma, Shreya, and Menma (from cassava plants), should be field tested on large scale for their efficacy against tuber crop pests by including them under AICRP (TC). There is a need for working out the active ingredients, mode of action, etc. for these biopesticides.
- Emergence of biotypes of *Bemisia tabaci* which transmits CMD should be monitored.

1.5.4 Crop Utilization

- Development of health-oriented food products (low glycemic) from tuber crops needs to be taken up.
- There is a need for development of edible biofilms from tuber starches and their blends with hydrocolloids.
- Isolation and characterization of bioactive principles from tuber crops (*Amorphophallus* sp., *Coleus forskohlii*, etc.) need to be intensified.
- Cassava processing should be scientifically studied and improved. The process of manufacture should be standardized and improved.
- The enzyme modification process developed by CTCRI has made ethanol production from cassava faster and easier. It has to be scaled up and CTCRI shall work in collaboration with the industry to increase its adoption. Its economics should be worked out at the earliest.
- The main problem faced by the industries is high cost of production which shall be addressed in order to prevent the import of starch from Thailand as it adversely affects the small-scale manufacturers of cassava starch at the Salem belt. Besides, the sago industry has been moved from orange to red category by Central Pollution Control Board. These environmental aspects should be studied by CTCRI, and a solution should be worked out to resolve the environmental issues of sago manufacture.
- The polymers and hydrogels from cassava starch developed by CTCRI should be refined and improved further so that the industry finds it worth adopting by keeping a liaison with the corporate sector. The potential applications of sago starch in industrial and other sectors such as cement industries need to be explored.
- Since elephant foot yam, taro, and yams are used only as vegetables at present, their use is restricted. Hence, there is a need to develop diversified products from these crops to make them more popular and enhance their value.
- Arrowroot starch fetches a very high price of over Rs. 400 per kg. There is a need to study the reasons for not spreading its cultivation to more areas in the country and also the poor utilization of tubers for starch extraction.
- Extruded products from yams and elephant foot yam have large potential. The dried and packed (dehydrated) taro tubers also have good market in the towns and cities. Hence, there is a need to popularize these products.
- The economics of production of ethanol, wine, etc. from sweet potato (developed by the Regional Centre of CTCRI, Bhubaneswar) should be worked out for further scaling up.
- Sweet potato with high vitamin A content can be a good source of nutrition for children and women. Hence, new products which are ready to cook, ready to eat, etc. should be developed and commercialized by having a collaborative program with leading food industries. The antioxidant properties of sweet potato should be studied in depth.

1.5.5 Extension and Social Sciences

- Participatory technology should be developed with regard to nutrient management and integrated pest/disease management in tuber crops.

- Technology transfer, commercialization, and utilization by clientele system need to be emphasized.
- The impact of new technologies shall be studied along with the economic gains realized by the farmers.
- Success stories and also the experience of progressive farmers should be documented and popularized among other farmers.
- The statistical information (average production, productivity, demand-supply, marketing, etc.) on tuber crops, particularly aroids and yams, is lacking. There is a need to generate this information by CTCRI in collaboration with AICRP (TC) centers.
- Information and communication technologies (ICTs) should be exploited fully in technology transfer activities. The website of CTCRI should be revamped to serve all stakeholders including the public. It shall be made dynamic and rich with content that includes the food value and nutraceutical importance of tuber crops and value-added products meant for all stakeholders. The extension material including brochures, handouts, posters, and videos in the local languages of major producing areas shall be made available on the website for free access. CTCRI should create a robust database of farmers, traders, input suppliers, industry, etc.
- There appears to be rich traditional indigenous technical knowledge (ITK) on tuber crops available with the farmers particularly with tribals on an array of topics. This traditional knowledge shall be collected and documented. If need be, a special project/program should be initiated under the Tribal Sub-Plan (TSP) or any other similar schemes.
- The economic feasibility of introducing cassava in Orissa for industrial use should be worked out. In this regard, farmers' visit to Salem and Tamil Nadu under Agricultural Technology Management Agency (ATMA) or similar such programs should be arranged. Possibilities of establishing a pilot-scale starch/sago factory shall be explored.
- Integrated disease management packages developed for collar rot and other diseases of elephant foot yam and taro leaf blight (specifically in the taro-growing areas of Eastern and NEH regions) should be popularized through demonstrations and awareness activities.
- The productivity of sweet potato in India is much below its reported potential. All-out efforts should be made to enhance productivity by popularizing technologies and information related to good quality planting material, nutrition management, control of weevil, etc. Area expansion in high-potential regions of Bihar, West Bengal, Rajasthan, and Eastern UP should be given priority.
- Swamp taro is considered to be the best for mitigating flood effects in Brahmaputra valley region. Hence the crop should be popularized in other flood-prone areas.
- Promotion of the less exploited tuber crop, yam bean should be taken up highlighting its nutritional components and its use as a raw salad with high dietary fiber.
- An ethnobotanic importance of tuber crops needs in-depth study for its better commercial exploitation.

1.6 Future Outlook

In the developing countries, tropical root and tuber crops have received far less attention with respect to research and development than other crops. Most of the basic and applied research in the area of crop protection of root and tuber crops has been conducted in the developed countries. Considerably more attention has to be given to IPM/IDM of tropical tuber crops in developing countries.

Effective plant quarantine is the first line of defense against the introduction of pests and diseases. It is obviously an important component of IPM/IDM. It requires coordination at national and international levels. With efficient and vigilant quarantine, the introduction and spread of exotic pests which are of serious nature can be avoided or greatly curtailed.

Use of pesticides to control the pests and diseases is generally not practical due to economic considerations; however, they can be economically used for the treatment of planting materials such as stem cuttings before planting. It is therefore necessary to use new technologies integrating flexible combinations with minimum use of pesticides. Such technologies may include use of host plant resistance, selection of disease-free planting material, field sanitation, improved cultural practices that reduce disease vulnerability, biological control, and developing disease diagnostic kits for major viruses. Many of the virus diseases need to be studied systematically for better understanding and formulation of an effective integrated disease management.

References

FAO, WFP, IFAD (2012) The state of food insecurity in the world 2012. Economic growth is necessary but not sufficient to accelerate reduction of hunger and malnutrition. FAO, Rome

FAOSTAT (2012) FAOSTAT data base. Food and Agriculture Organization of the United Nations, Rome

FAOSTAT (2013) FAOSTAT data base. Food and Agriculture Organization of the United Nations, Rome

Lebot V (2009) Tropical root and tuber crops: cassava, sweet potato, yams and aroids, Crop production science in horticulture no. 17. CABI Publishing, Wallingford, 413 pp

Scott G, Rosegrant MW, Ringler C (2000) Roots and tubers from the 21st century: trends, projections and policy options. Food, agriculture and the environment, discussion paper 31. International Food Policy Institute (IFPRI) and International Potato Institute (CIP), Washington DC, 64 pp

2 Cassava, *Manihot esculenta*

Abstract

Distribution, economic importance, symptoms/damage, host range, biology, survival and spread, and management of insect and mite pests and fungal, bacterial, viral and nematode diseases of cassava (*Manihot esculenta*) using regulatory, physical, cultural, chemical, biological, host resistance, and integrated pest management methods are discussed.

Keywords

Cassava • *Manihot esculenta* • Pests • Diseases • Nematodes

2.1 Introduction

Cassava (*Manihot esculenta*), popularly known as tapioca, belongs to the family Euphorbiaceae and is believed to have originated in South America, most probably Brazil, and was later carried to Africa (where it is now an important subsistence crop), Asia, and the Pacific Islands. Wild species of cassava are found in the natural habitat of Brazil. The genus *Manihot* consists of about 98 species, but none of the existing wild species can clearly be identified as the ancestor of cassava. Recently genetic variation at microsatellite loci has been used for the investigation on evolutionary and geographical origins of cassava from its wild relatives (Olsen and Schaal 2001). The evidence from molecular analysis points toward the possible origin of cassava from Brazilian wild species, viz., *Manihot tristis* and *M. esculenta* ssp. *flabellifolia*. The studies helped in uncovering the unique genetic diversity in *M. esculenta* ssp. *flabellifolia* and *M. esculenta* ssp. *peruviana* and their genetic similarity to cassava, supporting the hypothesis that these two species might be the ancestors of cassava. Genetic erosion of wild *Manihot* species is occurring at a fast rate in Brazil due to deforestation.

Cassava is native to South America and is grown throughout the tropics, with Brazil and Zaire being among the largest producers. It is a perennial shrub of approximately 2 m in height (Fig. 2.1). During the crop production process, it is grown as an annual. It is propagated vegetatively from stem cuttings of approximately 25 cm in length. Because the plants are heterozygous, seeds are only used in the breeding process. Planting is done by hand at densities of 10,000 plants/ha. The most common pests of cassava are weeds and systemic diseases, which are carried from one planting to the next in the cuttings.

P.P. Reddy, *Plant Protection in Tropical Root and Tuber Crops*,
DOI 10.1007/978-81-322-2389-4_2

Fig. 2.1 Cassava plant and tubers

Starch is stored in enlarged roots with 25–40 % starch being common. Harvesting begins 8–14 months after planting with the entire plant being uprooted by hand. Yields can range from 7 to 30 t/ha. Plants can be left unharvested for more than one season, with the roots becoming larger during the following season. This is not recommended for production in the United States, since root quality is reduced considerably during the regrowth and aging process. Cassava roots are very perishable with a shelf life of only a few days. Careful handling and storage under high humidity can prolong the shelf life by 1 or 2 weeks. Although it is not commonly practiced in the United States, young tender shoots are consumed as pot herbs in some countries.

The presence of hydrocyanic glucosides (HCN) in all plant parts presents some problems in marketing cassava. Selections have been made from both chance seedlings and in breeding programs which are low in HCN. These are the only types sold in US markets. More than 1 kg of unprocessed roots would have to be consumed before lethal doses of HCN would be reached. Peeling and boiling in water are common methods of removing a large proportion of the HCN in the roots. Other postharvest problems with cassava include proper handling and storage of cuttings under frost-free conditions.

Roots are usually peeled and boiled or baked. Commercial processing of cassava is limited to packers of frozen, peeled roots, which are marketed in the United States in packages like frozen French fries. This convenience pack may have the potential for expanded utilization of cassava in this country and Europe. Deep-fried chips, like potato chips, are produced and marketed in the Miami, Florida, area. Deep-fried root pieces are offered in Miami Latin restaurants under the name of "Miami fries." Cassava starch, known as tapioca, has limited potential for expansion. Even though cassava flour can be used as a partial substitute for wheat flour in the production of bread, market economics restrict this process to countries where wheat is an import commodity. The future of cassava production in the United States is limited by the bulky nature and perishability of the cuttings. The development of cassava which can be grown from seed has great potential. Being a diploid, it should not be difficult to develop material which breeds true to type and germinates uniformly.

Cassava is grown in an area of 19.641 million ha globally with an annual output of 252.204 million tons of tubers with productivity of 12.84 t/ha (Table 2.1) (Fig. 2.2). Here it is interesting to note that though India is not having a major area under this crop, its productivity is the highest in the world (38.58 t/ha in 2012).

The major problem in cassava cultivation is the lack of availability of disease-free planting materials of high yielding, high starch, and cassava mosaic disease (CMD)-resistant varieties. The old varieties are to be replaced with such novel potential lines in order to boost the per hectare yield of starch without escalation in the cost of production. Cassava has been identified as the only crop for industrial application and poverty alleviation in rural areas by providing job

Table 2.1 Area, production, and productivity of major cassava-producing countries in the world in 2011

Region	Country	Production ('000 t)	Area ('000 ha)	Productivity (t/ha)
Africa	Nigeria	52,404	3,737	14.02
	Congo (ex. Zairo)	15,569	2,171	7.17
	Ghana	14,334	1,072	13.36
	Angola	14,241	889	16.01
	Africa total	**140,966**	**13,047**	**10.80**
Asia	Indonesia	24,010	1,183	20.30
	Thailand	21,912	1,135	19.30
	Vietnam	9,876	560	17.63
	India	8,076	221	36.48
	Asia total	**76,681**	**3,913**	**19.60**
America	Brazil	25,442	1,741	14.61
	Paraguay	2,454	181	13.53
	Colombia	2,264	223	10.17
	Peru	1,117	94	11.83
	America total	**34,363**	**2,668**	**12.88**
	World total	**252,204**	**19,641**	**12.84**

Source: FAOSTAT (2012)

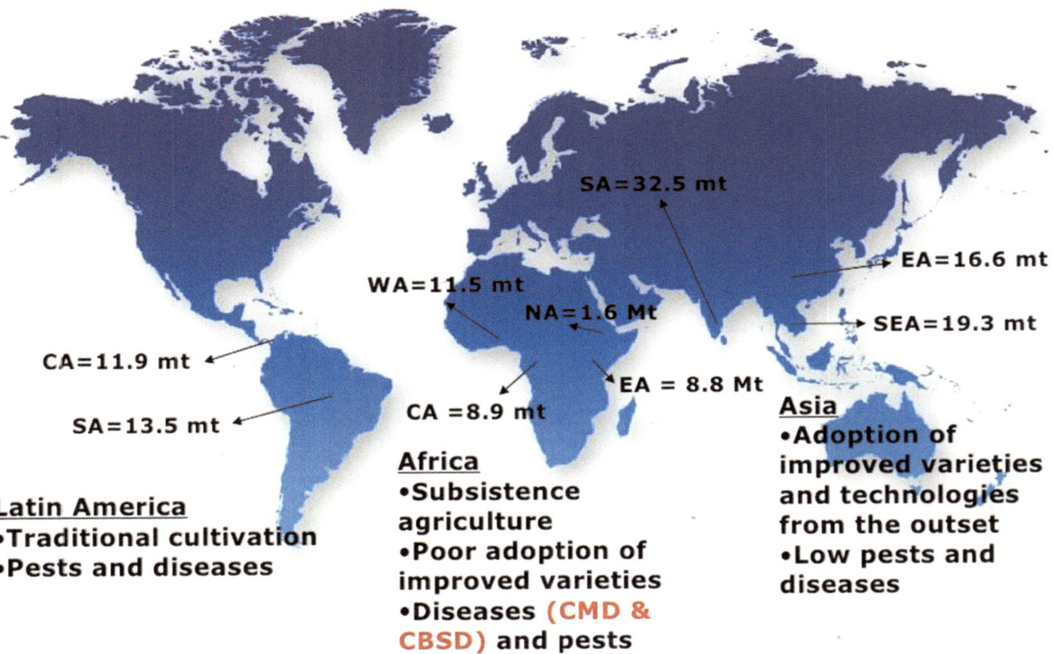

Fig. 2.2 Productivity of cassava in different countries

opportunities as well as food and feeds. Concerted effort initiated by researches, farmers, industrialists, and the processing agencies in this direction is expected to give fruitful results in cassava production in the country.

An economic crop of cassava can be obtained even under marginal conditions or on exhausted soils. The crop requires very little attention after planting. It has draught tolerance (mainly due to in-built mechanism to shed or drop the leaves

Table 2.2 Cassava-growing states, regions, and production systems in India

State	Production system	Area/region
Kerala	Upland rain-fed	Entire state
	Lowland irrigated	Entire state
Tamil Nadu	Upland rain-fed/lowland irrigated	Kanyakumari district
	Plains rain-fed	Salem, Erode, Namakkal, Dharmapuri, Villupuram, Cuddalore, Dindigul, Theni districts
	Plains irrigated	Salem, Erode, Namakkal, Dharmapuri, Villupuram, Cuddalore, Dindigul districts
	Hills rain-fed	Kolli, Pachamalai, Kalvarayan, Sidheri, Kadambur hills, etc.
Andhra Pradesh	Plains rain-fed	East Godavari, agency areas of Srikakulam, Visakhapatnam, Vijayanagaram districts
NEH region	Shifting (Jhum) cultivation, homestead-mixed stand rain-fed	Assam, Meghalaya, Tripura, Nagaland, Manipur
Maharashtra	Plain irrigated	Nasik, Beed, Latur, Sholapur, Pune, Sangli, Satara districts

under adverse moisture conditions) and produces more calories per unit area. The tubers contain 25–30 % starch, 1.5 % protein, and 2.1 % minerals. In India, cassava is cultivated in an area of 0.226 million ha, producing 8.746 million tons of tubers with an average yield of 38.58 t per hectare (Table 2.2). The area and production under cassava followed the global trends in the early nineties, and at present, stagnation is being observed. Kerala where the crop was first introduced in India accounted for 32 % of area under cassava (0.072 m ha). Tamil Nadu accounts for 62 % of area (0.140 m ha) and 1.36 % of area are in Andhra Pradesh (0.003 m ha). The Salem belt of Tamil Nadu and Samalkot belt of Andhra Pradesh are known for cassava as an industrial crop. The major cassava-producing states include Tamil Nadu, Kerala, Andhra Pradesh, Meghalaya, Pondicherry, Nagaland, Assam, and Karnataka (Table 2.3).

2.2 Insect and Mite Pests

2.2.1 Tobacco Whitefly, *Bemisia tabaci*

Among the 37 species of whiteflies identified, *B. tabaci* is the most important and ubiquitous pest inflicting directly and indirectly to the plants. Storey (1936) reported the role of *B. tabaci* as a vector of African mosaic disease in cassava, and it was confirmed by Chant (1958). There is a cassava-specific population of *B. tabaci* christened as "the Indian cassava biotype" that acts as a vector for Indian cassava mosaic disease (ICMV), but the sweet potato biotype is a non-vector of mosaic disease in cassava. Climatic factors have a pivotal role in their population buildup. Higher temperature and lower humidity are the congenial conditions for their fast multiplication. Pillai and Daniel (1979) recorded lower population of whitefly during August–September, with the lowest in August and higher during February and March with highest in March.

2.2.1.1 Damage

The whitefly is a major pest of cassava, particularly in Africa where it is responsible both for the transmission of plant viruses and, increasingly, for direct damage due to feeding by high populations. They are probably the most damaging insect pest in all cassava-producing regions. Whiteflies suck the sap and devitalize the plant (Fig. 2.3). They cause yellowish speckles with crinkling and curling of leaves. In severe infestation, black sooty mold develops. Adults and nymphs are covered with white waxy and powdery scale.

Table 2.3 State-wise area, production, and productivity of cassava in India during 2012

State	Area ('000 ha)	% share of area in India	Production ('000 t)	% share of area in India	Productivity (t/ha)
Tamil Nadu	140.00	61.76	5,964.49	68.19	42.604
Kerala	72.47	31.97	2,637.20	30.15	36.390
Andhra Pradesh	3.09	1.36	61.80	0.71	20.000
Assam	4.26	1.88	33.50	0.38	7.864
Meghalaya	5.30	2.34	28.18	0.32	5.317
Karnataka	1.00	0.44	12.80	0.15	12.800
Pondicherry	0.22	0.10	5.47	0.06	24.864
Andaman and Nicobar Islands	0.27	0.12	2.05	0.02	7.593
Mizoram	0.07	0.03	1.01	0.01	14.429
Lakshadweep	0.02	0.01	0.06	0.00	3.000
India	**226.70**		**8,746.56**		**38.582**
World	**20,385.206**		**262,585.741**		**12.881**
% share of India in world	**1.11**		**3.33**		

Source: FAOSTAT (2012)

Fig. 2.3 *Left*, whiteflies on leaves of cassava. *Right*, pupa and adults of *Bemisia tabaci*

2.2.1.2 Biology

Eggs are tiny, oval-shaped, about 0.25 mm in diameter, and stand vertically on the leaf surface. Newly laid eggs are white and then turn brownish. They are deposited on the underside of leaves, sometimes in a circle or oval-shaped patterns.

Larvae are transparent, ovate, and about 0.3–0.7 mm in size, and they move around on the plants looking for a feeding site upon hatching.

Pupae are dirty white and surrounded by wax and honeydews. During this stage, the red eyes of the emerging adults are visible.

Adults are about 1 mm long with two pairs of white wings and light yellow bodies. Their bodies are covered with waxy powdery materials. They are found feeding on top of the plants. A female can produce as many as 200 eggs in her lifetime and mating is not necessary. It takes about 40 days to develop from egg to adult.

2.2.1.3 Management

High reproductive potential, broad host range, and resistance to insecticides are some of the constraints faced in their management. The most feasible method for the management of whitefly is proper field sanitation and cultivation of resistant varieties. Removal of crop residues and rouging of infested plants also check the risk of carryover population.

1. *Physical Methods*
   (a) *Yellow Sticky Traps*: Place 1–4 yellow sticky cards per 300 m^2 field area. Replace traps at least once a week. It is difficult to determine the population of newly trapped whiteflies on a sticky card to those previously trapped ones.

   To make your own sticky trap, spread petroleum jelly or used motor oil on yellow painted plywood, 6 cm × 15 cm in size or above. Place traps near the plants, preferably 25 cm away from the plant, to ensure that the leaves will not stick to the board. Alternatively, the traps can be hung and positioned at 61 cm zone above the plants.
2. *Cultural Methods*
   (a) *Intercropping*: A 2-year experiment in Cameroon found that intercropping cassava with maize and cowpeas was associated with a drop of 50 % in the adult whitefly population and a 20 % reduction in the incidence of cassava mosaic disease.
   (b) *Botanicals*: The use of CTCRI bioformulation "Nanma" (prepared from cassava plants) is very effective against this pest.
   Neem Leaf Extract: Pound 1–2 kg of neem leaves gently. Place in a pot. Add 2–4 l of water. Cover the mouth of the pot securely with the cloth and leave it as such for 3 days. Strain to get clear extract. Dilute 1 l of neem leaf extract with 9 l of water. Add 100 ml of soap. Stir well and spray.
   Neem Oil Extract: Put 5 kg of finely ground seeds into a basin. Add hot water little by little until it is possible to knead the mixture. Knead and press the mixture to get the oil. Approximately 650–750 ml oil is extracted from this mixture. For the spray material, add 15 ml of neem oil to 1 l of water and 5 ml of soap.
   Neem Seed Powder Extract: Remove shells and pulps from dried and matured seeds. Pound the seeds gently in such a way that no oil comes out. Once done, add 50 g of powder in a pail of 1 l of water. Stir well for about 10 min and steep for at least 6 h but not more than 16 h. Stir it again for another 10 min. Strain, add 5 ml of soap, stir well, and spray.
   Neem Oil and Soap Spray: Add 15 ml of neem oil into 1 l of soapy water. Constantly shake the container or stir the extract while in the process of application to prevent oil from separating.
3. *Biological Methods*
   (a) *Parasitoids*: *Encarsia transvena* is the most dominant and potential parasitoid of *Bemisia tabaci* on cassava. The field release of endoparasitoid, *Encarsia formosa*, gave 69.53 % parasitization during August and 76.22 % during September.

   Two species of aphelinid parasitoids, *Eretmocerus mundus* and *Encarsia sophia*, were identified on *Bemisia tabaci* on cassava.
   (b) *Pathogens*: *Beauveria bassiana* attacks whiteflies and can be a useful biocontrol agent in situations where the crop is grown in high humidity conditions.

2.2.2 Spiraling Whitefly, *Aleurodicus dispersus*

This sap-feeding insect is a native of Caribbean Islands and Central America. *A. dispersus* is a polyphagous pest infesting on a wide range of plants. Incidence of spiraling whitefly in India was first reported in 1994 at Thiruvananthapuram and subsequently from five states in peninsular India on a variety of crops. A survey conducted in Kerala and Tamil Nadu revealed that the whitefly incidence was prevalent from September to May and its intensity of infestation was high during January to April. Maximum infestation was observed in February (400 insects/leaf) and

Fig. 2.4 Spiraling whitefly infestation on cassava leaf

minimum in September (average 3 insects/leaf); however, infestation found declined on the onset of monsoon (June). The team reported a total of 56 plant species belonging to 35 families including fruit crops, horticultural crops and garden plants, and plantation crops as its alternate hosts.

2.2.2.1 Damage

The spiraling whitefly is snow-white, seen particularly on the lower surface of the leaves *en masse* (Fig. 2.4). Female lays smooth yellow to tan oval-shaped eggs, 0.3 mm in length. Adult lays eggs in an irregularly spiraling pattern that is associated with a trail of white wax. It is from this that the whitefly derives its common name, spiraling whitefly. Upon hatching, crawlers (first instar larvae) move to find an appropriate leaf vein for feeding and settle. As they mature, they develop waxy tuffs. The first instars are mobile, while the latter three are permanently attached to the leaf (sessile). Immature and adult stages suck sap, causing yellowish specks on the leaves and damage the host plant. Severely infested leaves are covered with black sooty mold due to the secretion of honeydew by whitefly, and that hinders the photosynthesis of the host plant. Premature fall of leaf is one of the symptoms of its severe infestation. The insects are numerous mainly in the dry season.

2.2.2.2 Biology

After mating, females of the spiraling whitefly lay eggs on the undersurface of leaves. The eggs occur in spiral patterns (like fingerprints) of white material secreted by the insect on the leaves (Fig. 2.4). Adult is 2 mm long, white in color due to the powdery or waxy scales all over the body and wings. They congregate on both upper and lower surfaces, but preferred to feed and lay eggs on the lower surface of the leaf. Male is bigger than female. Eggs are pale yellow, laid singly in a spiraling manner, and are covered with white waxy substances. It is firmly attached on the leaf by a minute stalk. Egg laying starts 2–3 days after emergence, and on an average, 110 eggs are laid by a female. Longevity of adults varies from 18 to 26 days with a mean of 22 days. Number of eggs per egg mass ranges from 5 to 22 with a mean of 10.6. Adults are free-moving and able to fly.

2.2.2.3 Spread

The spiraling whitefly spreads by active flight and by being transported on stem of planting materials.

2.2.2.4 Host Range

There are 72 host plants that have been reported in Kerala. A total of 56 plant species belonging to 35 families including fruit crops, horticultural crops and garden plants [fruit trees (citrus, banana, plantains, papaya), vegetables, and ornamental plants], and plantation crops have been reported as its alternate hosts.

2.2.2.5 Management

The following control measures have been recommended for the management of spiraling whitefly:

- Spraying of dimethoate 30 EC at 0.05 % or biomite or profenofos.
- Neem oil and cotton seed oil at 0.01 % were also found to cause considerable mortality of the different stages of the pest.
- Insecticidal soap at 2.5 % also deters the adults of the spiraling whitefly.
- Colored sticky traps can be effectively used for controlling this notorious pest.

1. *Biological Methods*
   (a) *Predators*: Surveys throughout Tamil Nadu during 1997–2000 revealed the occurrence of 14 species of predatory insects, among them a nitidulid predator, *Cybocephalus* sp., and coccinellid, *Axinoscymnus puttarudriahi*, which were found to be predominant and reduced the population of spiraling whitefly in Tamil Nadu. The same species have been reported earlier to attack *A. dispersus* on guava in Indonesia. *A. puttarudriahi* has been recorded as a potential predator on *A. dispersus* in Sri Lanka.

   The important predators recorded from the field are *Menochilus sexmaculatus*, *Cybocephalus* sp., *Scymnus quadrillus*, *Axinoscymnus puttarudriahi*, *Pharascymnus hornii*, and *Oligota* sp.

   The predators like *Cybocephalus* sp., *Cryptolaemus montrouzieri*, and *Mallada astur* can also be effectively used for the biocontrol of spiraling whitefly.

   (b) *Parasitoids*: Several potential parasitoids have been reported suppressing spiraling whitefly population. Among them, *Encarsia* sp. and *Eretmocerus* sp. are very common in India. Although chemical pesticides are in use to manage the pest, this has been highly discouraged due to an array of reasons, including the deterioration of the natural enemies.

   In Minicoy Islands, the parasitization of spiraling whitefly by *Encarsia* spp. has been found to the extent of 59.4 % on tapioca during March 2000 (Ramani and Bhumannavar 2002).

   A new exotic, aphelinid, solitary, endoparasitoid, *Encarsia* sp., near *meritoria* was found parasitizing the nymphs of *A. dispersus* in Tamil Nadu. The parasitoid identified was closely related to *E. meritoria* and *E. haitiensis*. The occurrence of this new parasitoid showed the establishment of the exotic parasitoid which might have been introduced along with the pest *A. dispersus*. Establishment of *E. haitiensis* parasitoid on *A. dispersus* in new locations and higher parasitism of more than 80 % were found by several workers (Fig. 2.5).

   The preliminary studies on artificial release of the parasitoid, *Encarsia meritoria*, at orchard and insectary in Tamil Nadu Agricultural University showed the establishment of the parasitoid in new locations within a period of 3 months of initial release with the maximum parasitization of 70 %. The parasitization level in location-I was 69.53 % during August 2000 which ultimately resulted in population reduction of *A. dispersus* to nearly 70 %. In location II, the parasitism rate during second month after release was 26.14 %, and a maximum of 76.22 % parasitism was recorded during September 2000 (Table 2.4). Further establishment of *E. meritoria* to a widespread area in India may yield a successful management of this exotic pest.

   (c) *Integrated Methods*
   A. dispersus can be successfully managed by integrated management practices using sticky light traps, botanical insecticides, and biocontrol agents.

Fig. 2.5 *Left*, parasitoid *Encarsia haitiensis* of spiraling whitefly. *Right*, mummies of the spiraling whitefly

Table 2.4 Inundative release of parasitoid, *Encarsia meritoria*, for the management of spiraling whitefly on cassava

Month/fortnight	Per cent parasitism Location I (orchard)	Location II (insectary)
July 2000 I	16.14	19.10
II	39.20	26.14
August 2000 I	51.08	35.19
II	69.53	47.05
September I	62.88	60.15
II	43.72	76.22
October 2000 I	57.67	59.20
II	48.00	52.78

2.2.3 Stem Mussel Scale, *Aonidomytilus albus* (Hemiptera: Diaspididae)

2.2.3.1 Distribution

The pest is distributed in West and East Africa, Argentina, Brazil, India, Madagascar, Mexico, Taiwan, West Indies, and the United States.

2.2.3.2 Damage

The cassava white scale is found mainly on cassava stem surfaces. The females are wingless, firmly attached to the stems, and covered with white material. The males have wings.

Gray, elongate adult scales, 2.5–3.0 mm long and pale nymphs, infest the lower stems of the plant (Fig. 2.6). Infestation may become severe in cassava cuttings kept for planting; the consequent drying up reduces germination success. The insect sucks sap from cassava stems. This causes the stems to lose a lot of water and die.

This mussel-shaped soft scale is a serious problem in the field crop as well as during the storage of planting materials. The scales encrust onto the stem and multiply rapidly. Due to sucking of sap, the stems dry up and could not be used as planting material.

2.2.3.3 Biology

Eggs are deposited under the upper scale covering of the adult. After oviposition, body of the female shrinks and shrivels up. Incubation period is around 4 days. The first instar larvae crawl over the stem and later lead a sedentary life. Wind current facilitates the distribution of crawlers to the nearby plants. The cast skins of each instars deposit over its body. Only females are sedentary while males are good fliers. The life cycle is completed in 20–25 days. Infestation occurs in dry

Fig. 2.6 White scale infestation on stem of cassava

situations, and it aggravates during prolonged moisture stress.

2.2.3.4 Spread

Males of the cassava white scale can fly. However, the pest spreads mainly by wind and the transport and planting of infested stem cuttings.

2.2.3.5 Management

1. *Cultural Methods*: Clean healthy cuttings should be used for planting. Storing the cuttings in hygienic conditions and vertically in shade (Fig. 2.7) is also effective.
   (a) *Botanicals*: The use of CTCRI bioformulation "Nanma" is very effective against this pest.
2. *Chemical Methods*: Spraying during storage or dipping the cuttings in 0.1 % demeton-s-methyl, 0.5 % parathion, 0.05 % dimethoate, or 0.1 % malathion is recommended. Dipping the sets in dimethoate, monocrotophos (0.05 %), or methyl demeton (0.05 %) for 10 min is also effective.
3. *Biological Methods*
   (a) *Predators*: The scale insects are heavily preyed upon by Coccinellids *Chilocorus nigrita*, *Scymnus guimeti*, *Horniolus guimeti*, *Cheilomenes sexmaculata*, and *Stethorus parcepunctatus*.

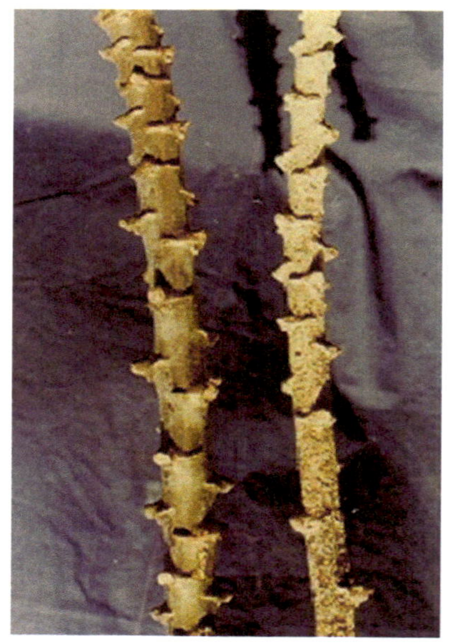

Fig. 2.7 Storing of cassava stems vertically in shade

   (b) *Parasitoids*: The scale insects are parasitized by two Hymenopteran parasitoids *Aspidiophagus* sp. and *Signiphora* sp.
4. *Host Resistance*: Cassava cvs. Sree Harsha, Sree Apporva, and Sree Atulya were found field resistant, while CO 1 and Indira Cassava were tolerant, and Sree Prabha exhibited field-tolerant reaction.

Fig. 2.8 White peach scale infestation on cassava stems

2.2.4 White Peach Scale, *Pseudaulacaspis pentagona*

2.2.4.1 Distribution

It is cosmopolitan in distribution.

2.2.4.2 Damage

The scale is most often seen in large numbers on the bottom of stems; there are often so many that farmers think a white fungus affects the plants (Fig. 2.8). Circular papery female scales (2 mm in diameter) and long (0.75 mm) male scales can be seen on the stem and leaf bases. Weak or old plants are particularly susceptible. The scale feeds on plant sap, and infestations cause leaf yellowing with a loss of healthy growth. Heavy infestations can result in stunting. Severe infestations on cassava have been reported from the western coast of Guadalcanal.

2.2.4.3 Biology

The nymphs emerge from eggs beneath the female scale and begin to feed. They secrete cottony threads, but these disappear when a thin skin is secreted over the nymph. At the first molt, the insects become disk-like. Cast skins at each molt are added to the scale. Males molt 3 times and then pupate. Females mate after the second molt and begin egg laying.

2.2.4.4 Host Range

Host range of white peach scale includes chili, bell pepper, and cassava in Solomon Islands.

2.2.4.5 Management

1. *Cultural Methods*
    - Cut out stems of plants infested by white peach scale and burn them.
    - Do not plant cuttings from plants infested with white peach scale.
    - Avoid planting new crops downwind from those infested with white peach scale, as the crawlers will spread to the new crop through the wind.
2. *Chemical Methods*: Insecticides should be avoided, unless control by natural enemies is ineffective. Insecticides are not always effective against females; they live under a protective scale under which they live.
    - Use soap sprays (5 tablespoons of soap in 4 l water) or white oil (petroleum jelly). These sprays work by blocking the breathing holes causing suffocation and death.

Spray the stems: the oils must contact the scales. White oil can be obtained as a commercial product or made by mixing together 1 cup cooking oil, 2 cups water, and 1 teaspoon dishwashing liquid. Dilute the mixture at 6 teaspoons per liter of water and spray on the infested parts. The addition of malathion (2 ml/l water) is useful.

- Use synthetic pyrethroids (e.g., lambda cyhalothrin or cypermethrin). These are useful against crawlers.

3. *Biological Methods*

Species of ladybird beetles and lacewings are known to feed on white peach scale in other parts of the world, and probably these insects do the same in Solomon Islands. *Encarsia diaspidicola*, a wasp, has been successfully released in Samoa and more recently in Hawaii with good results.

2.2.5 Black Scale, *Parasaissetia nigra*

Besides *Parasaissetia nigra*, *Saissetia coffeae*, *S. miranda*, and *Ceroplastes cirripediformis* have also been reported on cassava stems and are similar in appearance to black scale, i.e., hemispherical and waxy in appearance.

2.2.5.1 Distribution

It is distributed in most warm countries.

2.2.5.2 Damage

Colonies of dark-brown to black-domed soft scales 2–4 mm long can be seen on the stems. *P. nigra* infests twigs and branches (Fig. 2.9). The scales produce copious, sticky honeydew on which sooty molds develop, coating the plant and nearby surfaces; new growth is sometimes stunted and defoliation may be evident. The honeydew produced sometimes attracts ants.

2.2.5.3 Biology

The scale is the female insect; beneath the scale, she lays eggs which hatch into crawlers that migrate over the plant before settling down to feed. Although females do not move around, they retain their legs. Female crawlers molt twice before becoming adult, and males, which have two wings, molt 3 times.

2.2.5.4 Management

Cassava cuttings must not be allowed to develop heavy infestations of scale insects since they will produce poor plants when planted or do not germinate. Healthy plants will not suffer heavy yield loss due to scale insects. Kerosene emulsion and parathion have been recommended for soft scale control.

Orange-spotted ladybird, *Orcus australasiae*, seems to feed readily on the soft scale insects such as black scale, *P. nigra* (Fig. 2.10).

Fig. 2.9 Cassava black scale infestation on leaf and stem

Fig. 2.10 Orange-spotted ladybird predating on black scales (encountered an ant)

Fig. 2.11 Cassava root scale on underground cassava stem

Metaphycus helvolus (Hymenoptera: Encyrtidae) was more effective against *P. nigra* and reduced the population to commercially unimportant levels, eradicating the scale in many areas.

2.2.6 Root Scale, *Stictococcus vayssierrei*

2.2.6.1 Distribution

The cassava root scale seems to be restricted to parts of Central Africa.

2.2.6.2 Damage

The cassava root scale lives underground on the storage roots, feeder roots, and submerged stems of cassava. The insects are reddish-purple or brown in color, oval in shape, and look like ticks on cassava (Fig. 2.11). They lack wings and are attached firmly to the plant roots.

Young plants affected by the insects in Cameroon showed stunted growth, extensive leaf-fall, wilting, and tip dieback. Cassava root scale attack causes the storage roots to be smaller than normal and deformed. The insect attacks were more severe when cassava was planted on the flat surface rather than on ridges. There were more severe infestations when cassava was intercropped with other plants such as maize and groundnuts than when planted alone. Tubers were covered with the scale insect making them unattractive at market.

2.2.6.3 Host Range

In addition to cassava, the cassava root scale attacks yam, cocoyam, and groundnut.

2.2.6.4 Spread

In Southern Cameroon, Dejean and Matile-Ferrero (1996) reported that the ant *Anoplolepis tenella* was the principal agent for the dispersal of *S. vayssierei*. It has been demonstrated the active role of *A. tenella* workers in the transport and dissemination of scale crawlers within and between host plants.

2.2.6.5 Management

In an effort to control ant colonies, boric acid-sucrose solution showed a significant reduction of *A. tenella* population within 6 weeks. The control of *S. vayssierei* by spraying boric acid-sucrose solution in crop field is contingent to the reduction of the density of the associated ant *A. tenella*.

The bitter cassava genotype "Sadisa" can be used to reduce root scale infestation and damage.

2.2.7 Mealybugs, *Phenacoccus marginatus*, *P. manihoti*, *P. herreni*

2.2.7.1 Distribution

P. manihoti is found throughout most of the cassava-growing regions of Africa, while in the Americas, it is confined to Bolivia, southwestern

Fig. 2.12 Mealybug infestation on cassava leaf and stem

Brazil, and Paraguay. *P. herreni* is reported from northeastern Brazil, Colombia, the Guyanas, and Venezuela. Neither species is present in Asia.

2.2.7.2 Damage

Papaya mealybug (*P. marginatus*) is a very noxious insect pest on cassava, and this soft-bodied insect can easily be identified by the presence of white powdery waxy substances all over its body. They are seen in clusters on the stem, petiole, and leaf, particularly on the ventral side. Infestation is very high during warm and dry periods. Females are slightly yellow, measuring approximately 2.2 mm long and 1.4 mm wide, wingless, and can move by crawling or through air currents. A series of very short waxy filaments are seen over its body. Males are winged and can fly, but do not feed. Adults and nymphs suck juice from their host plant.

It feeds on the sap of plants and injects a toxic substance into its host, resulting in chlorosis (yellowing), plant stunting, leaf deformation, early leaf drop, etc. They produce honeydew, and this sticky layer is a perfect growth medium for a black fungus commonly known as sooty mold, which cover the entire leaf and cut the light available for photosynthesis. This will adversely affect the tuber yield. Mealybug feeding is restricted to the aerial part of the plant only.

The cassava mealybugs suck sap from cassava leaves and shoot tips (Fig. 2.12). The pest reduces the lengths of the internodes and causes the leaves to clump together into "bunchy tops." Leaves and shoots become distorted and yellow due to mealybugs feeding on them. The pest also distorts the stems, dries up the leaves, and eventually, if the attack is particularly severe, defoliates the plants. The damage is more severe in the dry than in the wet season.

2.2.7.3 Biology

Eggs are oblong, golden-yellow, and enclosed in woolly ovisacs located at the posterior end of the adult females. The eggs may hatch as soon as they are laid, giving the impression that young are born rather than hatched. The crawlers are flattened, oval, light yellow, six-legged insects, and with smooth bodies. Soon after feeding, they emit a white waxy covering over their bodies giving them a mealy appearance. The female body is oval, rose-pink, and covered with white waxy filaments. The tiny-winged male is needed for reproduction but only lives for a few days.

Populations of the cassava mealybug are all females. The insect lays eggs without mating. A single insect can therefore start a severe infestation. You may notice masses of golden-yellow eggs within the colonies of the pest. The pest is more abundant in the dry than in the wet season.

Female lays 100–600 eggs in a cluster which are greenish yellow in color. Egg laying is completed over a period of 1–2 weeks, and hatching occurs in about 5–10 days. On hatching the crawlers come out and move around in search of suitable feeding sites in sheltered areas. The young stages are very minute and often go unnoticed, but they are highly infective. Life cycle is completed in about 1 month. Adult females live for about 30 days.

2.2.7.4 Host Range

It is a polyphagous pest and has been recorded over 55 host plants. Some of them are papaya, cassava, mulberry, hibiscus, citrus, cotton, tomato, eggplant, peppers, beans, peas, sweet potato, mango, cherry, pomegranate, rubber, etc.

2.2.7.5 Spread

Newly hatched cassava mealybugs are tiny, light, and easily blown by wind from plant to plant. Also, the pest survives on stem surfaces and is spread by being carried by farmers on cassava stem planting materials.

The spread of infestation is achieved through planting materials, wind, water, rain, clothing, vehicle, etc. Passive transport of the pest is also possible through the field equipments, animals, or people moving during field operations. Persons who enter in the mealybug-infested field will be unknowingly carrying scores of crawlers through their dress and other exposed body parts. Since their size is too small to be identified by the naked eye, they often go unnoticed and may contaminate when the same person enters into uninfested field. Under favorable environmental conditions, mealybug can build up huge numbers in a very short time and cause considerable damage to its host.

2.2.7.6 Management

1. *Cultural Methods*: Select pest-free sets for planting. Burn the severely infested plants.
   (a) *Neem Oil and Soap*: Use a mixture of neem oil and soap solution for spraying. Add 20 ml of neem oil and 2–5 ml of soap solution in a plastic bucket and make up to one liter. Vigorously shake the solution till it looks milky white with foam. Dip the sets of cassava for 5 min in this solution, so larval stages of the mealybugs will be washed off and killed.

   Spray the same solution in mealybug-infested cassava field also. The nozzle of the spray should be turned toward the lower side of the leaf and ensure full coverage by the spray fluid. A second spray after 15 days may ensure the death of residual population.

   If the infestation is mild, dip a small cotton ball in mild soap solution and swab on the infested area. This can be done twice at an interval of 5 days.

   (b) *Cassava Biopesticide*: The biopesticides formulated at the Central Tuber Crops Research Institute, Thiruvananthapuram, Kerala, from cassava leaves such as "Shreya" and "Nanma" are very effective against cassava mealybug.

   Petroleum ether extracts of seeds of cassava, yam bean, and neem oil are also effective to control mealybugs (Table 2.5).

   Planting of flowering herbs as intercrops and/or perimeter crops helps in the buildup of the population of natural enemies.

   Control of ants helps in the management of mealybugs as they aid mealybugs to have access to the plant hosts.
2. *Chemical Methods*: Spraying of synthetic insecticides will be done only if really needed. There are a number of natural enemies in the field to check the population of mealybugs,

Table 2.5 Mortality of *Phenacoccus manihoti* in different botanicals

	Mortality of *Phenacoccus manihoti* (%)		
Concentration (%)	Cassava extract	Yam bean extract	Neem oil
5	100	100	100
3	100	100	100
1	100	100	100
0.5	70	87	70
0.05	60	70	70
Control	0	0	0

but the indiscriminate use of chemical insecticides will adversely affect its natural enemies and pave the way for pest resurgence.

Infected cuttings should be dipped or dusted with aldrin, BHC, or DDT. Mephosfolan and parathion sprays or dusts may be applied to the growing point if it is seriously affected by mealybugs and should control the pest.

Insecticides like profenofos 50 EC-2 ml/l, chlorpyrifos 20 EC 4 ml/l, and dimethoate 30 EC 2 ml/l are reported to be effective against mealybugs.

3. *Biological Methods*
   (a) *Predators*: A coccinellid *Cryptolaemus montrouzieri* successfully controlled mealybugs in storage. The growth and reproduction of mealybugs were favored by hot and humid conditions during February to April during the storage (Fig. 2.13). The optimum temperature for *C. montrouzieri* to control mealybugs was found to be between 25 and 30 °C. In storage, it was successfully controlled by two numbers of *C. montrouzieri*/kg tuber coupled with natural parasitization by *Anomalicornia tenuicornis*. During the months of low parasitic activity, four coccinellids were found sufficient for their control. Both these bioagents consumed the mealybugs within 3–4 days.
   (b) *Parasitoids*: Outstanding control of *P. manihotis* was obtained with the encyrtid parasitoid, *Apoanygyrus* (=*Epidinocarsis*) *lopezi*. The wasp (Fig. 2.6) is the most effective natural enemy against the cassava mealybug, and it has controlled the pest in most of Africa (Table 2.6).

Fig. 2.13 Adults of ladybird beetles feeding on cassava mealybugs

Cassava mealybug is also parasitized by encyrtid parasitoid *Anomalicornia tenuicornis* (Encyrtidae) up to 15–20 %, and the activity of the biocontrol agent was higher during December to March.

The wasp used in the biological control of the cassava mealybug prefers mealybugs that are large in size. Large mealybugs are found on vigorously growing cassava plants. Soil improvement practices which promote vigorous cassava plant growth will therefore improve biological control of the mealybug by the wasp.

The mealybug is successfully controlled by the parasitoid, *Acerophagus papayae* (Fig. 2.14), which is mass released to various infested areas. The *Apoanygyrus lopezi* is a promising biocontrol agent.

2.2.8 Striped Mealybug, *Ferrisia virgata*

This is a whitish mealybug with two longitudinal dark stripes, long glassy wax threads, and two long tails.

Table 2.6 Economic impact assessment of biological control of cassava mealybug in Africa

Pest species	Loss (%)	Successful biocontrol agent	% reduction in loss	Savings (US$ million)	Discount rate (%)
Cassava mealybug	40	The encyrtid wasp *Apoanygyrus lopezi*	90–95	7,971–20,226[a]	6

Neuenschwander (2004)

[a]Based on 27 countries in Africa, depending on scenario

Fig. 2.14 *Left*, a parasitoid wasp (*Apoanygyrus lopezi*) on cassava mealybugs. *Right*, parasitoid *Acerophagus coccois* on mealybug

Fig. 2.15 The striped mealybug

2.2.8.1 Damage

It sucks sap, but unlike the cassava mealybug, it does not inject any toxins into the plants; therefore, the attack is not as severe. The striped mealybug occurs on the underside of the leaves near the petioles and on the stems (Fig. 2.15).

Infestations of *F. virgata* remain clustered around the terminal shoots, leaves, and fruits, sucking the sap which results in yellowing, withering, and drying of plants and shedding of leaves and fruits. The foliage and fruit also become covered with large quantities of sticky honeydew which serves as a medium for the growth of black sooty molds. The sooty molds and waxy deposits result in a reduction of photosynthetic area.

2.2.8.2 Management

At the beginning of a local outbreak, severely infested branches should be cut and burnt immediately. It has been controlled by the coccinellid *Cryptolaemus montrouzieri* together with the syrphid *Allograpta obliqua*.

2.2.9 Hornworm, *Erinnyis ello*

2.2.9.1 Distribution

The pest is distributed in North, Central, and South America, Antigua, Barbados, British Virgin Islands, Dominica, Dominican Republic, Grenada, Guadeloupe, Jamaica, Martinique, Montserrat, St. Kitts, St. Lucia, St. Vincent, and Trinidad and Tobago.

2.2.9.2 Damage

Hornworms cause defoliation in cassava. The effect would be more severe under poor soil conditions. If attacked by a heavy infestation during the first 2 months of growth, some plants may die and the crop yields will suffer. The tuber quality is lost, the starch content is reduced, and the tubers are watery, thus increasing the cost of processing.

Hornworm eggs when deposited on the upper or lower leaves hatch within 7 days. The cassava hornworm larvae (Fig. 2.16) feed for 3–4 weeks until they mature on young leaves and growing shoot of the cassava plant by stripping away the foliage. In severe cases, the larva can defoliate the whole plant and the entire crop (Fig. 2.16).

2.2.9.3 Biology

The eggs are very tiny, shiny black, and are found in the crevices of bud, stems, and barks of the plant. A newly hatched larva is pale green with a black horn and is about 6 mm long. The matured larva is whitish-green with yellowish body spots and more pronounced black and red spot on its abdominal segment. The thoracic feet are pinkish and thrice annulated. The abdominal feet are pale reddish with a black band and whitish claspers. A larva can reach a length of 8–10 cm. The pupa is shiny and has pale-orange streaks on the wing cases, feet, and around each abdominal segment. There is also a curved line of the same color at the rear of the thorax. It is usually found on the ground among plant residues. An adult female has pale-gray forewings, while those of male are darker. Both hind wings are red with dark gray marginal bands. The adults fly only at night time, and they tend to migrate in swarms and lay eggs en masse.

2.2.9.4 Management

1. *Cultural Methods*: Handpicking of hornworms and then squashing them is very effective.
2. *Biological Methods*
   (a) *Predators*: In Colombia, a predatory wasp, *Polistes* sp., usually gives excellent control. *Polistes canadensis* predation has been increased by placing 2 m-high "wasp houses" covered in palm leaves in the cassava field.
   (b) *Parasitoids*: In Colombia, *Trichogramma fasciatum* kills 60–100 % eggs of *E. ello* on cassava. Braconid wasp parasitizes cassava hornworm and other insects, thereby reducing the population.
   (c) *Pathogens*: If possible, *Bacillus thuringiensis* sprays should be given to reduce *E. ello* larval numbers without damaging the predator and parasite population. Treatment of the first three instars is the most effective. Controlling breakout of larva can be achieved by spraying

Fig. 2.16 Larva of hornworm and defoliated cassava plants

B. thuringiensis (Bt) on susceptible plants before they become a major problem.

Spray plants thoroughly with Bt when larvae are less than 5 mm long. The larvae are not killed instantly. The pests have to ingest Bt first. It takes about 2–3 days for the pests to die. In the hot tropics, it is more effective to spray Bt in the late afternoon as there are longer and cooler hours ahead. Bt remains longer on the leaf surfaces and survives better in cooler temperature. Spraying in the morning provides a shorter and hotter environment. Do not spray Bt on wet days. Keep Bt formulations in cool storage places.

Baculovirus erinnyis is a hornworm larva virus. The larvae are collected in the field, liquefied in blenders, and mixed with water. The same mixture is then sprayed to the infested plants. Upon application, the virus can kill 90–100 % hornworm larvae. This virus is used to control cassava hornworm larvae in Latin America.

In Brazil, the granulosis virus (GV) is being used on over 2,000 ha to control *Erinnyis ello* on cassava. A standardized formulation consists of 20 ml (18 g) of crude, filtered macerate which for spraying is diluted in 200 l of water/ha. In Columbia, 50–70 ml larval macerate in 200 l of water with 0.2 ml Triton-ACT is sprayed per hectare. The speed of action in Columbia is phenomenal, as over 80 % mortality is recorded in 48 h. In Southern Brazil, a mortality of 90 % is attained by 4 days after spraying of 20 ml of larval macerate. An increase in numbers of natural enemies following GV application is thought to explain the observed spread of infection from sprayed areas.

2.2.10 Thrips, *Frankliniella williamsi* (Thysanoptera: Thripidae)

2.2.10.1 Damage

Thrips are yellow, tiny, elongated insects that are found on the upper and lower surfaces of leaves (Fig. 2.17). Infestations are more severe in the dry season. Both young and adult suck the sap from leaves and cause them to lose their color. If attack occurs early, the young leaves become distorted. Older tissues become blotched and appear silvery or leathery in affected areas, thus hindering photosynthesis. Some species damage growing points by sucking the sap of the plant resulting in loss of vigor and causes reduced yields. Thrips are also vectors of major viral diseases.

2.2.10.2 Biology

Larvae and adults of *F. williamsi* live in the growing points and on young leaves. Thrips insert their eggs in the midrib of the leaf undersurface. The greenish-colored nymphs live near the veins where they undergo two nymphal and two pupal stages. Thrips attack is most frequent during dry periods, and plants recover with the initiation of the rainy season.

2.2.10.3 Management

1. *Cultural Methods*
   - Good field sanitation—removal of weeds and residues of all previous crops.
   - Crop rotation—cultivation of crops (vegetables) which are not host to the pest.
   - Overhead irrigation will help in reducing the pest population during the dry season.
   - An integrated approach is recommended for the management of thrips.
   - Before planting, cuttings can be treated by immersing them in hot water (mixing equal volumes of boiling and cold water) for 5–10 min.
   - Planting the cuttings horizontally and completely underground.
   - Soaking the cuttings in an insecticide solution.
2. *Chemical Methods*: Insecticides such as Regent (Fipronil), Admire, Abamectin, and Vydate L at 5 ml to 4.5 l of water may be applied to both upper and lower leaf surfaces for effective control.

2.2.11 Aphids, *Aphis gossypii* (Homoptera: Aphididae)

2.2.11.1 Damage

Aphids attack the cassava plants at all stages of growth and are usually found in dense clusters on the undersurface of the young leaves, thus causing leaf distortion (Fig. 2.18). In severe cases, aphids

Fig. 2.17 Immature nymph of thrips and damage to leaf

Fig. 2.18 Adults and nymphs of aphids

cause the leaves to become chlorotic, followed by wilting. On young tender stems and growing point, sucking of plant sap from the tissues causes the plant to become weak. Seedlings are weakened and killed when infestation is high, and growth of older infested plants is retarded. Infested leaves curl, shrivel, and may turn brown and die.

Aphids secrete a sweet substance known as "honeydew" while they feed. This substance attracts ants and serves as a substrate for sooty mold (black fungus), thus impairing photosynthesis. Aphids also serve as vector for viruses.

2.2.11.2 Management

1. *Cultural Methods*: Good field sanitation—removal of weeds and plant residues from previous crops.
2. *Biological Methods*: The natural predator, ladybird beetle frequently feeds on aphids. When aphid population is low and ladybird beetles are present, there is no need for chemical control.
3. *Chemical Methods*: This may be applied when the population is high. A contact or stomach insecticide may be used such as Fastac, Decis, or Karate at 6 ml to 4.5 l of water, Sevin 85 % W.P. (carbaryl) at 6 g to 4.5 l of water, or malathion 57 % EC at 15 ml to 4.5 l of water.

2.2.12 Variegated Grasshopper, *Zonocerus variegatus*

2.2.12.1 Damage

The variegated grasshopper feed on cassava plants, chewing leaves, petioles, and green stems. It defoliates the plants and debarks the stems (Fig. 2.19). Groups of adult and nymphal grasshoppers defoliate and sometimes strip the bark of the cassava. The pest damage is more common

Fig. 2.19 *Left*, adult of the variegated grasshopper on cassava. *Right*, cassava plants defoliated and debarked by the variegated grasshopper

on older than on younger cassava plants and is more severe in the dry than in the wet season. In Nigeria the pest occurs during November–March. Repeated defoliation reduces yield.

2.2.12.2 Biology

Adults are usually dark-green with yellow, black, and orange markings on their bodies (Fig. 2.19), and the nymphs are black with yellow markings on the body, legs, antennae, and wing pads. The young nymphs gather in large numbers on weeds and low-growing crops. The female lays their eggs just below the surface of the soil under the shade of the cassava tree. Eggs start to hatch at the beginning of the main dry season.

After mating, female-variegated grasshoppers lay many egg pods just below the surface of the soil. The egg pods look like tiny groundnut pods. The egg-laying sites always have vegetation which casts shade on the ground and keeps it moist, soft, and suitable for egg laying. These sites are usually close to cassava fields and small in surface area. In most of West and Central Africa, adult grasshoppers can be seen in large numbers at such sites, usually between March and May. The eggs start to hatch at the beginning of the main dry season, usually in October and November.

2.2.12.3 Spread

The variegated grasshopper spreads by flying from farm to farm. However, the insect does not fly over long distances. It spreads faster in areas where the forest has been cleared than in thick vegetation.

2.2.12.4 Host Range

In addition to cassava, the variegated grasshopper also feeds on citrus, cashew, cowpea, plantain, vegetables, and many other crops.

2.2.12.5 Management

1. *Cultural Methods*: In any year, the abundance of the variegated grasshopper depends largely on the number of egg pods that survive in the soil during the wet season. The destruction of egg pods will therefore reduce the numbers of the pest. Farmers can locate and mark egg-laying sites early in the wet season. At a later stage, they can then dig up the soil at the sites to expose and destroy the egg pods. The digging up of eggs should be done before the eggs start to hatch early in the dry season, for example, in October, in most of West Africa.

   The variegated grasshopper does not lay egg pods deep in the soil. Therefore, it is easy to dig out the egg pods. However, egg pod destruction needs to be carried out over a wide area in the wet season in order to control the pest effectively. This will require the participation of many farmers on many neighboring cassava farms. If only one farmer destroys the eggs in and around his/her farm, the pest will later invade the farm from the neighboring farms and bushes. Extension agents can organize the community of villages to dig up and destroy the egg pods on as many farms as possible.

Certain weeds, for example, the Siam weed, *Chromolaena odorata*, harbors immature stages of the variegated grasshopper. From the weeds, the pest will move onto cassava plants. You can therefore discourage the pests from gathering in your farm by removing these weeds in your farm.

2. *Biological Methods*: Fungi have been found that kill the variegated grasshopper. The fungi are spread as "spores" which are like tiny seeds. The spores land on a pest, germinate, and the fungus then penetrates the body of the pest, growing and killing it within a few days. When a diseased grasshopper dies, its dead body may remain firmly gripped to the plant (Fig. 2.20) or drop to the ground. "Biopesticides" consisting of fungus spores mixed in oil are being prepared by scientists as commercial products against the grasshopper. The product can be sprayed on weeds such as the Siam weed, *Chromolaena odorata*, to kill newly hatched nymphs which gather in large numbers on the weed. The product can also be sprayed directly on cassava to kill nymphs and adults of the grasshopper on the plant.

2.2.13 Shoot Fly, *Siblua pendula*

2.2.13.1 Damage

This pest (Fig. 2.21) damages the tips of the young plant, causing growth to be stunted. The growing point becomes brown and deformed due to white larvae burrowing inside it. There is a whitish-yellow exudate from the damaged shoot. The growing point may die and secondary shoots develop, which may also be attacked. Heavy infestation takes place at the beginning of the rainy season. Mature plants do not suffer as much as the younger ones.

Fig. 2.20 Nymph of the variegated grasshopper killed by fungal disease

2.2.13.2 Biology

The adult is a shiny blue fly. Eggs are laid among the folded leaves of the growing plant, and after 4 days, the larvae emerge and bore into the plant to feed. The larval stage lasts 20–23 days and pupation, which takes place in the soil, lasts 25 days.

2.2.13.3 Management

1. *Cultural Methods*: Planting when the shoot fly population is low is recommended.
2. *Chemical Methods*: Trials in which the growing point was removed to simulate shoot fly damage have indicated that older plants do not suffer economic losses from shoot fly. Therefore, pesticides should be used to protect young plants. The larvae are reported to be difficult to control, and some attempts have been made to bait the adults with mixtures of malathion (0.15 %) and molasses (5 %); diazinon, dicrotophos, dimethoate, methamidophos, or monocrotophos have been recommended for larval control.

Fig. 2.21 Cassava shoot fly

3. *Host Resistance*: The use of resistant cultivars is recommended. In Brazil, the varieties 14181AC and Ouro do Vale were found to be highly resistant to this pest.

2.2.14 Fruit Flies, *Anastrepha manihoti*, *A. pickeli*

2.2.14.1 Distribution

Anastrepha manihoti is distributed in Brazil, Panama, and Peru, while *A. pickeli* is present in Argentina, Brazil, Panama, and Peru.

2.2.14.2 Damage

The female which is yellow oviposits in the stem tissue, and upon hatching the yellowish-white larvae bore their way into the pith region. Stems bleed a white exudate due to yellowish-white legless larvae boring in the pith 10–20 cm below the growing point. Infection with bacterial stem rot is sometimes associated with fruit fly attack. Heavy attacks and combined effect of insect and pathogen can lead to the death of the main growing point, causing formation of lateral shoots and reduced yields. Yields have been reduced from 4 up to 33 % when infected materials were planted. This pest also reduces the quality of the propagation material.

2.2.14.3 Biology

The flies are yellowish-brown with distinct dark markings on their wings. When at rest, they commonly extend the wings and raise and lower them (Fig. 2.22). They lay eggs in the stem. The larvae are about 5 mm long, wider at the posterior end than at the head, where the sclerotized mouth parts may be seen. Pupation takes place in the soil.

2.2.14.4 Management

1. *Chemical Methods*: Chemical control is probably only necessary when young plants (2–5 months) are severely attacked. The larvae can be killed in the stem by spraying with fenthion at the recommended rates.

2.2.15 Leaf-Cutting Ants, *Atta* sp., *Acromyrmex* sp.

2.2.15.1 Damage

The pest cuts semicircular pieces of leaves and removes the buds and carries them back to their nest (Fig. 2.23). Whole plots of cassava can be

Fig. 2.22 Fruit fly on cassava

Fig. 2.23 Leaf cutting ants on cassava

defoliated. These ants' nests can be identified by the piles of dirt around the entrance holes. Outbreaks usually occur during the first month of crop growth.

2.2.15.2 Biology

The ants live in nests up to 10 m in diameter extending several meters below ground level. A nest is made up of brood chambers in which fungus gardens are maintained. The leaves are chewed and used as soil for the fungus on which the ants feed. The ant colony consists of queen, various casts and the brood.

2.2.15.3 Management

1. *Chemical Methods*: Insecticides are the most effective means of control. Baiting is the recommended method for ant control. Poisoned baits applied over a wide area during dry season are the most satisfactory method of control. The foraging ants carry the poison back to the nest where the rest of the colony is affected and dies. The bait must be attractive and remain so for 2–4 weeks, and the poison must be slow acting, to give the foragers time to carry it back to the nest. If it is necessary to apply bait in the wet season, then it should be water-proofed to prevent decay; 2 % methyl-trichloro-silane is recommended for this purpose. Baits consisting of 0.4 % insecticide (mirex or aldrin are the preferred insecticides)+8.5 % soybean oil+91.1 % dried citrus meal are recommended. The insecticide should be dissolved in the soybean oil and then well mixed with the citrus meal. Commercially prepared baits are available. The farmer should protect his hands while applying bait.

   Fumigating the nests with carbon disulfide, sulfur smoke, or arsenates will also destroy the colony, or alternatively the nest entrance may be treated with aldrin or dieldrin aqueous solution.
2. *Integrated Methods*: This may include digging, removing, and flooding nests, using metal or grease to seal the nests, or painting white bands around trees.

2.2.16 Lacewing Bugs, *Croythaica cyathicollis* (Hemiptera: Tingidae)

2.2.16.1 Damage

Lacewing bugs are usually found in clusters on the undersurface of leaves (Fig. 2.24). Their sucking action results in leaf mottling. The mottled areas eventually become necrotic, and this leads to premature abscission of leaves.

Lace bugs attack mainly during the dry season and are severe during prolonged droughts. Adults are gray and nymphs are white, both found on the underside of the leaves. Damaged leaves have small yellow spots that later turn reddish-brown,

Fig. 2.24 Nymphs and adult lace wing bug

resembling mite damage. The rate of photosynthesis decreases and lower leaves fall.

2.2.16.2 Biology

Eggs are laid on the middle and lower leaves and hatch after 8 days. Nymphal instars last 3, 2, 5, 3, 3, and 5 days approximately, and the adults live for about 50 days. Adults are gray and about 3 mm long; the nymphs are paler and smaller.

2.2.16.3 Management

1. *Cultural Methods*: Good field sanitation—removal of weeds and plant residues from previous crops.
2. *Chemical Methods*: Decis, Karate, or Fastac at 6 ml to 4.5 l of water and Sevin or Padan at 5 g to 4.5 l of water are recommended for chemical control of lacewing bugs. Fenitrothion, applied at recommended rate, should control this insect.

2.2.17 Cutworms

2.2.17.1 Damage

1. *Surface Cutworms*: The black cutworm feeds on the basal part of the stem near the surface of the soil. The larvae are greasy gray to brown with faint lighter strips.
2. *Climbing Cutworms*: These climb the stem feeding on the buds, foliage, and stem causing the upper part of the plant to wilt and die. The larva is dark gray to black with lateral yellow strips (Fig. 2.25).

Fig. 2.25 Cut worms on cassava

3. *Subterranean Cutworms*: These remain in the soil feeding on the roots and underground parts of the stem. They cause damage to young plants killing up to 50 %.
4. *Crickets*: The black cricket or the mole cricket causes most of the damage by cutting the young shoots right after germination.

2.2.17.2 Management

Soil treatment with aldrin/carbofuran or poison baiting (consisting of 10 kg sawdust + 8–10 l of water + 500 g sugar/molasses + 100 g trichlorfon to be scattered over 0.5–1.0 ha) is effective for the management of cutworms.

2.2.18 White Grubs, *Leucopholis coneophora*

It is primarily a pest of coconut, but infestation is also noticed on cassava cultivated in coconut garden. The pest infestation is severe in light sandy soil area where rain fall is high.

2.2.18.1 Damage

The grubs, whose adult stage is a beetle, attacks the cuttings and roots of cassava plant (Fig. 2.26). Damages are characterized by the destructions of the bark of the cuttings and the presence of tunnels in the woody part. These cuttings may rot and die. Young plants often suddenly wilt and die. White grub attacks plants which become pale-yellowish and slowly wither and dry. The grubs feed on developing roots and commonly in young plants.

2.2.18.2 Biology

Female lays eggs during June–July in soil to a depth of 7.5–15.0 cm very close to the root of grass. Incubation period is about 20 days. Newly emerged grub feeds on the tip of the root, and later instars feed even on mature roots of coconut or cassava. Infested plant looks pale and dries up later. The grubs can be collected from the root zone of the dried plant. By October–November, the grubs mature, and they move deeper into the soil for pupation. Pupation takes place in an earthen cocoon, and it lasts for almost 30 days. During the onset of monsoon (June), the fully developed adults emerge out.

Fig. 2.26 White grub larva, adult beetle, and damage to cassava tuber

2.2.18.3 Management

Cultural Methods: Deep plowing exposes early stages of the grubs to sun light and to their natural enemies. This pest can also be managed by collecting and destroying the adults during monsoon by setting light trap.

Chemical Methods: Application of Chlorpyrifos 0.04 % around the base of the plant kills the pest. The pest can also be managed by spraying with endosulfan or quinalphos as the adults emerge after the first monsoon shower (April to June) when they feed on the leaves and die. Soil application of carbaryl (10 %) dust is useful for controlling this pest.

2.2.19 Lace Bug, *Vatiga manihotae*

2.2.19.1 Distribution

The pest is distributed in South America.

2.2.19.2 Damage

Early defoliation is caused by large numbers of lace bug adults and nymphs feeding on the underside of the leaves. The leaves develop yellow spots which later turn brown. Lace bugs are recognized by the lacy reticulate pattern on their pronotum and hemelytra; the prothorax is expanded.

2.2.19.3 Biology

Eggs are laid on the middle and lower leaves and hatch after 8 days. Nymphal instars last 3, 2, 5, 3, and 5 days approximately, and the adults live for about 50 days. Adults are gray and about 3 mm long; the nymphs are paler and smaller.

2.2.19.4 Management

Spraying fenitrothion is effective for the management of lace bugs.

2.2.20 Termites, *Odontotermes obesus*

2.2.20.1 Damage

They damage cassava stems and roots late or in the dry season. In newly planted cassava farms, termites chew and eat stem cuttings (Fig. 2.27) resulting in poor growth, death, and rot. In older cassava plants, termites chew and enter the stem which weakens them and cause breaking. Termites cause poor establishment of sets/seedlings, resulting in drying of the plants. Termite damage occurs mostly in the dry season.

Many different kinds of termites damage cassava stems and storage roots. Termites live in soil or in nests above the ground. They can also be found in tunnels on the surface of cassava stems. Termite nests contain worker, soldier, queen, and

Fig. 2.27 Termite nest

king termites. Worker and soldier termites are the ones you normally see when you break open the nests. The workers and soldiers are small insects with white or brown bodies and brown heads. They may or may not have wings.

Worker termites cause all the damage to crops and feed all the other members of the nest. The soldier termites fight off other creatures which may enter or destroy the nest.

2.2.20.2 Biology

In a new established colony, the queen lays about five eggs. When these hatched, the larvae are fed by the gastric juices of the queen. As the colony expands, the eggs are laid singly or in double rows of 16–24 eggs glued together by a gelatinous secretion, depending on the species. Incubation takes from 24 to 90 days depending on the climatic conditions.

The larva undergoes seven nymphal instars, but this again varies according to their species and the environmental conditions. These are fed and cared for by the workers. They mature over a period of 2–6 months, depending on their species.

Adult termites are small, 4–15 mm long, and vary in color from white to tan and even black, depending on the species.

Termites can be identified according to the caste they belong. There are four castes in the colony, namely, the queen, king, soldier, and the worker. The queen is the largest termite in the colony. Her role is to lay eggs to increase the size of the colony. She can lay thousands of eggs in a day. The king is always beside her ready to mate her. The soldiers have large onion-like heads and powerful jaws. They emit liquid when disturbed. The workers are the majority in a termite colony. They gather the food, feed the queen, take care of and feed the larvae, as well as build and maintain the nest. Unlike ants, the male and female termites can be workers. The nymphs with fully developed wings will be the future kings or queens.

2.2.20.3 Management

1. *Physical Methods*: Burning plant residues on top of termites' mound would suffocate them.
2. *Cultural Methods*
   - Destroy the termites' nest, locate and kill the queen, and pour boiling water or burn dried grass straws (any plant debris) to kill the rest.
   - For the "worker termites," locate their soil runways/tunnels and destroy these either by hand tilling or by flooding.
     (a) *Botanicals*: Spray Spanish needle seed extract (*Bidens pilosa*): Bring to boil 1 teacup of mature seeds for 5 min. Cool and then strain to get a clear filtrate. Add 1 l of water and spray.
3. *Chemical Methods*: In areas where termites are particularly a problem, the cut ends of cassava stem cuttings should be smeared with a watery paste of soil mixed with kerosene. This can limit termite damage.

   In case of severe incidence, soil application of carbaryl (10 %) dust or spray of chlorpyrifos (0.05 %) is recommended.

2.2.21 Gall Midge, *Latrophobia brasiliensis* (Diptera: Cecidomyiidae)

2.2.21.1 Damage

The gall midges are tiny flying insects. They are known to incite gall formation on plants during feeding. They feed on flowers/flower buds, fruit, developing seed, and decaying material. Galls are mostly found on leaves (Fig. 2.28), buds, and stems. A large number also feed on fungi (and is sometimes beneficial to agriculture).

Fig. 2.28 Galls on leaves and adult gall midge

Flies usually induce galls on cassava and are generally found in the upper surface of the leaf where they lay their eggs. The larvae cause abnormal cell growth, forming galls (Fig. 2.28). The galls are yellowish green to red in color, and when they open, the larvae can be seen inside. Severe outbreaks usually retard growth.

2.2.21.2 Management

1. *Cultural Methods*
   - It is advisable to plant in dry areas.
   - Establish fields in open locations and space plants to allow adequate ventilation.
   - Control weeds around plants.
   - Affected leaves should be removed and destroyed once a week to prevent the pest from becoming a serious problem.
   - Remove all fallen leaves from fields and discard by burying or burning.
   - Color traps present a potential tactic for monitoring or mating disruption. Along with biocontrol and other tactics, they form the basis of an IPM program.
2. *Chemical Methods*: The following chemicals can be sprayed to manage gall midge:

   Insecticides: Sevin, Trigard, diazinon, Admire, malathion

   Fungicides: Manzeb, Maneb, Ridomil

   Herbicides: Fusillade, Roundup, Gramoxone, Nabu-S

2.2.22 Cinch Bugs, *Blissus* sp.

2.2.22.1 Damage

The cinch bugs are sucking insects that operate underground. They suck juices from the root area and wound the roots of the plant with their mouth parts. This pest lives in the soil and punctures the tubers making them unmarketable. Soil microorganisms (*Fusarium*, *Aspergillus*, *Genicularia*, *Pythium*, *Phytophthora*, and *Diplodia*) can penetrate through these wounds and cause rots. The symptoms may not be seen until the cassava tuber is cut.

2.2.22.2 Management

Cinch bugs are best managed through the use of cultural practices. The use of trap crops, such as *Crotalaria* spp. and sunflower interplanted in affected cassava fields, can reduce infestations. However, the best control method is crop rotation, which breaks the life cycle of the pest.

2.2.23 Storage Pests

2.2.23.1 Damage

1. **Cassava Chips Borer or Coffee Bean Weevil**, *Araecerus fasciculatus*

   It is a polyphagous-stored product pest. Adult beetle is grayish-brown in color, antenna with 3 apical segments forming a long loose antennal club. It is 3–4.5 mm long. Life cycle takes about 50 days. Female lays on an average 50 eggs. Adults live for 70 days. Adults and grubs feed on the cassava chips and convert them into powdery mass. Infestation causes weight and quality loss.

2. **Cigarette Beetle**, *Lasioderma serricorne*

   It is a small oval or globular beetle, 2–2.5 mm long, with prothorax covering the deflexed head. Elytra cover the abdomen. Antenna is serrate. Adult lives for about 2 weeks. Life cycle is completed in 40 days. It damages cassava chips and flour that leads to weight and quality loss of the products.

3. **Flour Beetles**, *Sitotroga cerealella*, *Tribolium castaneum*

   Causes flour to be webbed together and grubs feed within, causing weight and quality loss.

4. *Other Storage Pests*

   Lesser grain borer (*Rhyzopertha dominica*), black borer (*Dinoderus minutus*), rice weevil (*Sitophilus oryzae*), and gray moth (*Ephestia cautella*) also cause damage to cassava chips.

   These insects bore into chips, and the shape of the chips is lost and reduced to a powdery mass. These borers cause weight and quality loss (Fig. 2.29).

2.2.23.2 Management

Drying cassava chips to a moisture level below 10 % is recommended to reduce insect infestation. The chips should be stored using suitable packing material, which can prevent reabsorption of moisture. Polythene impregnated jute bag, closely woven plastic bags, metal bins, etc., can be used for long-time storage. The store should be cleaned before storing chips. Infested cassava should be fumigated for disinfestation before storing. Methyl bromide and aluminum phosphide (0.6 g/50 kg) are the fumigants most widely accepted for effective and safe disinfestation. Impregnating the bags using malathion (0.5 %), fenvalerate (0.1 %), or nimbicidin (2 %) before storing the chips also reduces insect infestation. Admixture of salt at 3–5 g/100 g chips before drying reduces insect infestation considerably during storage.

The predator *Cheyletus eruditus* appears to be the most widely distributed and devours 3–5 eggs of *Rhyzopertha dominica* beetle per day. *Lyctocoris* sp. is the most successful predator of all, destroying 8–12 eggs of *R. dominica* per day. It is polyphagous and attacks other stored product pests.

The parasitoid *Anisopteromalus* sp. is found in a few areas and parasitize *R. dominica* up to 13.3 to −17.1 % of beetle population.

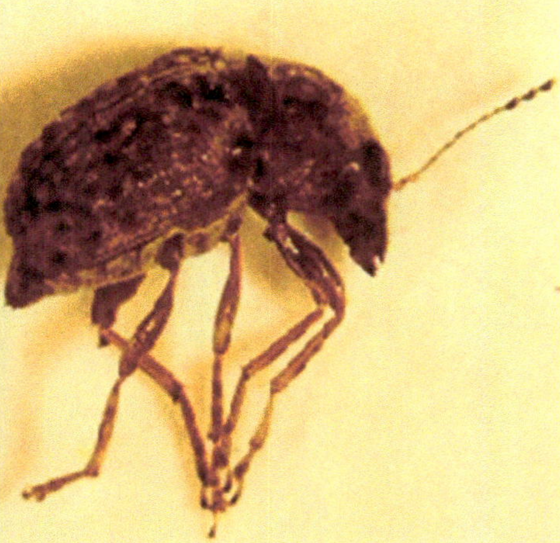

Fig. 2.29 Storage pests of cassava

Rice weevil, *Sitophilus oryzae*, larvae are parasitized by ectoparasitic wasps *Anisopteromalus calandrae* and *Lariophagus distinguendus*, and their potential for controlling rice weevil has already been demonstrated.

2.2.24 Red Spider Mite, *Tetranychus* spp.

Tetranychus telarius and *T. noecaledonicus*, commonly called red mites, feed on undersurface of the leaf.

2.2.24.1 Distribution

The pest is distributed throughout the semitropics and in protected cultivation in temperate zones.

2.2.24.2 Damage

The infestation starts from the mature lower leaf. Symptoms of the infested leaf show yellow specks along the main leaf vein, and during severe infestation, leaf withers off. Feeding by the mite on the lower surface of the leaves causes characteristic blotching.

Lower leaves develop yellow dots along the veins; these dots become reddish, and if the dry season is prolonged, the infestation spreads to higher leaves (Fig. 2.30). Leaves dry up and fall and some plants die. The mites responsible are brownish-red with dark lateral internal markings; they produce webs.

2.2.24.3 Biology

Reproduction takes place during dry weather; the optimum temperature for development is 32 °C. Single eggs are laid on the undersurface of the leaves or on the webbing. The development of egg to adult takes about 17 days, and there may be 19 broods per year.

2.2.24.4 Management

1. *Chemical Methods*: Foliar application of urea followed by spraying water at the beginning of mite infestation or spraying the insecticide, dimethoate (0.05 %), at the peak period of infestation can be adopted to control the menace.

   Spraying of 0.04 % binapacryl, 0.05 % chlorobenzilate, 0.03 % dimethoate or parathion, 0.01 % lime sulfur, 0.16 % mancozeb, 0.001 % naled, or 0.07 % phoslane is recommended for the management of spider mites.
2. *Biological Methods*
   (a) *Predators*: Coccinellid *Stethorus gilvifrons* and thrips *Scolothrips indicus* are the most important predators.

The most successful biocontrol agent, *Typhlodromalus aripo* (Acari: Phytoseiidae), is

Fig. 2.30 Red spider mite infestation on leaves of cassava

confined to shoot tips and so allows persistence of the host population and is also better able to survive on alternate sources of food when the host is scarce. It is now established in some 20 countries and has reduced mite damage by more than 50 %. *Aceria lopeji* was rapidly successful throughout the range of climates in the infested areas of Africa.

2.2.25 Green Spider Mite, *Mononychellus tanajoa*

2.2.25.1 Damage

The active stages feed on the bottom surface of leaves by sucking fluids from cells. This causes yellow spotting of leaves (chlorosis) which can increase from a few spots to complete loss of chlorophyll (Fig. 2.31). Most cassava green mite (CGM) individuals are generally found on the upper third of the cassava plant. Leaves damaged by CGM may also show mottled symptoms which can confuse it with symptoms of cassava mosaic virus disease (CMD). Damage is most severe during the dry season.

Severely damaged leaves dry out and fall off, which can cause a characteristic candle stick appearance. Because of reduced plant growth, accumulation of starch in the storage roots is slowed, sometimes even reversed, and root yield losses in the absence of any control measures can reach up to 50 %. Where leaves are eaten as vegetables by farmers, a corresponding loss ensues. Reduced growth and stunting of the tips is also responsible for contorted and thin stems, thus affecting the planting material to be used for the next season.

The size of CGM populations, and hence yield losses, is generally influenced by several factors including the (1) age of the host plant, young plants are more exposed and susceptible to CGM attacks than older plants; (2) season, damage severity is greater during dry than wet season, and heavy rainfall can reduce CGM populations; (3) temperature, populations increase with increasing temperature leading at times to very rapid increase in populations and damage; and (4) poor agronomic practices, plants grown in poor soils are more susceptible to mite attacks.

2.2.25.2 Biology

Eggs are tiny, spherical, pale-white, and are laid on the undersides of leaves often under the webbings. Eggs hatch in 4 or 5 days. Nymph looks similar to the adult but is only the size of an egg. It has only 6 legs. It molts 3 times before becoming an adult.

The adult is also very tiny, maybe yellowish, greenish, pinkish, or reddish, depending on the species. It looks like a tiny moving dot. It has an

Fig. 2.31 Green mite on cassava

oval body with 8 legs and with 2 red eyespots near the head of the body.

The male is smaller than the female with a more pointed abdomen. A female usually has a large, dark blotch on each side with numerous bristles covering her legs and body.

In the field, eggs are laid singly on the lower surface of leaves. Apart from the egg which is inactive, there are four active stages including larva, protonymph, deutonymph, and adult male and female. In the laboratory, egg, larval, protonymphal, and deutonymphal stages last 3–4, 1–2, 1–2, and 2–3 days, respectively. CGM development occurs in about 10 days, and a female can live up to 30 days and lay more than 60 eggs during her life span. Peak CGM densities occur during the first half of the dry season, with a smaller peak occurring within about a month of the start of the long rainy season.

During heavy rainfall, all stages of the mite may be washed off the plant. Females disperse in the wind after lowering themselves on silken threads. This mite does not construct webbing.

2.2.25.3 Management

1. *Cultural Methods*: Several cultural methods, such as adjusting planting time for the crop to escape severe damage at young age, mixing varieties to avoid genetic uniformity, and removing infested tips, have been tried but without much success, primarily because these practices were not well suited for traditional farming systems.

   Good field sanitation—removal of weed and plant residues from previous crops helps in reducing the mite population. The following cultural methods are recommended:

   - Plant on the onset of the rainy season to encourage vigorous growth.
   - Horizontal planting of cuttings is recommended rather than the slanting position.
   - Intercrop cassava with pigeon pea in triple and double rows. Cowpea and peanut are also good cassava intercrops.
   - Apply 5–7 t of manure per hectare to maintain the soil fertility.
   - Canavalia or crotalaria are good companion crops. Sow 1 row of Canavalia or crotalaria between rows of cassava immediately after planting cassava. Let these grow until harvest. Plow under them to incorporate into the soil.
2. *Chemical Methods*: Although several insecticides and acaricides (such as dimethoate and dicofol) can control CGM, they are often too expensive for farmers and may be dangerous if not well applied and in countries where leaves are consumed as vegetables. During severe infestations, chemical control may become necessary. Miticides such as Abamectin, Newmectin, or Vertimec may be used at 5 to 4,500 ml of water for controlling the mites.
3. *Biological Methods*: Because CGM was introduced into Africa from South America, where it is widespread and generally is not a serious cassava pest, biological control was chosen as the most appropriate method of control. Several predatory mites of the family Phytoseiidae that generally keep the mites under control in the South America were introduced into Africa from Colombia and Brazil. All Colombian species failed to establish. However, three predators from Brazil, *Neoseiulus idaeus*, *Typhlodromalus manihoti*, and *T. aripo*, became established in several countries (Fig. 2.32). Of the three predators,

Fig. 2.32 Predatory mite, *Typhlodromalus aripo* (*left*) feeds on cassava green mite (*right*)

Table 2.7 Economic impact assessment of biological control of cassava pests in Africa

Pest species	Loss (%)	Successful biocontrol agent	% reduction in loss	Savings (US$ million)	Discount rate (%)
Cassava green mite	35	The phytoseiid mite *Typhlodromalus aripo*	80–95	2,157[a]	10

Neuenschwander (2004)
[a]For Nigeria, Ghana, Benin

T. aripo is the most widely distributed. The predator occurs mainly on young leaves at cassava shoot tips. It spreads by wind and by being carried on stem cuttings. In Africa, *T. aripo* was the most successful in reducing the populations of the green mite by 35–60 % with a parallel increase in fresh root yield by 30–37 % (Bellotti 2002).

Classical biological control of cassava green mite was achieved by identifying, introducing, and establishing predatory mites (*Typhlodromalus aripo* and *T. manihoti*) and later an acaropathogenic fungus (*Neozygites tanajoae*) from climatically similar areas of Brazil (Table 2.7).

Farmers can increase the spread of predatory mites by plucking and carrying cassava shoot tips with the predator from one field to another. Farmers can also increase the survival and spread of these predators by growing cassava varieties whose new leaves clump together at the shoot tip. These will attract the predators better than varieties whose young leaves are widely spread. Even though farmers may not grow such varieties for food or sale, they can grow a few plants in farms to attract the predators.

Farmers can also leave certain weeds such as *Euphorbia heterophylla* and *Mallotus oppositifolius* on cassava farms to attract phytoseiids. The predators can live on these weeds when their food is scarce on cassava, ensuring that they are there to provide biological control when the cassava green mite comes up again. During weeding, farmers can leave the weeds to grow along the margins or in other parts of cassava farms—but, of course, not so many that they compete with the crop. This cultural practice will be especially useful at sites where cassava is grown continuously, with little or no fallow.

Predatory mites from Brazil, *Neoseiulus idaeus*, *Typhlodromalus manihoti*, and *Typhlodromalus aripo*, were effective. Where *T. aripo* has been present for three or more years, it can reduce CGM by an average of 50 % and increase root yield by an average of 30 %. Control is greater where both *T. aripo* and *T. manihoti* are present together.

4. *Host Resistance*: The IITA variety TMS 30572 and the national varieties 8017 and 8034 in Cameroon, and MS6 and NR 8082 in Nigeria are good against the cassava green mite.

2.2.26 White Spot Spider Mite, *Oligonychus (Homonychus) peruvianus*

2.2.26.1 Distribution

The pest is distributed in Guatemala, Mexico, Peru, Trinidad, the United States (California and Texas), and Venezuela.

2.2.26.2 Damage

Old leaves are attacked and whitish spots (webbing) develop on the underside close to the veins; these spots are yellow-brown on the upper surface. The mites feed under the webbing.

2.2.26.3 Biology

Egg laying and nymphal development take place beneath the webbing.

2.2.26.4 Management

Same as for green spider mite.

2.2.27 Rodents

2.2.27.1 Damage

Production of tuber crops such as cassava in the state has gone down steeply during the past decade. The fall is mainly due to the rat problem, specifically of Bandicoots. Rodents are common vertebrate pests of cassava. Rodent attack may become more serious under changing climatic conditions or due to a sudden increase in impact of other factors.

Rodents are common vertebrate pests of cassava. The most important vertebrate pest of cassava is considered to be the cane rat, and large groups can completely destroy the entire crop in a single visit. In cassava fields, cut stems are chopped into bits, and partly dug and eaten roots with teeth marks are usually visible at feeding sites.

2.2.27.2 Management

The following measures are recommended for rodent control:

- Set traps in the fence.
- Cover exposed roots with soil.
- Weed your cassava farm to discourage rodent pests.
- Harvest roots as soon as they are mature (James et al. 2000).

Direct control is by using poisonous baits. Placement of maize meal containing 3 % zinc phosphide at 5-m intervals around the edges of fields gave a kill of 93 % rodents (Taylor 1968).

Burrow fumigation with aluminum phosphide is also effective. Phostoxin generates gas when it comes into contact with moisture in the atmosphere, leading to hydrolysis and production of phosphine gas which is poisonous to rats.

Indirect control suppresses rodents through habitat management and their use as food. Simple husbandry practices such as field sanitation, clearing of farm borders and removal of potential nest sites, continuous inspection of farms to identify and check damage, and prompt harvesting can eliminate many rodents.

Smoke generated from fires set at the entrance of burrows and blown into them is widely used to kill the giant rat and other mouselike rodents.

Artificial flooding of burrows with irrigation water can also flush out most rodents. Digging or plowing drives out rodents from their burrows.

2.2.28 Beneficial Insects

The following are some of the beneficial insects that are indigenous on fields and gardens or can be introduced for the control of pests. The following methods can be adopted to conserve these beneficial insects:

- Provide hiding sites and alternative habitats such as mulches and other ground covers.
- Plant small flowering plants on borders, hedges, and other perennial habitats as source of food and shelter.
- No indiscriminate use of synthetic pesticides.

2.2.28.1 Ground Beetle

Ground beetles (Fig. 2.33) feed on slugs, snails, cutworms, cabbage root maggots, grubs and insect pupae, and small caterpillars. Most species are nocturnal, and they hide during the day in soil crevices, under rocks and stones, decaying logs, leaf litter, or composting materials. When disturbed or when other vertebrates prey upon them, they emit an odor or gas, as a type of defense mechanism, preventing them from being eaten by

Fig. 2.33 Ground beetle feeding on larva

other predators. Ground beetles live on or below the ground, hence the name. Development from the egg to the adult stage takes about a year, although adults may live for 2–3 years or longer.

1. *Conservation*
   - Practice mulching in some sections of your field to provide a habitat for the ground beetles.
   - Provide permanent beds and perennial plantings to protect population.
   - Plant white clover and/or amaranth as ground covers.

2.2.28.2 Lacewing

Lacewings (Fig. 2.34) feed on aphids, leaf miner, mealybugs, thrips, whitefly, armyworms, bollworms, cabbage worm, codling moths, corn borer, cutworm, diamondback moth (DBM), fruit worm, leafhopper nymphs and eggs, potato beetle, scale insects, spider mites, and caterpillars of most pest moths. If given the chance, they can also prey on adult pests.

Larvae are known as aphid lions. They attack their prey by taking them with their large sucking jaws and injecting paralyzing poison and then sucking out the body fluids of the pest. A larva can eat 200 or more pests or pest eggs a week. An older larva can consume 30–50 aphids per day. It can consume more than 400 aphids during its development. The larvae resemble alligators with pincers like jaw. However, they become cannibalistic if no other prey is available. They feed for 3–4 weeks and molt three times before pupation. Each adult female may deposit more than 100 eggs. Many species of adult lacewings do not prey on pests. They feed on nectar, pollen, and honeydew. An adult will live for about 4–6 weeks depending on the climatic conditions.

Fig. 2.34 Adult lacewing

1. *Conservation*
   - Flowering plants such as dill, cosmos, sunflower, carrots, and dandelions are good source of pollen and nectar for adults.
   - Provide source of water during dry season.

2.2.28.3 Ladybird Beetles

Ladybird beetles (Fig. 2.35) feed on aphids, mealybugs, scale insects, spider mites, and whiteflies. When disturbed, some of them emit a strong smelling yellow liquid as a protection against other predators. Their colors vary from red, orange, steel blue, yellow-brown, or yellow elytra, frequently spotted or striped with black. They feed on pollen, nectar, water, and honeydew, but aphids or other prey are necessary for egg production. They are the best-known predators of aphids and are capable of eating up to 50–60 per day and about 5,000 aphids in their lifetime.

1. *Conservation*
   - Ladybird beetles are found in most agricultural and garden habitats. Their presence indicates that natural biological control is occurring. It is important to maintain habitats planted with several flowering crops. These give the ladybird beetles varied food sources.

Fig. 2.35 Ladybird beetles

- When food is not available, they tend to eat each other. Their beneficial predatory behavior and activities are continuous when there is no indiscriminate use of synthetic pesticides.

2.2.28.4 Mealybug Destroyer, *Cryptolaemus montrouzieri*

Mealybug destroyer (Fig. 2.36) feed on mealybugs. The larva feeds on mealybug eggs, young crawlers, and the honeydew produced by mealybugs. It can consume up to 250 mealybugs. The pupa is found in sheltered stems. Adult mealybug destroyer is dark-brown or blackish beetle. A female can lay up to 10 eggs a day in a mealybug colony or in a group of mealybug eggs. It may live up to 2 months.

(a) *Conservation*: Mealybug destroyers only thrive when there are mealybugs. They feed on mealybugs, which are necessary for their reproduction. Members of carrot (fennel, dill, angelica, tansy) and sunflower families (goldenrod, coreopsis, sunflower, and yarrow) are good habitats for adult mealybug destroyers.

1. *Spider*

   Spiders (Fig. 2.37) feed on moths and caterpillars. A female can produce 200–400 eggs, but only 60–80 spiderlings can hatch from these. Females can survive 2–3 months. All spiders are poisonous to insects, but only a few species are poisonous to humans, like the black widow and the brown recluse.

(b) *Conservation*

- Mulching along some sections in dikes of rice paddies, in field corners, or a portion of the fields can increase the number of spiders. They can hide in the layer of mulch that serves as their alternate habitat. They can also prey on other small insects inside the mulch.
- Remember that the more food the spiders can eat, the faster their population buildup will become. Cover crops are also important to provide overwintering sites of spiders' sacs.
- A spider population depends on the availability of food, the habitat, and the environmental conditions. Avoid use of pesticides as much as possible, for broad-spectrum insecticides can easily kill them.

2.3 Diseases

Among the factors that affect cassava production, diseases still remain the major constraint that can bring Africa's cassava production to a halt. The recent East African Cassava Mosaic pandemic and the food shortages that resulted from it add value to the above statement. African cassava mosaic disease is still widespread and causes severe yield losses in production systems that

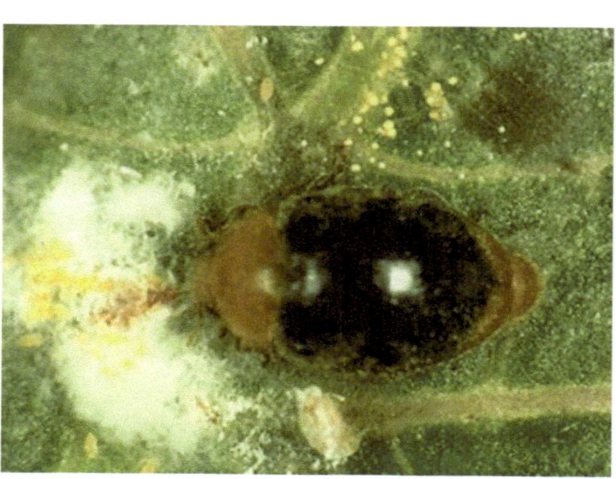

Fig. 2.36 *Cryptolaemus montrouzieri*

Fig. 2.37 Spider

depend on susceptible cultivars. Cassava bacterial blight, anthracnose, bud necrosis, leaf spots, and root rot diseases affect yields of cassava in almost all producing countries in Africa. Information on yield losses due to diseases is often based on estimates, but observations indicate that losses are significant in most of the cassava-growing areas of Ghana.

The importance of cassava diseases include:

- Diseases cause low yields of edible roots which affect income of farmers.
- Food security is reduced by diseases.
- Severe outbreaks of diseases such as cassava bacterial blight can result in famine (in whole communities or countries).
- Cassava diseases that affect stems can lead to loss or shortages in the supply of planting materials.
- Loss of leaves through diseases can affect the availability of leafy vegetables.
- Loss of leaves and poor yield of storage roots can affect livestock production in communities that use cassava as animal feed.

Even though cassava is a hardy crop, it has been shown that diseases can cause serious losses. Cassava bacterial blight is the most important disease of cassava, and virus diseases are also of considerable threat to the crop. As cassava is propagated vegetatively, the use of disease-free planting material is the most important single factor in disease control. Under traditional farming systems, yields have been more than doubled by using disease-free planting material and improving cultural practices.

2.3.1 Tuber Rot, *Phytophthora palmivora*

Tuber rot of cassava is a serious problem in Tamil Nadu. The disease is primarily caused by *Phytophthora palmivora*, and many secondary invaders also get associated with the disease at later stage. *Phytophthora* root rot is prevalent in waterlogged soils. The disease causes yield losses up to 80 %.

2.3.1.1 Symptoms

The pathogen attacks both mature and young plants, causing sudden wilting and severe soft rotting of the swollen roots. The affected plants give rise to rotten tubers (Fig. 2.38). The plants

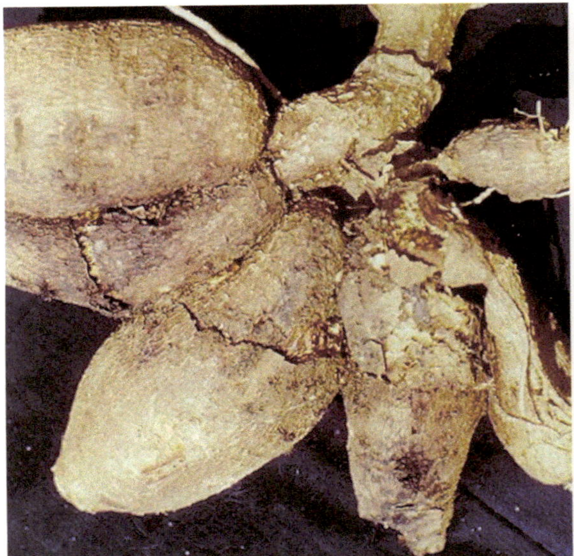

Fig. 2.38 Cassava tuber rot

can be easily pulled without pressure from the ground. Two-third reductions in tuber yield have commonly been observed in affected fields in Tamil Nadu. The disease is characterized by the appearance of dark-colored round to irregular-shaped water-soaked lesions (15–30 mm in diameter) on mature tubers in the field. White mycelial mats of the fungus develop around these lesions. On advancement of infection, the lesions enlarge causing internal browning, oozing of internal fluids, and shriveling of the tubers. The infected tubers emit a characteristic foul smell and rot within 5–7 days depending on the soil conditions. However, no external symptoms are visible on leaves and stem of infected plants. Hence, early detection of the disease is very difficult.

The pathogen *P. palmivora* produces uniform white mycelial colony on agar medium. The mycelium is coenocytic and hyaline. It produces apical, pear-shaped papillate sporangia freely liberated on sterile water. The pathogen can be isolated from freshly infected brown and marginal areas of internal tuber tissues.

2.3.1.2 Survival

The main mode of survival of the pathogen in the soil is through infected tubers left over after harvest, and this serves as the source of inoculum for the next crop. The fungus also survives in the form of dormant mycelium and chlamydospores. The survival period in the soil is enhanced with high soil moisture. In the absence of any infected materials, the survival of the pathogen in air-dry soil was only up to 6 weeks as against 24 weeks in soil with 40 % soil moisture.

2.3.1.3 Management

The disease can be controlled by cultural practices such as providing good drainage, use of lighter soils, early harvest, crop rotation with cereals, and fallowing. Wide spacing tends to discourage the spread of root pathogen. Crop residues should be removed and burnt, and healthy planting material should be used for planting.

Breaking the hard pan by deep plowing (45–60 cm) (Fig. 2.39), ridge planting, regulating irrigation, improvement of drainage condition, strict adherence to sanitation, and use of *Trichoderma* sp. showed substantial reduction of cassava tuber rot and increased the quantity of marketable tubers. The sett treatment (30 g/l for 15 min.) and soil application (1 kg/ha) of *Trichoderma viride* recorded the lowest tuber rot incidence (7.33 %) and highest harvest index (66.9 %). *Trichoderma viride*, *T. harzianum*, and *Pseudomonas fluorescens* are found effective against cassava tuber rot.

Fig. 2.39 Breaking the hard pan by deep plowing

2.3.2 Brown Leaf Spot, *Cercosporidium henningsii*

Brown leaf spot is a major fungal disease in heavy rainfall areas. It causes considerable premature defoliation and loss in tuber yield up to 30 % in susceptible variety like Malayan-4.

2.3.2.1 Distribution

Brown leaf spot disease is a widespread fungal leaf disease in all cassava-growing areas.

Fig. 2.40 Brown leaf spot on tapioca leaves

2.3.2.2 Symptoms

Among the leaf spot diseases, brown leaf spot caused by *Cercospora henningsii* causes substantial yield losses in areas with high and extended rainfall. Older leaves tend to be more susceptible than younger ones, and the disease tend to be more widespread and severe in crops over 5 months old. The disease causes characteristic brown spots (3–12 mm in diameter) on both sides of the leaves (Fig. 2.40). Spots on the upper surface are angular and uniformly brown with a distinct dark border. Spots on the lower surface have less distinct margins and appear grayish in the center because of the presence of fruiting bodies of the fungus. The lesions expand and become irregular and angular in shape as they are limited by leaf margins and veins. Commonly an indefinite halo appears round the lesions. As the disease develops, the infected leaves turn yellow, become dry, and fall off. Susceptible varieties can be severely defoliated during the rainy season.

2.3.2.3 Development and Spread

Warm, wet conditions favor the development and spread of *C. henningsii*. Conidia of the fungus on older or fallen leaves are spread by rain splash and wind to younger leaves and to new plantings. The conidia need moisture to germinate on the leaf surface. The fungus penetrates through stomata and grows trough the intercellular spaces. The fungus survives during the dry season in old lesions, often on fallen leaves.

2.3.2.4 Management

1. *Cultural Methods*: Measures designed to reduce the humidity within the stand of cassava, such as wider spacing and the varieties with a low leaf area index, are recommended. The severity of the disease can be reduced if the planting is timed so that the most susceptible growth stage (over 5 months old) does not coincide with the wet season.
2. *Chemical Methods*: Spraying of copper fungicides, Thiophanate, Benomyl, and Carbendazim at monthly intervals starting from third month, was effective in controlling the disease.
3. *Host Resistance*: Cassava cvs. Sri Prakash and Sri Visakam released from Central Tuber Crops Research Institute, Thiruvananthapuram, Kerala, India, are resistant to brown leaf spot disease.

2.3.3 White Leaf Spot, *Cercospora caribaea*

2.3.3.1 Distribution

White leaf spot (*C. caribaea*) is commonly found in the humid but cooler cassava-growing regions and has been reported in certain areas of Asia, North America, tropical Africa, and Latin America. In these areas, this pathogen may cause considerable defoliation of susceptible varieties of *M. esculenta*, the only reported host species.

Fig. 2.41 Cassava plant with white leaf spot symptoms

2.3.3.2 Symptoms

Lesions caused by *C. caribaea* are smaller and different in color to those induced by *C. henningsii*. They are circular to angular, usually 1–7 mm in diameter, and white or rarely yellowish-brown (Fig. 2.41). The lesions are sunken from both sides to about one-half the thickness of the healthy leaf blade. While the white spots remain distinct, the lesions frequently have a diffuse colored border on the lower leaf surface. The border sometimes appears as an irregular violet-brown line surrounded by a yellow or brownish halo. The center of the spots is given a grayish velvety aspect during the fructification of the pathogen which occurs predominantly on the underside of the leaf.

Penetration occurs through stomatal cavities and invasion of the tissues through intercellular spaces. When the leaf spots thus produced reach about 5–7 mm, a stroma is formed from which the conidiophores are later produced.

2.3.3.3 Spread and Survival

Secondary disease cycles are repeated throughout the rainy season when the conidia are dispersed by rain splash and wind. The fungus survives the dry season in old, infected tissues and renews its activity with the coming of the rainy season and the renewed growth of the host. Weeds also serve as alternative hosts for leaf spot fungi.

2.3.3.4 Management

1. *Cultural Methods*
   - Field sanitation—Removing leaves, crop debris, and weeds from around plants to prevent disease spreading.
   - Crop rotation.
   - Use of disease-free planting material.
2. *Chemical Methods*
   - Spray a fungicide if necessary.
   - Apply herbicide, e.g., Gramoxone, Roundup, or Karmex, for annual and perennial weed control.

2.3.4 Phomopsis Blight/Concentric Ring Leaf Spot, *Phomopsis manihotis* (*Diaporthe manihotis*)

2.3.4.1 Distribution

The disease is widespread throughout the cooler cassava-growing areas at high altitudes or in the lowlands during the rainy season. The disease occurs in Africa, Brazil, Colombia, India, and Philippines.

2.3.4.2 Symptoms

The disease occurs in cooler cassava-growing areas during the rainy season when temperatures are below 22 °C. Susceptible varieties may be severely defoliated and occasionally dieback or death may occur. The disease is characterized by large brown spots with indefinite margins. Lesions are usually found at the tips or edges of the leaf lobes or along the main veins or midrib. At first the upper surface of the lesions consists of concentric rings formed by the brown pycnidia (fruiting bodies) of the fungus (Fig. 2.42). These rings are often absent on the older lesions as the epidermis disintegrates, allowing mature pycnidia to be washed off by rain so that the lesions resemble those caused by *Cercospora viscosae*. Few pycnidia are produced on the lower surface of the lesions, which appear uniformly brown.

When humidity is high grayish-brown, hyphae can be observed on the surface of the lesions. On

Fig. 2.42 Phomopsis blight—*brown spots* with concentric rings of pycnidia on the upper surface of the leaf

the underside of leaves, veins and veinlets become necrotic in the region of the lesions. The lesions expand, causing leaf blight. Eventually the necrosis spreads to the whole leaf and petiole. Leaves then wilt and drop, sometimes resulting in extensive defoliation. Dieback of young shoots can also be caused, and in extreme cases, the plant may die. Diseased stems become brown and are often covered with pycnidia.

2.3.4.3 Host Range

Tests have shown that *Solanum melongena* is susceptible to *P. manihotis*.

1. *Development and Spread*

   The young upper leaves are more susceptible to the disease. The disease is more prevalent at high altitudes or in lowland areas during the rainy season, because conditions are best for spore germination (20–25 °C).

   Conidia formed on the leaves and stems are dispersed by watersplash, with maximum germination between 20 and 25 °C. Survival between crops is not understood, but probably occurs as pycnidia or perithecia in leaf and stem debris.

2. *Management*

   Field-resistant cultivars have been observed, and these may be used in a breeding program for resistance.

   Meristem-tip culture eliminates the disease.

2.3.5 Powdery Mildew, *Oidium manihotis* (*Erysiphe manihotis*)

2.3.5.1 Distribution

The disease is distributed worldwide and endemic in almost all cassava plantations.

2.3.5.2 Symptoms

The first sign of the disease is a white fungal mycelium growing over the surface of the leaf. Yellowish lesions are formed where the fungus penetrates the leaf. The lesions are indefinite and contain pale brown angular spots of necrotic tissue (Fig. 2.43).

The disease occurs most commonly during the dry season in warmer cassava-growing areas. Mature leaves appear to be the most susceptible.

2.3.5.3 Management

Resistance has been observed in some varieties.

2.3.6 Anthracnose, *Colletotrichum gloeosporioides* (*Glomerella manihotis*)

Cassava anthracnose disease (CAD) is widespread in most of the cassava-growing regions of Africa. It is estimated that CAD causes yield losses in the neighborhood of 30 % or more in susceptible cultivars. The disease affects both

Fig. 2.43 Powdery mildew symptoms on cassava leaf

leaf and stem production. Severe anthracnose attacks can cause death of stems which can affect the availability of planting materials especially in large-scale production systems.

2.3.6.1 Distribution

The disease is widespread in distribution and appears to be serious in Southeast Asia, Africa, and Brazil.

2.3.6.2 Symptoms

Anthracnose tends to appear after long periods of rain. Young shoots seem to be the most susceptible. Leaf border blight and spots appear on young leaves, which become distorted. Infected tissue becomes necrotic and whole leaves may die. The disease also attacks stems, causing a dieback of young stems and the production of cankers on older stems (Fig. 2.44). In the center of the cankers, pinkish areas of the fungal fruiting bodies can be observed. Anthracnose tends to disappear at the onset of the dry season.

Cassava anthracnose disease appears as cankers ("sores") on the stems and bases of leaf petioles (Fig. 2.44). Cankers weaken the petioles so that the leaf droops downward and wilts. The wilted leaves die and fall, causing defoliation and shoot tip dieback or complete death of the shoot. Soft parts of cassava stems become twisted under severe attack by the disease.

2.3.6.3 Spread

The disease usually starts at the beginning of the rains and worsens as the wet season progresses.

The main sources of the fungus that causes cassava anthracnose disease are cassava plants with the disease. The fungus spreads by wind-carrying spores from cankers on the stems or by planting stem cuttings with cankers. The fungus enters cassava plants through wounds and feeding punctures made by the bug *Pseudotheraptus devastans*. Dead cassava stems and leaves with the fungus also serve as sources of the disease if they are not destroyed after root harvest.

2.3.6.4 Host Range

The fungus that causes cassava anthracnose can also cause diseases on other crops such as coffee, pepper, avocado, banana, papaya, and yam.

2.3.6.5 Management

1. *Cultural Methods*
   (a) *Healthy Planting Material*: In situations where desired varieties are susceptible to

Fig. 2.44 *Left*, cankers at the bases of leaf petioles. *Right*, cassava shoot with wilted leaves

anthracnose, healthy stems free of cankers and dieback tissues must be selected to provide cuttings for new plantings.

Healthy cuttings sprout well and give rise to young vigorous growing plants that are likely to escape early infections. Where canker-bearing stems are the only available sources of planting materials, stem cuttings must be dipped in appropriate fungicides recommended by extension agents in their operational areas. Spores and other fungal materials of disease-causing fungi can be destroyed by a few minutes of dipping in suitable fungicides.

(b) *Farm Sanitation*: Stems and leaves from infected plants after harvest must be destroyed by burning to reduce the amount of fungal spores and other infective structures that can cause infections in the next generation of plants.

The fungus that causes cassava anthracnose can also cause diseases on other food crops such as pepper, avocado, banana, pawpaw, and yam. Spores therefore can be transferred from these plants to cause infections in cassava. Plant debris from plants suspected to be harboring the anthracnose causing fungus must be destroyed by burning, particularly during land preparation before cassava is planted.

2.3.7 Root Rot, *Macrophomina phaseolina, Botryodiplodia theobromae, Fusarium* spp.

2.3.7.1 Symptoms

Leaves on affected plants turn brown and wilt and the plant has a scorched appearance. Leaves may remain attached to the plant or drop to the ground. Plant death will occur. Examination of roots reveals root dieback and swelling of tubers. Tubers may have light brown, dark gray, blue, or pink discoloration (Fig. 2.45). Rotting roots may be soft and produce a foul odor. Infection by *Botryodiplodia* fungi may cause the appearance of white fungal structures at the base of the stem, particularly during the wet season. Root rot disease emergence is often favored by waterlogged and poorly drained soils.

Fig. 2.45 Cassava tubers destroyed by root rot disease

2.3.7.2 Spread

The important sources of cassava root rot fungi are soils, cassava root, and stem debris contaminated with the fungi. The fungi enter cassava plants through wounds caused by pests or farming tools or by piercing the roots by themselves. Farm tillage tools used in cassava farms with the disease should be cleaned after use to prevent the fungi on them from spreading to other areas. Similarly, cassava plant debris in farms with the disease serves as source of root rot fungi and should be destroyed by burning.

2.3.7.3 Host Range

Cassava root rot fungi attack a wide range of other crops including maize, soybean, sunflower, and coffee.

2.3.7.4 Management

Plant cassava in well-draining soils; remove and destroy all crop debris by burning; sanitize all tools after use.

Avoid farming very close to rivers and streams that are likely to overflow their banks in the rainy season. It is common to find farmers who have limited access to land cultivating cassava in valleys that are subjected to flooding in good rainy days.

Waterlogged soils are poorly aerated, and cassava roots developing in such soils will suffer from rots not necessarily due to diseases. Roots of plants growing in waterlogged soils are not healthy and can easily be attacked by diseases. A sandy loamy soil that is well drained is a good soil type for cassava.

Some of the organisms or agents that cause root rot diseases of cassava such as certain types of fungi require some amount of water to allow them to swim to effect new infections. This is one of the reasons why root rot diseases of cassava are common in poorly drained soil.

The soil should be fertile enough to give a good healthy crop. Avoid planting in soils that are poor in nutrients. Efforts must be made to improve soil fertility by the addition of inorganic or organic fertilizers whenever possible. A good example of organic fertilizer is poultry manure. What must be kept in mind is that a healthy plant is not easily attacked by diseases, but even when attacked, it is able to fight the disease better.

Do not cultivate cassava on land that has a history of root rots and other major diseases. This information can be obtained from neighbors farming close to the land. If the only available land has a history of rots, then good disease management practices must be maintained if good yields are expected.

Avoid using planting materials from fields with visible signs of root rot diseases even if the stems look healthy.

Collect plant debris (stems and roots) with fungi together and burn to destroy particularly after harvest.

Some root rot-causing organisms actively degrade storage roots when they are well developed and harvesting is delayed. Early harvesting, therefore, prevents or reduces the incidence of rots of this nature on farms.

Continuous cropping of cassava on the same piece of land contributes to increase in incidence and severity of diseases including root rots. It is generally effective to rotate cassava with cereals to help reduce the levels of inoculum (spores, etc.) on fields. It is advisable not to plant cassava continuously for 3 years on the same piece of land particularly in localities where high cassava disease pressures exist.

Fallowing of land after five or more years of continuous cassava cultivation for a period of 3–5 years is a good measure that can reduce the incidence and severity of diseases. When practiced properly, it can help eliminate a disease from a locality or reduce its incidence and severity significantly. The absence of host plants from a field deprives pathogens or disease-causing organisms of their nutrient resulting in the decline of pathogen populations.

2.3.8 *Polyporus* Root Rot, *Polyporus sulphureus*

A parasitic mushroom (*Polyporus sulphureus*), originally suspected to be pathogenic on woody plants, has been found attacking cassava plants and causing severe root rots in Ghana. This parasitic mushroom is capable of causing 100 % yield loss on farms where susceptible cultivars are planted.

2.3.8.1 Symptoms

General symptoms of root rot diseases include wilting of leaves, which in most cases is accompanied by defoliation. Other symptoms include swollen roots with colored inner tissues (Fig. 2.46). Rotten roots may be soft and produce an offensive odor (this is often associated with rots caused by bacteria). Shoot or stem dieback is a feature of plants with underground rotten roots. Root rot diseases may lead finally to death of infected plants.

2.3.8.2 Host Range

The *Polyporus* root rot mushroom can attack a number of crops including yam and citrus. Care should therefore be taken in the selection of crops for rotation or intercropping with cassava.

2.3.8.3 Management

1. *Regulatory Methods*: Cassava stems from fields with visible signs of root rot diseases must not be used as planting materials, even if they look healthy. Stems from such fields are likely to carry spores of root rot fungi.

   Localized or national quarantine measures to check movement of planting materials from

Fig. 2.46 *Left*, fruiting bodies of the parasitic mushroom *Polyporus sulphureus* at the base of collar. *Right*, rotten root of cassava harvested 24 months after planting (note the inner colored rotten tissue)

root rot endemic areas into new localities may be necessary to check the spread of root rot diseases.

2. *Cultural Methods*
   (a) *Good Farm Sanitation*: Destruction of plant debris including rotten roots and stems bearing fruiting bodies immediately after harvest is a good measure that destroys spores of pathogenic fungi. The leathery fruiting body after harvest can persist for over a year and is a good source of spores for new attacks in the subsequent season's crops. Disease severity on farms can be reduced through destruction of debris that carry spores into the next planting season.

   Fruiting bodies of parasitic mushrooms such as *Polyporus sulphureus* start their development on cassava plants after the first few rains following the dry season. Harvesting and destroying the fruiting bodies soon after they appear prevent them from producing spores required for the spread of the disease.

   (b) *Early Harvesting*: Some root rot-causing organisms actively invade and degrade storage roots when they are well developed, and harvesting is delayed. Early harvesting therefore prevents or reduces incidence of rots in some varieties of cassava.

   (c) *Crop Rotation*: The *Polyporus* root rot mushroom can attack a number of crops including yam and citrus. Care should therefore be taken in the selection of crops for rotation or intercropping with cassava.

   (d) *Fallow*: Fallowing of land after five or more years of continuous cassava cultivation for a period of 3–5 years is a good measure that can reduce incidence and severity of cassava diseases. When practiced properly, fallowing can help eliminate a disease completely from a locality or reduce its incidence significantly. In localities where pressure on available land is high, crop rotation may be the better of the two options.

   (e) *Clean Farm Tools*: Farm tools such as cutlasses, hoes, and plows used on fields with a root rot history must be cleaned immediately after use before being used on a second farm. This reduces the spread of diseases from farm to farm.

   (f) *Avoid Planting Cassava as The First Crop After Clearing Woodlands or Forests*: Certain parasitic mushrooms can grow on woody trees or cassava depending on which one is available or preferred. It is therefore not advisable to cultivate cassava as the first crop just after clearing forests or woodlands. In the absence of woody trees, introduced cassava plants will be attacked. Woody stumps left on farms after clearing woodlands may serve as secondary hosts or reservoirs to fungal pathogens that attack cassava. It is therefore a good practice to remove stumps from cleared fields.

3. *Host Resistance*: Results of work in *Polyporus* endemic areas in Ghana indicate that some varieties tested gave yields two times higher than the susceptible local cultivars.

2.3.9 Cassava Bud Necrosis

Bud necrosis is a fungal disease usually found in cassava-growing areas with humid environments. Incidence of the disease is therefore higher in the humid forest zones compared to the drier savanna areas. Observations indicate that poor farm sanitation (weedy fields) in high relative humidity zones promotes high incidence and severity of the disease.

2.3.9.1 Symptoms

The main symptom of the disease is the appearance of dark or gray patches or necrotic lesions on stem surfaces of susceptible cultivars (Fig. 2.47). The necrotic areas are made up of fungal tissue of the causal organism. Necrotic areas often cover buds on the stem, giving the disease its name. The fungal matter sometimes covers buds ("eyes") on cassava stem cuttings (Fig. 2.47). The affected buds die, which reduces the sprouting ability of stem cuttings.

Fig. 2.47 Cassava plants with bud necrosis lesions on stem

2.3.9.2 Host Range

The fungus that causes bud necrosis can also live on other crops such as yam, banana, and mango. Debris from these plants suspected to be harboring disease-causing organisms must be destroyed by burning, particularly during land preparation.

2.3.9.3 Spread

The main sources of the fungus that causes bud necrosis are cassava plants with the disease. Dead cassava stems and leaves with the fungus also serve as sources of the inoculums if they are not destroyed after root harvest. The disease spreads through spores that are carried by wind from plant to plant or from farm to farm. Bud necrosis of cassava, however, largely spreads through the use of infected stem cuttings bearing necrotic lesions for planting.

2.3.9.4 Management

1. *Cultural Methods*
   (a) *Healthy Planting Materials*: In farming communities where the desired varieties grown are susceptible to the disease, farmers must use cuttings derived from healthy stems completely free from necrotic lesions.

   Stem cuttings with infected buds often fail to sprout when planted, resulting in poor plant establishment on farms. In large-scale commercial production farms, refilling or replacement of cuttings that fail to sprout increases production cost to farmers.

   (b) *Good Farm Practices*: Incidence and severity of bud necrosis is significantly reduced on farms when good planting distances that allow free movement of air around plants are maintained. The disease is better controlled when weeds are also well managed. These good practices contribute effectively to disease control when healthy planting materials are always used to start farms.

   Plant debris, especially stem pieces bearing necrotic lesions, must be destroyed by burning immediately after harvest to reduce sources of infective fungal spores.

2.3.10 Rust, *Uromyces manihotis*

2.3.10.1 Symptoms

This disease is characterized by the formation of pustules on the veins, petioles, or green stems.

Fig. 2.48 Rust symptoms on leaves and stems of cassava

These pustules are orange or light to dark-brown depending upon the age of the pustule or the type of fungal fructification. They usually induce distortion of the affected plant (Fig. 2.48).

2.3.10.2 Management

Biological control is suggested by means of aspersions with *Darluca* spp. suspensions.

2.3.11 Super Elongation Disease, *Sphaceloma manihoticola* (*Elsinoe brasiliensis*)

2.3.11.1 Distribution and Economic Importance

The disease is distributed in Brazil, the Caribbean, Central America, Colombia, and Venezuela. *Sphaceloma manihoticola* has also been recorded from Cook Islands associated with scab-like lesions and leaf distortion, but without stem elongation.

Sphaceloma manihoticola is considered a major pathogen of cassava that can drastically reduce yields (Alvarez and Molina 2000). Early infection of super elongation disease has been reported to result in up to 80 % yield loss. Late infection does not result in significant losses.

2.3.11.2 Symptoms

Field symptoms included raised, pale, tan-colored corky lesions on the petioles and midribs of leaves and stems (Fig. 2.49). Infrequently, chlorotic spots with necrotic centers were seen on the leaf lamina, which dried and fell out giving a "shot hole" appearance. Leaves were often twisted and curled.

Besides the typical leaf and petiole lesions, severe infection of young plants can result in the elongation of internodes resulting in thin and weak stems (due to the hormone gibberellin), which accounts for the common name super elongation disease. This symptom is frequently seen in severely infected young plants, but was not observed in Trinidad. However, it is not unusual for plants suffering from this disease not to exhibit elongation of the internodes, especially during the dry season (Krausz 1976). Tubers are produced, but they are smaller and thinner than normal.

2.3.11.3 Host Range

Besides *Manihot* spp., *Euphorbia brasiliensis*, *Jatropha curcas*, and *J. aconitifolia* var *papaya* (Zeigler et al. 1984) are also reported as hosts.

2.3.11.4 Development and Spread

Super elongation disease is most severe during the rainy season and is spread by the use of infected planting material and wind flow. Volunteer plants are a major source of infection. Conidiophores are produced in the cankers which are acervuli. Spores are believed to be spread by means of rain or wind. At the start of the dry season, infection and spread of the disease decline.

Fig. 2.49 Symptoms of super elongation disease (leaf, stem, and petiole lesions)

Temperatures of 20–23 °C and relative humidity of 90–100 % increase the disease incidence. The disease increases at the resumption of wet conditions as free moisture is necessary for conidia to germinate. Initially the fungus grows on the epidermis and following direct penetration grows in the intercellular spaces of the epidermis and cortex. The cankers are formed when mycelia aggregates, push up the cortex, and cause epidermis to rupture.

2.3.11.5 Management

The key for managing super elongation disease is to ensure that the planting material used to establish the crop is clean. Farmers must therefore inspect fields from which they obtain planting material before the leaves fall.

Fungicide trials with copper hydroxide, Benomyl, and Maneb have shown promise. The disease can be eradicated from infected cuttings by treating them with captafol (80 %) dip at 4,000 ppm a.i. for 3 min.

The use of resistant varieties where possible is the best method of control. Meristem-tip culture also eliminates the disease.

2.3.12 Bacterial Blight, *Xanthomonas campestris* pv. *manihotis*

Cassava bacterial blight (CBB) can be found in all cassava-producing countries and is the most devastating of all-known diseases of the crop in Africa. In susceptible cultivars, CBB can cause 100 % yield loss when environmental conditions optimum for disease outbreaks occur. CBB causes death of leaves and stems, and as such severe yield losses are often encountered when outbreaks of the disease occur. Severe outbreaks of CBB in epidemic proportions caused famine in Zaire (now Congo DR) and Nigeria in the early 1970s. In severe outbreaks of the disease, shortages in the supply of planting materials may result.

2.3.12.1 Distribution

The bacterial blight is widely distributed in Argentina, Brazil, Cameroon, Colombia, Ghana, India, Indonesia, Madagascar, Malawi, Malaysia, Mauritius, Mexico, Nicaragua, Nigeria, Panama, Philippines, Taiwan, Thailand, Togo, Uganda, Venezuela, and Zaire.

2.3.12.2 Symptoms

Symptoms of secondary infection are angular leaf spotting, followed by blight (Fig. 2.50), defoliation, wilting, and dieback. At first the leaf spots are small, angular, and water-soaked, but they enlarge, covering all or part of the leaf, and turn brown or dark-brown. The initial spots are clearly visible on the undersides of leaves. The necrotic leaves roll and dry up, and after remaining attached to the stem for a short while, they fall off. Gum is often exuded from young infected

Fig. 2.50 Bacterial blight on tapioca leaf

stems, petioles, and leaf spots and collects as droplets. On leaves these droplets occur mainly on undersides along veins or veinlets. On petioles and stems, the gum is exuded from cracks. The gum forms a yellowish glistening scab when dried (Fig. 2.50).

Vascular tissue of infected petioles and stems becomes necrotic and appears as brown strings. Leaves supplied by these vascular tissues wilt and young stem tissues rot. The young green stems rot faster than the older greenish-brown ones which maintain a healthy appearance. The rotting of the young stems causes characteristic dieback symptoms.

Root rot may appear when susceptible cultivars with a good water balance are infected. Roots have dry vascular necrosis. On young immature plants, complete destruction of the aerial portions may be caused. New shoots are usually produced, but these rapidly become infected, especially during the rainy season. All symptoms appear mainly during the rainy season and vary according to the susceptibility of the variety and the period of infection.

2.3.12.3 Development and Spread

The spread of the disease occurs if the bacterial causal organism is transferred from an infected plant to a new susceptible plant. Often the bacterium gains entry into plants through wounds or openings or scratches on leaves and stems.

Farm tools particularly cutlasses can easily be used to transmit the disease from farm to farm if tools are not cleaned after work in an infected field.

Fig. 2.51 A bacterial blight-infected plant showing wilting and grasshoppers. The grasshoppers are believed to be involved in spreading the disease from plant to plant

Stems of infected plants are sources of the bacterium that causes the disease. The spread of the disease is largely through the use of cuttings derived from infected stems in starting new farms.

It is common to find grasshoppers on plants showing symptoms of CBB compared to healthy plants in the same area (Fig. 2.51). This observation has been made in several disease documentation surveys. Grasshoppers and other insects that feed on cassava may therefore be involved in the transmission of the disease from plant to plant on the same farm or over long distances.

2.3.12.4 Management

1. *Quarantine Measures*: In cases of localized outbreaks of CBB, it may be necessary for a localized quarantine measure to be implemented that ensures that planting materials from CBB outbreak zones are not moved into other communities. Farmers must be educated to appreciate the importance of such a measure.

   Quarantine authorities must monitor movement of planting materials across borders and ensure that uncertified planting materials are quarantined to be sure of their safety. Tissue culture materials from unreliable laboratories or sources must pass quarantine regulations before they are allowed into cropping systems of countries.
2. *Cultural Methods*
   (a) *Rouging of Plants*: Farmers cultivating susceptible varieties must regularly inspect their plants and remove and destroy the plants showing disease symptoms through burning. This practice of search and destroy must seriously be followed after the first few rains following the dry season. It is safe to eat storage roots from CBB-infected plants. Early detection and destruction of plants showing symptoms of the disease can prevent or slow down the spread of the disease on a farm.
   (b) *Fallow*: If complete outbreaks on whole farms are observed, it may be necessary to harvest roots immediately and destroy stems and leaves and other plant debris through burning. Infected plant debris may also be plowed deep into the soil. In both situations, the land must be allowed to fallow for few years (3 years minimum) before it is planted again to cassava. The bacterium responsible for the disease is unable to survive over a long period outside the host.
   (c) *Crop Rotation*: In communities where land for peasant farming is scarce, fallowing of land may not be a good disease control option. After destruction of infected plant debris as described in the preceding section, cereals, grains, or legumes can be planted on the same piece of land. After three seasons, the land can be returned to cassava.
   (d) *Use of Healthy Planting Material*: Farmers must consciously harvest stems from healthy plants completely free of any of the described symptoms of the disease. It is generally wise to avoid using planting materials from any farm suspected to be under CBB attack even if symptoms of the disease have not been observed on plants.
   (e) *Pest Control*: Grasshoppers and other insects that can serve as carriers of the bacterium from farm to farm must be controlled using environmentally friendly measures. Controlling grasshopper movement after the early first rains of the wet season can reduce the spread of CBB.
3. *Biological Methods*: Several bacterial antagonists such as *Bacillus cereus*, *B. subtilis*, and *Pseudomonas* spp. and some fungi such as *Trichoderma* spp. and *Gliocladium* spp. were applied successfully to control the bacterial blight (Amusa and Odunbaku 2007).

   Foliar application of *Pseudomonas fluorescens* and *P. putida* significantly reduced the number of angular leaf spots per leaf on susceptible clones in cassava (Hernandez et al. 1985).

2.3.13 Bacterial Stem Rot, *Erwinia cassava*

2.3.13.1 Symptoms

An infested stem rots internally resulting in the wilting of young shoots which is then followed by tip collapse and dieback. This is characterized by a pungent, soft stem rot, and discoloration of the woody portion of the plant (Fig. 2.52). Shoots of infected plants often wilt and die. Diseased cuttings used for planting do not always germi-

Fig. 2.52 Symptoms of bacterial stem rot on cassava

nate, and if they do, they are stunted with a small number of thick roots.

2.3.13.2 Management

Spray mint leaf extract. Grind 250 g of leaves to make into a paste. Add 2 l of water, stir well, strain it to have a clear extract, and spray.

2.3.14 Cassava Mosaic Disease

Cassava mosaic disease (CMD) is the most serious viral disease of cassava in India. Cassava mosaic disease is caused by the cassava mosaic geminivirus.

2.3.14.1 Distribution and Economic Importance

The disease occurs in East, West, and Central Africa and India. Estimate of losses caused by mosaic ranges from 30 to 80 %.

2.3.14.2 Symptoms

The first symptom appears on young leaves as chlorotic speck. Gradually they enlarge and intermix with green tissue to provide a mosaic pattern. The pale discoloration may be intensified to yellow color depending on the varieties. The leaf area is reduced, and in extreme cases leaf distortion and shoe string appearance are observed (Fig. 2.53). The intensity of symptoms varies with season, genotypes, and virus load present. There may be symptom variation in the same plant. Some plants show symptoms on the younger leaves and recover at the later stages. In serve case of infection, the growth of the plant is affected which ultimately leads to crop loss. The severity of the disease is assessed by using the following scale.

Grade 1—Only specks <100 mm^2
Grade 2—Wide areas of mosaic, no distortion
Grade 3—Distortion, leaf area reduction 25 %
Grade 4—Distortion, leaf area reduction 26–75 %
Grade 5—Distortion leaf area reduction >75 %

The disease is caused by single-stranded DNA virus of the genus *Begomovirus* under the family *Geminiviridae*. Two distinct virus species under the genus *Begomovirus*, viz., *Indian cassava mosaic virus* (ICMV) and *Sri Lankan cassava mosaic virus* (SLCMV), are known to cause cassava mosaic in India. They are serologically different from *African cassava mosaic virus* (ACMV).

Tuber yield of diseased plants varies with different varieties as well as time of infection. Up to 80 % crop losses have been recorded in highly sus-

Fig. 2.53 Symptoms of cassava mosaic disease (CMD) infection on leaves (mosaic and leaf distortion)

Fig. 2.54 Reduction in tuber size caused by CMD

ceptible varieties as against 18 % in field-tolerant varieties (Fig. 2.54). Crop loss will be maximum when the disease appeared at the time of planting. No reduction in yield has been recorded when the disease occurred after 5 months of planting.

2.3.14.3 Particle Morphology

The particles typically are bisegmented, approximately 30×20 nm (Fig. 2.55).

2.3.14.4 Transmission

A specific biotype of whitefly (cassava biotype) is known to transmit the virus. Insect transmission alone does not cause much spread of the disease—man appears to be the main factor. The virus could be easily transmitted through grafting and dodder. The disease is not transmitted through seeds.

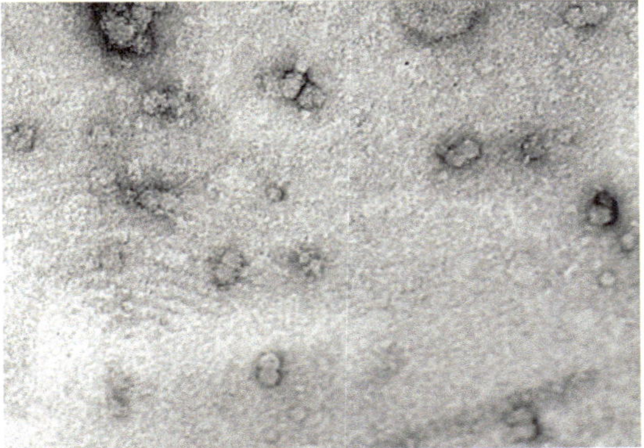

Fig. 2.55 Virus particles of Indian cassava mosaic virus

2.3.14.5 Spread

Primary spread of the disease is through the infected planting material. Secondary spread of the disease is through the insect vector, whitefly *Bemisia tabaci*. The virus occurs in the saliva of the whitefly. During feeding, the insect injects saliva containing the virus into cassava leaves. The virus multiplies and occurs in large numbers in the leaves and stems. Cassava mosaic disease is also spread by planting stem cuttings from plants infected with the disease.

2.3.14.6 Host Range

Cassava mosaic disease attacks only cassava.

2.3.14.7 Management

1. *Cultural Methods*
   (a) *Use of Disease-Free Planting Material*: Meristem culture combined with thermo- and chemotherapy will help virus elimination from the infected plant. However, it is recommended that meristem-raised tissue-cultured plants (Fig. 2.56) should be indexed for the absence of virus through molecular diagnostic technique like PCR or NASH. Wherever possible, the virus-free planting materials may initially be multiplied in the vector-free zones or disease-free areas, and later on the affected areas may be replaced by healthy plants. It should be remembered that the tissue-cultured plants are not resistant to disease; hence, they are likely to get infected under field conditions. However, it has been demonstrated that loss due to secondary spread of the disease is minimum compared to primary spread through infected cuttings.

Like other viral diseases, it is very difficult to control CMD with any one control measure. An integrated approach for the management of the disease has been recommended. The following practices are advocated for effective management of the disease.

- Selection of disease-free meristem-derived planting material, followed by clonal multiplication with periodical screening and rouging of freshly infected plants, will be useful to raise a good disease-free crop.
- Use disease-free healthy planting materials by selection.
- Production of virus-free planting material in vector-free locations (high altitude).
- Spray 0.05 % dimethoate for control of vector.
- Aqueous leaf and root extracts of *Boerhaavia diffusa* and *Mirabilis jalapa* were found to inhibit the ICMV under glasshouse conditions.
- Disease-free planting material can be multiplied on a large scale at higher altitude where the whitefly population is low or nil.

Fig. 2.56 Production of CMD-free plants through meristem culture and hardening of plants

- Breeding for resistance to CMD.
- Use field-resistant or field-tolerant cultivars like Sree Padmanabha, Sree Vijaya, H-97, H 165, Sree Visakham, Sree Sahya, etc.
- Strict field sanitation and rouge of infected plants and burn.
- Use of true cassava seeds.
- Planting of border rows with crops like maize or sorghum.
- Raising the plants in the nursery at closer spacing prior to transplantation into the main field is a useful step to prevent the primary spread of the disease in the main field.
- Adherence to strict sanitary practices such as timely harvest and prompt disposal of crop residues, and eradication of self-sown plants and weeds which may harbor both the disease and vectors.

(b) *Cassava True Seed Program*: Cassava true seed program (Fig. 2.57) was useful in popularizing cassava in nontraditional areas where the crop was grown mainly for starch purpose. The male sterile line "Ambakkadan" which is a popular cultivar grown in Kottayam and Idukki districts of Kerala had recorded the highest mean number of fruits per plant (168). The weight of 100 seeds ranged from 9 to 12.9 g. Seeds could be stored under ambient conditions up to 6 months without loss of viability. Male sterile lines recorded very good germination (80–90 %). Field establishment when the seedlings were transplanted from primary nursery ranged from 75 to 91 %. The seedling progenies were free from CMD, compared to the clones. Progenies of CE-647, S-1, and Ambakkadan were good yielders at closer spacing of 70 × 70 cm, and the yields were at par with that of the clones of promising varieties. Ambakkadan seedlings recorded higher tuber yield of 2.2 kg plant^{-1}, and the cooking quality was very good in both seasons. In the first clonal evaluation of seedling progeny lines, the progeny lines of CE-639, CE-630, CI-453, S2, and Ambakkadan have recorded distinctly higher yield than the control. The clonal

Fig. 2.57 Production of true cassava seeds for managing CMD

selections of Ambakkadan × TCH-2 recorded maximum tuber yield (29.38 t ha^{-1}) followed by Ambakkadan × S-856 (27.80 t ha^{-1}), and the selections were free from CMD.

2. *Biological Methods*: Among 4 rhizobacteria [*Pseudomonas aeruginosa* (RB9), *Bacillus pumilus* (RB26), *B. subtilis* (EN16), and *Enterobacter cloacae* (EN22)] and 11 different combinations, the incidence of CMV was less in all the treated plants compared to control. About 53 % CMV incidence was noticed in the case of *E. cloacae*-treated plots compared to 90 % in control. There was significant increase in plant height, girth, and tuber yield in all the treated plants compared to control. Among the combination treatments, *B. subtilis* and *E. cloacae* showed significant reduction of CMV (31 %), but it was less than *E. cloacae*.
3. *Host Resistance*: Triploid cassava cvs. Sree Apoorva, and Sree Athulya are tolerant to cassava mosaic disease (CMD). Growing field-tolerant varieties—H97, H165, Sree Visakham and Sree Sahya—are also helpful for managing the disease.
   (a) *GM Cassava*: The recent progress in the biotechnology of genetic modification and the continuing importance of cassava in the welfare of impoverished people in Africa make GM orphan crops such as cassava promising tools to significantly reduce poverty and improve food security in the least-developed countries (LDCs) in sub-Saharan Africa.

Transgenic cassava resistant to cassava mosaic disease (CMD) is of particular interest to many African countries due to CMD's prevalence in the continent and the relatively advanced knowledge of the disease in American and European research institutions.

Exploitation of the plant immune system against viruses (Ding and Voinnet 2007) by expression of hairpin RNA homologous to viral sequences has proven effective to generate virus-resistant crops including cassava (Yadav et al. 2011). The modularity of the hairpin RNA technology is suitable to generate plants resistant to multiple viral species. Stretches of viral sequences can be combined into a single hairpin construct (Bucher et al. 2006). Alternatively DNA sequences highly conserved between viral species can be targeted to generate transgenic plants with broad-spectrum resistance. Analysis of the CBSV and UCBSV isolates revealed high DNA sequence conservation in the 3′ region of the viral genome, where the gene for the coat protein (CP) sequence is located. The stable resistance to both CBSV and UCBSV can be engineered by using hairpin RNA homologous to the 3′-end of the CBSV-CP sequence in the cassava model cultivar 60444 (previously referred to as TMS 60444). Using an improved cassava transformation protocol, the technology was transferred to a selected Nigerian cassava landrace previously identified to be CMD resistant (Raji et al. 2008) (Figs. 2.58 and 2.59). The combination of natural and engineered virus resistance is therefore a promising approach to combat multiple cassava viral diseases in Africa.

2.3.15 Cassava Brown Streak Disease

Cassava brown streak disease (CBSD) is the most important of all cassava diseases known in Africa. Almost all traditional cultivars of cassava cultivated in different countries on the African continent are susceptible to this disease.

2.3.15.1 Distribution and Economic Importance

The disease is prevalent in Africa, particularly in coastal cassava-growing regions in East Africa: Kenya, Malawi, Tanzania, and Uganda (Lennon et al. 1986).

Yield losses due to CBSD range between 20 and 95 % in susceptible cultivars. On a continent that experiences food insecurity, losses of this magnitude in yield of such an important food crop cannot be accepted. Most traditional farmers who constitute the majority of food producers in Africa do not control CBSD disease. In Ghana, for example, yields of cultivars susceptible to CBSD disease are under 10 t/ha in most farming

Fig. 2.58 Geminivirus infected cassava (*Left*: wild-type cassava, *Right*: transgenic cassava) (Vanderschuren et al. 2009)

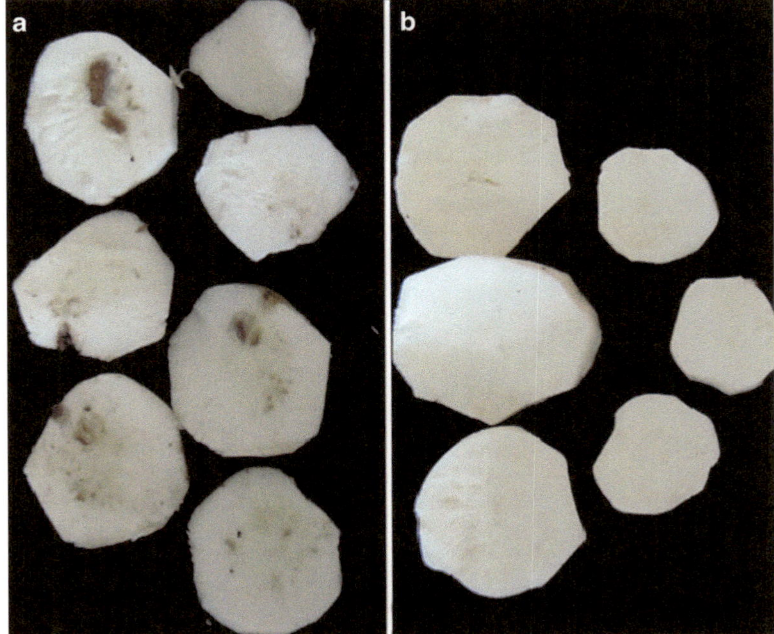

Fig. 2.59 CBSV—infected cassava roots (**a** wild type, **b** transgenic)

communities compared to yields of 30 t/ha or more that can be achieved with improved varieties resistant to CBSD disease.

2.3.15.2 Symptoms

1. *Symptoms on Leaves*: It appears as patches of yellow areas mixed with normal green color commonly referred to as chlorosis. It produces characteristic yellow or necrotic vein banding on leaves which may enlarge and join to form comparatively large yellow or necrotic patches (Fig. 2.60). The infected leaves do not become distorted in shape as occurs with leaves infected by cassava mosaic disease.

2. *Symptoms on Stem*: The disease appears as dark-brown "streaks" and "spots" on stems, with dead spots on leaf scars. These streaks are most prominent on upper green portions of the stem. "Streaks" may appear as scratch-like wounds on stems (Fig. 2.61). The diseased plants may show shoot tip death, which may progress into cassava stem dieback. In severe cases, there may be leaf drying and shoot dieback.
3. *Symptoms on Tubers*: The disease may cause cracks and discoloration in the storage roots and often causes root constriction and malformation. The harvested roots have corky, yellow-brown necrotic spots and root rot becomes evident (Fig. 2.62). In the roots of a susceptible variety, the disease causes a dry, hard rot which is irregular, yellow blotchy chlorosis. Necrosis is most pronounced in the edges of the root when the root/tuber is cut across with a knife.

Fig. 2.60 Symptoms of cassava brown streak disease on the leaves

2.3.15.3 Particle Morphology

It is not clear whether 1 or 2 viruses with filamentous particles are involved. In pure virus preparations, two particle lengths are sometimes observed: 650–700 nm (carlavirus-like) and a longer, more fragile flexuous (potyvirus-like) particle (distant serological relationships to both a carlavirus and potyviruses have been observed) (Fig. 2.63). Infected test plants contain pinwheel inclusions.

2.3.15.4 Transmission and Spread

Infected plants are the main sources of the virus responsible for the disease. The disease is spread through the planting of stem cuttings from diseased plants. The whitefly *Bemisia tabaci* is the vector in the transmission of the disease from plant to plant through its feeding on cassava

Fig. 2.61 Symptoms of cassava brown streak disease on stem

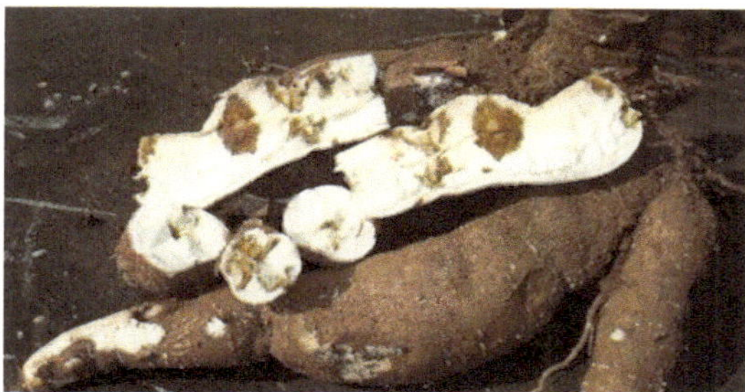

Fig. 2.62 Symptoms on cassava tubers

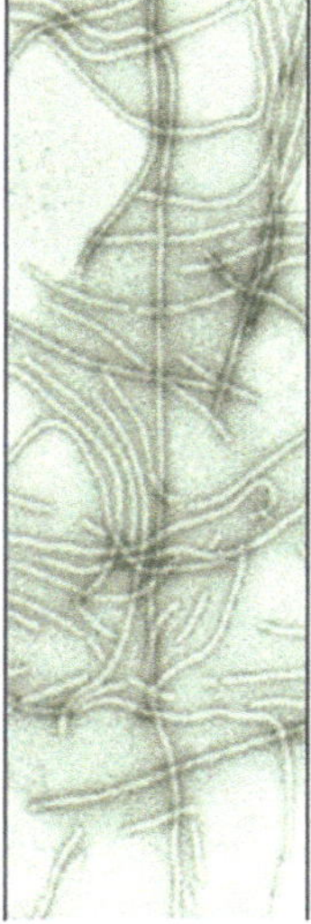

Fig. 2.63 Filamentous virus particles

leaves. Virus numbers increase in leaves and stems of infected plants. The disease spreads largely through the use of infected stem cuttings as planting materials.

2.3.15.5 Host Range

Brown streak disease is not known to attack other crops.

2.3.15.6 Management

1. *Cultural Methods*
   (a) *Use of Healthy Planting Materials*: The spread of the disease as mentioned earlier is largely due to the use of infected cuttings as propagating materials. Mosaic disease can be controlled effectively through the use of healthy or virus-free planting materials. Farmers must select stems from plants with no mosaic symptoms to produce cuttings.

   If the source of planting materials is from a branching susceptible variety, spread of the disease can be reduced if cuttings for planting are derived from the branches rather than the main stem. Cuttings from the main stem often sprout with the disease. In nonbranching varieties, it is preferable to avoid using cuttings from the basal portion of stems as planting material. Cuttings from the basal portions of stems give rise to young diseased plants at sprouting.

   Plants that develop from healthy cuttings often grow vigorously to escape early infections and give good leaf and root yields. Programs of planting material multiplication that supply stems to large numbers of farmers must ensure that disease-free stems are distributed to prevent large-scale spread of diseases especially into new communities.

   (b) *Rouging*: The practice of removing the few diseased plants showing symptoms of the disease from a population of healthy ones and destroying them (rouging) reduces spread of diseases particularly on farms. Regular inspection of plants on a farm is necessary if effective disease control is to be achieved.

   Young plants from sprouted cassava stem cuttings showing mosaic symptoms can be replaced with new healthy ones to maintain high planting densities.

   (c) *Thermotherapy and Meristem Tip Culture*: Thermotherapy followed by meristem-tip culture gives virus-free plants.

2.3.16 Witches' Broom

Cassava witches' broom is caused by phytoplasma.

2.3.16.1 Distribution

Disease has been detected in Vietnam, Thailand, Cambodia, Laos, Indonesia, and the Philippines. The disease is also locally important in Brazil

Fig. 2.64 Cassava witches' broom—excessive proliferation of branches is one of the most typical symptoms

(states of Ceara, Pernambuco, and Sao Paula) and southern Mexico (Costa and Kitajima 1972).

2.3.16.2 Symptoms

Three different types of symptoms have been reported: (1) stunting and excessive proliferation of branches; shoots have small leaves and shortened internodes, without distortion or chlorosis (Fig. 2.64); (2) proliferation of shoots from the cutting with generally weak growth; (3) a few weak, stunted shoots germinate from the cutting which never reach normal size (Kitajima and Costa 1979). Moderate temperatures (between 13 and 20 °C) favor the disease, and at higher temperatures the symptoms disappear.

2.3.16.3 Transmission

The disease is transmitted by vegetative propagation. Mechanical transmission has also been reported, but insect transmission is unknown.

1. *Management*

Remove and destroy plants suspected of being infected; remove all cassava debris from field after harvest; disinfect all tools and equipment between cuttings.

The disease can also be managed by thermotherapy followed by meristem-tip culture.

2.3.17 Frog Skin Disease, Phytoplasma

Cassava frog skin disease (CFSD) is an important disease affecting cassava roots caused by the phytoplasma. Symptoms disappearance in leaves was obtained by growing cassava-infected plantlets in a chloroxytetracycline solution producing further evidence for the phytoplasma etiology

2.3.17.1 Distribution

FSD is found in the Andean regions of central and southern Colombia. It is also endemic in the Amazon regions of Colombia. CMD is found in the northern areas of Colombia near the Caribbean (Anon 1990; Lozano and Nolt 1989).

Recently, it has been reported with increasing frequency in Colombia, Brazil, Venezuela, Peru, Costa Rica, and Panama, with an incidence of up to 90 % recorded in Colombian commercial fields in the production areas of Valle del Cauca, Cauca, Meta, and the North Coast.

2.3.17.2 Symptoms

This disease is not common but can cause tremendous losses (50–100 %) when it occurs. It is characterized by the thickening of the root

Fig. 2.65 Symptoms of frog skin disease on cassava tubers

epidermis, thin tubers, the irregular swelling (Fig. 2.65), and easy breakage of the periderm and thick, cracked, wrinkled skin. CFSD causes deep lesions in roots, eventually reducing their diameter; therefore, in many cassava cultivars, symptoms are observed only when the plants are harvested. Some cultivars may also show leaf symptoms such as mosaic, chlorosis, and curling or curvature in leaf margins.

2.3.17.3 Transmission

The disease spreads rapidly. The whitefly *Bemisia tuberculata* is suspected to be the vector of both FSD and CMD. This disease is also transmitted by cuttings, grafting, and dodder.

2.3.17.4 Management

Healthy planting material should be used for planting. Thermotherapy followed by meristem-tip culture also eliminates the disease.

2.4 Nematodes

Important nematodes parasitizing cassava are root-knot and root lesion nematodes. Ray et al. (1990) reported *Helicotylenchus multicinctus* and *Tylenchorhynchus mashhoodi* from Orissa.

2.4.1 Root-Knot Nematodes, *Meloidogyne* spp.

Meloidogyne incognita and *M. javanica* are the most widely reported and economically important nematodes (Sorley et al. 1983).

2.4.1.1 Distribution and Economic Importance

Meloidogyne incognita, *M. javanica*, and *M. arenaria* have been reported from cassava from Brazil, Colombia, Dominican Republic, Hawaii, India, Ivory Coast, Nigeria, Rhodesia, Tanzania, Togo, Trinidad, and the United States. Nirula (1963) reported *M. javanica* for the first time from India.

Jatala (1988) reported considerable yield reduction in cassava due to *M. incognita* and *M. javanica* infestation. *M. incognita* race 2 reduced fresh storage root weight by 49 % and dry top weight by 22 %.

2.4.1.2 Symptoms

Meloidogyne spp. feeding produces physically damaging galls with eggs inside them (Fig. 2.66). Galls later merge as the females grow and enlarge, and they interfere with water and nutrient supply (Gapasin 1980). Cassava roots become tough

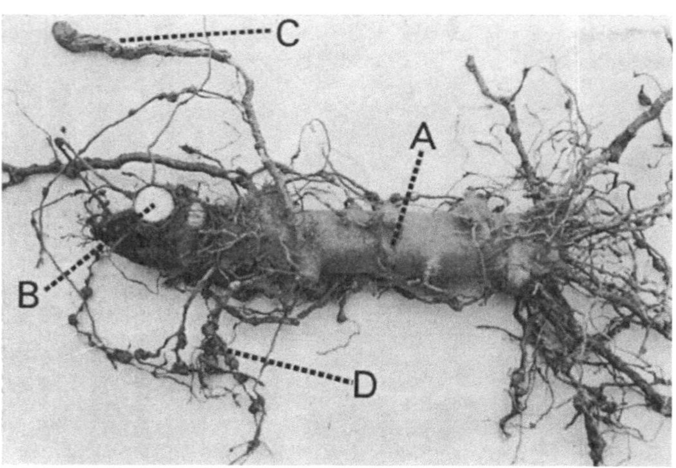

Fig. 2.66 Root-knot nematode on cassava

with age and restrict the movement of the juveniles and the egg release. It is therefore possible that extensive galling can be observed even at low densities following infection (Coyne 1994). Other pest and diseases can gain entry through the physical damage caused by gall formation, leading to rots. They have not been shown to cause direct damage to the enlarged storage roots, but plants can have reduced height if there was loss of enlarged root weight (Caveness 1982). Attack by root-knot nematodes was considered serious when cassava was intercropped with other susceptible hosts like brinjal and roselle. *M. incognita* race 2 and *M. javanica* significantly reduced stalk length, stalk weight, and storage root weight of cassava after 15.5 months growing period (Caveness 1981).

2.4.1.3 Management

1. *Cultural Methods*: The summer fallowing which is already being practiced by farmers in many parts of cassava-growing areas in Kerala due to nonavailability of water is an excellent means of controlling nematode population. It is also essential to avoid planting susceptible crops either in the previous season or as intercrops. Cassareep, a by-product of the cassava industry, was apparently effective in controlling *M. incognita* and *M. javanica* on cassava.
2. *Host Resistance*: Cassava cvs. Sree Sahya and Narayaniyakappa were found resistant to root-knot nematodes (Sreeja et al. 1998). Freitas and Moura (1986) reported cassava cv. Mandiocol to be resistant to *M. incognita* and *M. javanica*.

2.4.2 Lesion Nematodes, *Pratylenchus brachyurus*, *P. sefaensis*

Pratylenchus brachyurus and *P. sefaensis* were found attacking cassava. *P. brachyurus* is regarded as a serious pest of cassava causing damage to cortical tissues of roots and also predisposing plants to infection by secondary pathogens. *P. sefaensis* reduced fresh storage root weight by 45 % and dry top weight by 21 %.

2.4.2.1 Symptoms

The lesion nematode produces root lesions and subsequent rotting of roots. Affected plants are stunted with yellowing of leaves and dieback of twigs and produce small or no tubers. Leaves also show nutrient deficiency symptoms (Mohandas et al. 1990).

2.4.2.2 Management

1. *Chemical Methods*: A 7.2 % increase in tuber yield was recorded after application of DBCP at 50 l/ha.
2. *Host Resistance*: A number of cassava varieties (Atitogen, Ba Pou II, Agba Tiega, Agba

Boquia, Sodjievi) show some resistance to the lesion nematode.

2.5 Future Thrusts

- Emergence of biotypes of *Bemisia tabaci* which is a vector of CMD should be monitored.
- Multidisciplinary approach shall be adopted to tackle some of the problems limiting the productivity of cassava such as mosaic disease.
- Virus elimination through in vitro approaches along with thermotherapy and meristem culture for the production of quality planting material.
- Threats from established and emerging pests and diseases in traditional and new cassava niches need to be monitored.
- Strict quarantine should be followed for monitoring movement of cassava planting material in order to prevent the entry of cassava brown streak virus (CBSV) which is not reported in India till now but is widely prevalent in Africa and causes major havoc in tuber production.
- As the Cassava mosaic disease is one of the major diseases in India, studies are to be continued in order to monitor the development of new virus strains (as it has happened in Africa). This will help in keeping the disease under control by development of resistant varieties and vector control measures.
- Development of transgenic cassava resistant to cassava mosaic virus through marker-aided selection (MAS) and RNAi-mediated approach should be given due attention by increasing the transformation efficiency and developing more effective gene constructs.
- Detailed studies should be undertaken on cassava tuber rot and development of disease-resistant varieties using biotechnological methods (pathogen-derived and elicitor-induced methods).
- Quick diagnostic methods to be developed for important virus like Sri Lankan cassava mosaic virus (SLCMV).
- Use of biocontrol agents (*Trichoderma harzianum*, *Pseudomonas fluorescens*, etc.) should be intensified so as to formulate effective technology for the management of soilborne diseases in cassava.
- Disease-free cassava mother cultures should be established for all the released cultivars and maintained for further mass multiplication.
- Emerging pests and diseases due to climate change and introduction of invasive pests and pathogens are also main challenges. Appropriate technologies to contain such biotic stresses have to focus on refined eco-friendly methods. Availability of healthy planting material, particularly cassava free from mosaic disease, is a challenge to meet the emerging demands. In vitro elimination of cassava mosaic virus has to be standardized using meristem and nodal culture combined with chemotherapy and thermotherapy. Strategy for mass multiplication of disease-free planting materials has to be improvised. Production of large quantities of polyclonal antibodies and developing cost-effective, easy, and sensitive diagnostic kits for indexing the planting materials are necessary.
- Special emphasis has to be given for the identification and characterization of pathogens, development of diagnostic kits/techniques, and production of virus-free mother plants of cassava for mass propagation. Ready-to-use diagnostic kits have to be developed for important viral diseases with special emphasis on CMD and cassava tuber rot. New molecular detection techniques like microarray LAMP, utilization of nanotechnology for quick diagnosis, etc., should also be explored. Production of polyclonal antibody for detecting viral and fungal diseases in large quantities and development of ELISA kit are the prime target in this plan. Pathogen-specific primers are to be designed for PCR-based detection of these diseases.
- The technology developed in controlling the cassava tuber rot disease should be popularized widely in Tamil Nadu especially in the disease-prone areas in irrigated plains and hills.

References

Alvarez E, Molina ML (2000) Characterizing the *Sphaceloma* fungus, causal agent of super elongation disease in cassava. Plant Dis 84:423–428

Amusa NA, Odunbaku OA (2007) Biological control of bacterial diseases of plants in Nigeria: problems and prospects. Res J Agric Biol Sci 3:979–982

Anon (1990) Annual report of the Central Tuber Crops Research Institute, Thiruvananthapuram

Bellotti AC (2002) Arthropod pests. In: Hillocks RJ, Thresh JM, Bellotti AC (eds) Cassava: biology, production and utilization. CABI Publishing, Wallingford, pp 209–236

Bucher E, Lohuis D, van Poppel PM, Geerts-Dimitriadou C, Goldbach R et al (2006) Multiple virus resistance at a high frequency using a single transgene construct. J Gen Virol 87:3697–3701

Caveness FE (1981) Root-knot nematodes on cassava. Annual report IITA. IITA, Ibadan, Nigeria, pp 64–65

Caveness FE (1982) Root-knot nematode as parasites of cassava. IITA Res Brief 3(2):2–3

Chant SR (1958) Studies on the transmission of cassava mosaic virus by *Bemisia* spp. (Aleyrodidae). Ann Appl Biol 46:210–215

Costa AS, Kitajima EW (1972) Studies on virus and mycoplasma diseases of the cassava plant in Brazil. In: Proceedings IDRC/IITA cassava mosaic workshop. International Institute of Tropical Agriculture, Ibadan, Nigeria

Coyne DL (1994) Nematode pests of cassava. Afr Crop Sci J 2(4):355–359

Dejean A, Matile-Ferrero D (1996) How a ground-dwelling forest ant species favors the proliferation of an endemic scale insect (Hymenoptera: Formicidae; Homoptera: Stictococcidae). Sociobiology 28:83–195

Ding SW, Voinnet O (2007) Antiviral immunity directed by small RNAs. Cell 130(3):413–426

FAOSTAT (2012) FAO statistical yearbook 2012. Food & Agriculture Organization of the United Nations, Rome

Freitas OMBL, Moura RM (1986) Response of cultivars of cassava (*Manihot esculenta* Crantz) to *Meloidogyne incognita* and *M. javanica* (Nematoda: Heteroderidae) and comparisons with Hydrogen cyanide content. Nematologia Brasiliera 10:109–131

Gapasin RM (1980) Reaction of golden yellow cassava to *Meloidogyne* spp. inoculation. Ann Trop Res 2:49–53

Hernandez JM, Laberry R, Lozano JC (1985) Biocontrol of cassava (*Manihot esculenta*) root rot by fluorescent *Pseudomonas*. In: VIIth symposium of international society for tropical root crops, Gosier (Guadeloupe)

James B, Yaninek J, Neuenschwander P, Cudjoe A, Modder W, Echendu N, Toko M (2000) Pest control in cassava farms. International Institute of Tropical Agriculture (IITA). ISBN: 978-131-174-6. www.iita.org

Jatala P (1988) Nematodes in tuber and root crops and their management – improvement of sweet potato (*Ipomea batatas*) in East Africa with some references of other tuber and root crops. Report of the workshop on sweet potato improvement in Africa. ILRAD, Nairobi, Kenya, UNDP Project CIAT – CIP – IITA, pp 91–100

Kitajima EW, Costa AS (1979) Microorganismos do tipo micoplasma associados a molestias do tipo amarelo en algunas plantas cultivadas e ornamentais no Estado de Sao Paulo e no Distrito Federal. Fitopatol Bras 4:317–327

Krausz J (1976) The superelongation disease of cassava. Ph.D. thesis, Cornell University, Ithaca, 81 pp

Lennon AM, Aiton MM, Harrison BD (1986) Cassava viruses from Africa. In: Annual report 1985. Scottish Crop Research Institute, Dundee, 168 pp

Lozano JC, Nolt BL (1989) Pest and pathogens of cassava. In: Kahn RP (ed) Plant protection and quarantine: selected pests and pathogens of quarantine significance, vol 2. CRC Press, Pr Inc, Boca Raton, pp 174–175

Mohandas C, Palaniswami MS, Potty VP (1990) Survey, identification and pathogenicity of nematodes in tuber crops. Annual report of Central Tuber Crops Research Institute (1989–90), Thiruvananthapuram, pp. 72–73

Neuenschwander P (2004) Harnessing nature in Africa: biological pest control can benefit the pocket, health and the environment. Nature 432:801–802

Nirula KK (1963) Collateral host plants of root-knot nematodes. Curr Sci 32(25):221–222

Olsen K, Schaal B (2001) Microsatellite variation in cassava (*Manihot esculenta*, Euphorbiaceae) and its wild relatives: further evidence for a southern Amazonian origin of domestication. Am J Bot 88(1):131–142

Pillai KS, Daniel RS (1979) Monthly variations in the population of whitefly *Bemisia tabaci* the vector of cassava mosaic disease. J Root Crop 5:8–10

Raji A, Ladeinde O, Dixon A (2008) Screening land races for additional sources of field resistance to cassava mosaic disease and green mite for integration into the cassava improvement program. J Integr Plant Biol 50:311–318

Ramani S, Bhumannavar BS (2002) Interaction of two indigenous predators of the spiralling whitefly, *Aleurodicus dispersus* Russell (Homoptera: Aleyrodidae) with the introduced parasitoid, *Encarsia guadeloupe* Viggiani (Hymenoptera: Aphelinidae). Pest Manag Hortic Ecosyst 6:79–105

Ray S, Sahoo NK, Mohanty K (1990) Plant parasitic nematodes associated with tuber crops in Orissa (Abstract). National symposium on recent advances in the production and utilization of tropical tuber crops. ISRC, Central Tuber Crops Research Institute, Thiruvananthapuram

Sorley R, Ohair SK, Parrado JL (1983) Nematodes of cassava, *Manihot esculenta* Crantz. Nematropica 13:261–287

Sreeja P, Nageswari S, Mohandas C (1998) Reaction of cassava cultivars to root-knot nematode, *Meloidogyne incognita* (Kofoid and White, 1919) Chitwood, 1949. In: Mehta UK (ed) Nematology: challenges and opportunities in 21st century. Sugarcane Breeding Institute, Coimbatore, pp 137–139

Storey HH (1936) Virus diseases of East African plants VI – a progress report on the studies of the disease of cassava. East Afr Agric J 2:34–39

Taylor KD (1968) An outbreak of rats in agricultural areas of Kenya in 1962. East Afr Agric For J 34:68–71

Vanderschuren H, Alder A, Zhang P, Gruissem W (2009) Dose-dependent RNAi-mediated geminivirus resistance in the tropical root crop cassava. Plant Mol Biol 70:265–272

Yadav JS, Ogwok E, Wagaba H, Patil BL, Bagewadi B et al (2011) RNAi-mediated resistance to cassava brown streak Uganda virus in transgenic cassava. Mol Plant Pathol 12:677–687

Zeigler RS, Lozano JC, Alvarez E (1984) A summary of recent research on the super elongation disease of cassava. In: Proceedings of the sixth symposium of the International Society for Tropical Root Crops. International Potato Center, Lima, pp 363–370

3 Sweet Potato: *Ipomoea batatas*

Abstract

Distribution, economic importance, symptoms/damage, host range, biology, survival and spread, and management of insect and mite pests and fungal, bacterial, viral and nematode diseases of sweet potato (*Ipomoea batatas*) using regulatory, physical, cultural, chemical, biological, host resistance, and integrated pest management methods are discussed.

Keywords

Sweet potato • *Ipomoea batatas* • Pests • Diseases • Nematodes

3.1 Introduction

Sweet potato is an economically important member of the family Convolvulaceae (Fig. 3.1). It is one of the life-sustaining crops rich in carbohydrates that stand between man and starvation. Sweet potato ranks fourth and sixth on the list of dry matter production per hectare and edible energy production per hectare, respectively. It is also used for making starch, syrup, and alcohol. It occupies a prime place in terms of calories production/unit area and time. It is grown as a starchy food crop throughout the tropical, subtropical, and frost-free temperate climatic zone in the world.

The cultivated sweet potato is a hexaploid with $2n=90$. White-fleshed sweet potato is a trailing perennial vine, grown as an annual. Originating in South America, its production has spread throughout the tropics, and it is a staple food in countries of Africa and the South Pacific. The main differences between the white-fleshed and the familiar orange-fleshed sweet potatoes grown throughout the southern United States are that the white-fleshed types tend to have a higher dry matter content with 25–40 % starch and sugar content, are usually less sweet, generally are allowed to grow to a larger size, have variable root shape, and have a skin color that ranges from red to white. In addition the white-fleshed types are usually grown from stem tip cuttings of 30–40 cm in length, whereby little attention has been paid to the qualities and yield potential of the enlarged storage roots below the soil. Planting densities are approximately 30,000 plants/ha, and harvest begins from 4 to 6 months after planting. Although planting is not mechanized, vegetable transplanters could be modified to accommodate the cuttings. Harvesting aids used

P.P. Reddy, *Plant Protection in Tropical Root and Tuber Crops*,
DOI 10.1007/978-81-322-2389-4_3

Fig. 3.1 *Left* – orange-fleshed sweet potato is an important source of beta-carotene, the precursor to Vitamin A. *Right* – anthocyanin (powerful antioxidants)-rich variety of sweet potato

for cocoyam are also used for white-fleshed sweet potato harvest in Florida.

Unlike the other root and tuber crops, sweet potato importation into the United States is prohibited due to concern over importation of exotic diseases and insects. Thus, nearly all of the white-fleshed sweet potatoes marketed in this country are grown in southern Florida. Roots are either baked or boiled. There is little processing potential for white-fleshed sweet potato as it is currently known. Limited amounts of research efforts have been focused on selecting types with little or no sugar content. Such types could be grown in warm climates and used as potato substitutes.

Sweet potato is one of the world's highest-yielding crops in terms of production per unit area. It exceeds that of major cereals such as rice, maize, and wheat and has a higher food value. Sweet potato has been cultivated in Kenya since the end of the nineteenth century and is the second most important and widely distributed food security crop after maize. The crop is mainly grown by poor farmers and is cultivated on about 75,000 ha spread over various agroecological zones in the country. The ability of sweet potato to adapt to a wide range of growing conditions, in both fertile and marginal areas, makes it a versatile crop for Kenya's farming systems.

Approximately 80 % of the world sweet potato is grown in Asia, just under 15 % in Africa and about 5 % in rest of the world. Among the Asian countries, China is the largest producer. Sweet potato is cultivated in an area of 9.2 m ha with an annual production of 138.4 m tons in China which occupies first position in area, globally accounting for 68.3 % of the world area and producing 87 % of world sweet potato. Remaining area is under countries like Uganda, Vietnam, India, etc. Area, production, and productivity of sweet potato are showing an increasing trend globally.

Lack of stability in tuber production as seen in various seasons and locations is the major problem in the cultivation of sweet potato. Studies are initiated to assess the causes of nontuberization, uneconomical yield performance, as well as integrated nutrient management. Studies should be undertaken to identify the most compatible and profitable cropping systems involving sweet potato suiting to major sweet potato-growing states.

It is being cultivated in 0.110 m ha producing 1.040 million tons of sweet potatoes with an average yield of 9.50 t per hectare (Table 3.1). India ranks sixth in area but the productivity is very low (8.9 t/ha), lower than the world productivity level. Orissa, Uttar Pradesh, West Bengal,

Maharashtra, Karnataka, Chhattisgarh, Tamil Nadu, Assam, Madhya Pradesh, and Meghalaya are the leading states in sweet potato cultivation (Tables 3.1, 3.2, and 3.3).

3.2 Insect, Mite, and Vertebrate Pests

Many insect pests have the potential to reduce the quality and yield of sweet potatoes. Insects that damage the roots directly are the most troublesome and are referred to as soil insect pests. They can cause economic loss in relatively low numbers, and it is difficult to control them with insecticides because they live below the soil surface. Insects that injure the foliage reduce the yield of the plants indirectly and are referred to as foliage-feeding insects. These insects can cause economic loss, but usually only at very high numbers, and it is relatively easy to control them with insecticides because they are exposed on the plant.

The sweet potato root can be injured by several soil insects including the sweet potato weevil, rootworms, wireworms, white grubs, white-fringed beetles, and flea beetles. The sweet potato weevil larva is the only insect that tunnels throughout the root. Other soil insects feed on the surface of the developing root. The injury caused

Table 3.1 Area, production, and productivity of major sweet potato-producing countries in the world in 2006

Region	Country	Production ('000 t)	Area ('000 ha)	Productivity (t/ha)
Africa	Nigeria	3,205	989.0	3.2
	Uganda	2,604	590.0	4.4
	Tanzania	1,056	526.0	2.0
	Rwanda	886	148.5	6.0
	Burundi	835	125.0	6.7
	Angola	659	143.8	4.6
	Madagascar	526	123.9	4.3
	Ethiopia	409	50.3	8.1
	Egypt	300	10.0	30.0
	Kenya	231	24.3	9.5
Asia	China	100,215	4,629.0	21.7
	Indonesia	1,840	176.9	10.4
	Vietnam	1,461	188.4	7.8
	Japan	1,053	40.8	25.8
	India	933	107.2	8.7
	Philippines	575	120.6	4.8
	Bangladesh	311	35.1	8.9
	North Korea	392	31.5	12.4
	South Korea	283	17.2	16.4
	Solomon	89	6.2	14.4
America	USA	714	35.5	20.1
	Brazil	514	45.3	11.3
	Cuba	450	74.3	6.1
	Argentine	273	15.8	17.2
	Peru	185	11.3	16.4
	Haiti	161	55.5	2.9
	Mexico	69	3.2	21.5
	Dominican Republic	36	6.0	5.9
	Jamaica	25	1.4	18.5
	Venezuela	13	1.5	8.8

Source: www.fao.org (2007)

Table 3.2 Sweet potato-growing states, regions, and production systems in India

State	Production system	Area/region
Odisha	Plane rain fed Plane irrigated	Entire Orissa (Major districts Ganjam, Keonjhar, Bolangir, Sundargarh, Phulbani, Mayurbhanj, Sambalpur, Kalahandi)
Bihar/Jharkhand	Plane rain fed/irrigated In diara lands after receding of flood	Madhubani, Madhepur, Ranchi, Gumla, East Champaran, Vaishali, Sitamarhi, and Samastipur districts
Uttar Pradesh	Plane rain fed/irrigated	Etah, Mainpuri, Firozabad, Badaun, Fatehpur, Shahjahanpur, Farooqabad, Sultanpur, Kanpur rural, Allahabad, Hardoi, Unnao districts
Assam	Plane rain fed	Dhubri, Kamrup, Nagaon, Sonitpur, Kokrajhar, Lakhimpur, Morigon, Barpeta, Darrang, and Goalpara districts
Meghalaya	Plane rain fed	East Khasi Hills, West Khasi Hills, Jaintia Hills, West Garo Hills

Table 3.3 State-wise area, production, and productivity of sweet potato in India during 2012

State	Area ('000 ha)	% share of area by the state in India	Production ('000 t)	% share of area by the state in India	Productivity (t/ha)
Orissa	43.80	39.68	412.68	39.65	9.422
West Bengal	22.98	20.82	236.96	22.77	10.312
Uttar Pradesh	17.40	15.76	226.03	21.72	12.990
Assam	9.25	8.38	36.39	3.50	3.934
Chhattisgarh	3.55	3.21	35.61	3.42	10.031
Madhya Pradesh	2.18	1.97	24.07	2.31	11.041
Meghalaya	4.40	3.98	14.55	1.40	3.307
Tamil Nadu	0.83	0.75	13.37	1.28	16.108
Andhra Pradesh	0.48	0.43	9.64	0.93	20.083
Bihar	0.46	0.42	9.54	0.92	20.739
Nagaland	0.94	0.85	9.17	0.88	9.755
Kerala	0.37	0.34	4.57	0.44	12.351
Karnataka	2.70	2.45	3.5	0.34	1.296
Rajasthan	0.76	0.69	3.08	0.30	4.053
Andaman and Nicobar Islands	0.15	0.14	0.85	0.08	5.667
Mizoram	0.11	0.10	0.71	0.07	6.455
Lakshadweep	0.03	0.03	0.12	0.01	4.000
India	**110.39**		**1,040.84**		**9.50**
World	**8,087.116**		**103,145.500**		**12.754**
% share by India	**1.365**		**0.98**		

Source: FAOSTAT (2012)

Table 3.4 The sweet potato growth stages and its pests

Growth stages	Insects/mites
Seedling stage—leaves	Armyworm, cutworm, grasshoppers
Vegetative stage—leaves and vines	Armyworm, flea beetles, grasshoppers, stink bug, whitefly
	Sweet potato vine borer, sweet potato weevil
Reproductive—tubers and vines	White grub, wireworm
	Sweet potato vine borer, sweet potato weevil
Maturation—tubers	Sweet potato vine borer, sweet potato weevil, white grub, wireworm

by rootworms and wireworms is similar and cannot be separated easily. These insects chew small, shallow circular or oblong holes through the skin of the root. Flea beetles chew shallow, narrow, winding tunnels along the surface of the root. The injury caused by white grubs and white-fringed beetle larvae is similar to each other and different from the other soil insects. White grubs gouge broad, shallow to deep, irregular areas on the surface of the root. White-fringed beetle larvae chew comparatively narrow and irregular "channels" on the surface of the root.

The following Table 3.4 shows the pests that might damage the plant growth.

The most common pest of sweet potato is weevil, *Cylas formicarius*, which reduces tuber yield and quality, although there are simple cultural techniques that reduce the losses due to this pest. Young plants are susceptible to foliage feeders, such as hornworm, and it may be necessary to treat the unestablished plants with insecticide. *Cylas* spp. and *Alicodes* spp. continue to damage the tubers in storage.

3.2.1 Weevil: *Cylas* spp.

The main problem in production is sweet potato weevil (*Cylas formicarius*) infestation in the roots, which can result in total crop loss if left uncontrolled. Sweet potato weevil is the most serious pest of sweet potato in around the world. It causes damage in the field, in storage, and is of quarantine significance. It is inherently of interest to entomologists due to its strikingly colorful appearance and extremely long rostrum (beak).

3.2.1.1 Distribution and Economic Importance

Sweet potato weevil was first noted in the United States in Louisiana in 1875, and then in Florida in 1878 and Texas in 1890, probably entering by way of Cuba. It is now found throughout the coastal plain of the Southeast from North Carolina to Texas. It is also found in Hawaii and Puerto Rico and widely around the world in tropical regions.

Cylas formicarius from pantropical regions, *C. puncticollis* from Africa (countries south of Sahara), and *C. brunneus* from West Africa have been recorded on sweet potato.

Sweet potato weevil is responsible for losses ranging from 5 to 97 % in areas where the weevil occurs. There is a positive relationship between vine damage or weevil density and tuber damage. However, the plants exhibited some compensatory ability, with the relationship between vine damage and yield nonlinear and sometimes not significant.

3.2.1.2 Damage

A symptom of infestation by sweet potato weevil is yellowing of the vines, but a heavy infestation is usually necessary before this is apparent. Thus, incipient problems are easily overlooked, and damage is not apparent until tubers are harvested. The principal form of damage to sweet potato is mining of the tubers by larvae. The infested tuber is often riddled with cavities, spongy in appearance, and dark in color (Fig. 3.2). In addition to damage caused directly by tunneling, larvae cause damage indirectly by facilitating entry of soilborne pathogens. Even low levels of feeding induce a chemical reaction that imparts a bitter taste and terpene odor to the tubers. Larvae also mine the vine of the plant, causing it to darken, crack, or collapse. The adult may feed on the tubers, creating numerous small holes that measure about the length of its head. The adult generally has limited access to the tubers, however, so damage by this stage is less severe than by larvae.

Adult feeding on the foliage is seldom of consequence. Yield, storage life, and plant vigor are reduced.

3.2.1.3 Biology

Eggs are laid in holes in the vines or tubers, and hatching takes place within a week if the temperature is between 20 and 30 °C. The larvae tunnel in the vines and tubers for about 14 days. When fully developed (fourth instar), the larvae are 3 mm in diameter. Pupation takes place inside the plant and lasts about 1 week.

A complete life cycle requires 1–2 months, with 35–40 days being common during the summer months. The generations are indistinct, and the number of generations occurring annually is estimated to be five in Texas and at least eight in Louisiana. Adults do not undergo a period of diapause in the winter but seek shelter and remain inactive until the weather is favorable. All stages can be found throughout the year if suitable host material is available.

Fig 3.2 Larval damage on sweet potato tuber

1. *Egg*: Eggs are deposited in small cavities created by the female with her mouthparts in the sweet potato root or stem. The female deposits a single egg at a time and seals the egg within the oviposition cavity with a plug of fecal material, making it difficult to observe the egg. Most eggs tend to be deposited near the juncture of the stem and root (tuber). Sometimes the adult will crawl down through cracks in the soil to access tubers for oviposition, in preference to depositing eggs in stem tissue. The egg is oval in shape and creamy white in color. Its size is reported to be about 0.7 mm in length and 0.5 mm in width. The duration of the egg stage varies from about 5 to 6 days during the summer to about 11 to 12 days during colder weather. Females apparently produce two to four eggs per day or 75–90 eggs during their life span of about 30 days. Under laboratory conditions, however, mean fecundity of 122 and 50–250 eggs per female has been reported.
2. *Larva*: When the egg hatches, the larva usually burrows directly into the tuber or stem of the plant. Those hatching in the stem usually burrow down into the tuber. The larva is legless, is white in color, and displays three instars (Fig. 3.3). The mean head capsule widths of the instars are 0.29–0.32 mm, 0.43–0.49 mm, and 0.75–0.78 mm for instars 1–3, respectively. Duration of each instar is 8–16, 12–21, and 35–56 days, respectively.

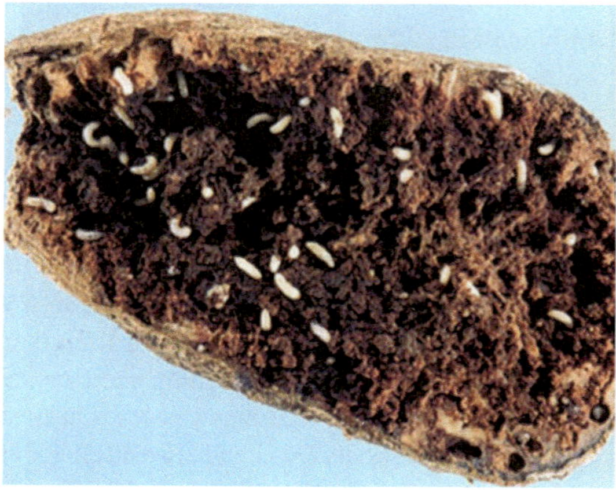

Fig. 3.3 Larvae of sweet potato weevil, *Cylas formicarius*

Fig. 3.4 Adult of sweet potato weevil, *Cylas formicarius*

Temperature is the principal factor affecting larval development rate, with larval development (not including the prepupal period) occurring in about 10 and 35 days at 30° and 24 ° C, respectively. The larva creates winding tunnels packed with fecal material as it feeds and grows.

3. *Pupa*: The mature larva creates a small pupal chamber in the tuber or stem. The pupa is similar to the adult in appearance, although the head and elytra are bent ventrally. The pupa measures about 6.5 mm in length. Initially the pupa is white, but with time this stage becomes grayish in color with darker eyes and legs. The duration of the pupal stage averages 7–10 days, but in cool weather it may be extended to up to 28 days.
4. *Adult*: Normally the adult emerges from the pupation site by chewing a hole through the exterior of the plant tissue, but sometimes it remains for a considerable period and feeds within the tuber. The adult is striking in form and color. The body, legs, and head are long and thin, giving it an ant-like appearance. The head is black; the antennae, thorax, and legs are orange to reddish brown; and the abdomen and elytra are metallic blue. The snout is slightly curved and about as long as the thorax; the antennae are attached at about the midpoint on the snout (Fig. 3.4). The beetle appears smooth and shiny, but close examination shows a layer of short hairs. The adult measures 5.5–8.0 mm in length. Under laboratory conditions at 15 °C, adults can live over 200 days if provided with food and about 30 days if starved. In contrast, their longevity decreases to about 3 months if held at 30 ° C with food and 8 days without food. Adults are secretive, often feeding on the lower surface of leaves, and are not readily noticed. The adult is quick to feign death if disturbed. Adults can fly but seem to do so rarely and in short, low flights. However, because they are active mostly at night, their dispersive abilities are probably underestimated. Females feed for a day or more before becoming sexually active but commence oviposition shortly after mating; the average preoviposition period is 7 days. A sex pheromone produced by females has been identified and synthesized.

3.2.1.4 Host Range

This weevil feeds on plants in the plant family Convolvulaceae. Although it has been found associated with several genera, its primary hosts are in the genus *Ipomoea*. Among vegetable crops, only sweet potato, *I. batatas*, is a suitable host. Native plants can be important hosts of sweet potato weevil. Railroad vine, *Ipomoea pes-caprae*, and morning glory, *I. pandurata*, are among the suitable wild hosts.

3.2.1.5 Management

The following management practices are recommended:

- Destruction of "nests" for egg laying.
- Careful ridging and regular hoeing.
- Regular irrigation to prevent the soil from cracking.
- Do not plant close to existing sweet potato fields.
- Use deep-rooted varieties and/or varieties with a short cycle.
- Do not leave damaged crops that may regrow in the field and become a source of infestation. Remove and destroy them (incinerate or bury at depth).
- Remove weeds belonging to the *Ipomoea* genus over a distance of at least 150 m around the field.

- For *Exophthalmus*: Do not grow other host plants in the immediate vicinity (avocado tree, citrus, *Gliricidia*, etc.).
- In Kilifi, Kenya, farmers create planting mounds or ridges incorporating a good amount of fresh leaves of *Lantana camera* before planting sweet potatoes. This improves soil organic matter and at the same time serves as a repellent of the sweet potato weevil, thus improving both yield and quality of harvested tubers.
- During storage, the sweet potato weevil can be controlled at ambient temperature by processing under low oxygen and high carbon anhydride atmospheres. At 25 °C, storage in 2–4 % oxygen and 40–60 % carbon anhydride destroyed adult weevils within 2–7 days.
- The fungus, *B. bassiana*, has been observed to inflict high levels of mortality under conditions of high humidity and high insect density, but field conditions are rarely conducive for disease epizootics.
- Mulches of plastic or rice straw have shown a reduction of weevil damage. The soil surface should be covered soon after planting, and the cover should be maintained until harvest. The mulch not only helps to retain soil moisture but also prevents the weevils from gaining access to roots through soil cracks.
- Avoid tubers with evidence of attack. Dispose of the affected tubers.
- Apply an insecticide to the cuttings. Spray a systemic insecticide in the field and a contact insecticide in the storage facility.
- Flood the field for 1–2 weeks before planting in order to destroy the larvae in the crop debris.
- In Uganda, sweet potato cvs. 'Araka Red' and 'Tedo Oloo Keren' are considered tolerant to *Cylas* spp.
- Practice proper field sanitation. After harvest remove all plant residues especially the discarded and unharvested tubers.
- Remove *Ipomea* weeds, especially the morning glory, as this weed is the weevils' alternate host.
- Plow the field after harvest to expose the pests to predators. If there is access to irrigation water, submerge the field for at least a day. The weevils do not like being in wet areas and on rotten and decayed materials.
- Practice crop rotation, if possible, to disrupt the life cycle of the pest. Rice, sorghum, yam, and edible sunflower are found to be good rotation crops.
- Properly select your planting material. The tender vine which is about 25–30 cm long is ideal and is usually free from eggs and larvae. The female lays its eggs on the vine near the base of the plant, and the larvae are found feeding on this plant part.
- Practice hilling-up to prevent soil from cracking. Cracked soil is an entry point for the weevils to feed and reproduce on the tubers.
- Storage containers and other facilities are possible source of new infestations. Clean and fumigate them. Complete eradication of the pest is a collective effort among the farmers in a given community (e.g., village, barangay) and involves not planting sweet potato and removing *Ipomea* weeds regularly for a period of at least 6 months.

1. *Cultural Methods*: Cultural practices are sometimes recommended to alleviate weevil problem. Isolation is frequently recommended, and it is advisable to locate new fields away from previous crops and distant from sweet potato storage facilities, because both can be a source of new infestations. However, despite the infrequency of flight by adults, dispersal can occur over considerable distances. Dispersal rates of 150 m per day have been observed, with dispersal more rapid in the absence of suitable hosts.

Sanitation is particularly important for weevil population management. Discarded tubers and unharvested tubers can support large population, and every effort should be made to remove such host material. Related to this, of course, is the destruction of alternate hosts; control of *Ipomoea* weeds is recommended.

Sweet potatoes should be planted only in fields that have had no *Cylas* infestation within the last 12 months and preferably more than 1 km away from any infested land, and only

Cylas-free planting material should be used. Earthing up of the plants, particularly those varieties with a tendency to push out of the ground, reduces attack. After harvest, the field should be cleared of trash, which should be burnt or fed to livestock.

Removal of alternate hosts, selection of weevil-free planting material, and mulching with *Eupatorium* or *Clerodendron* leaves at 3 t/ha were effective in reducing the weevil damage in Kerala (Gopalakrishnan et al. 2013).

(a) *Crop Rotation: Crop rotation (paddy-sweet potato-cowpea) can minimize weevil damage. Boemeryl acetate, a kairomone present in the periderm of sweet potato (Coleus), can attract both male and female weevils which offer immense scope for the management of weevils.*
(b) *Barrier Cropping*: Barrier cropping of sweet potato with yam bean at 2:1 ratio recorded lowest weevil-infested tubers (10.6 %) and lowest weevil population (22.2/kg tuber) and gave highest marketable tuber yield (18.8 t/ha) in Bihar.

Planting of sweet potato and marigold (barrier crop)/garlic in 1:1 ratio is recommended for the reduction of weevil infestation (16.13 %). Planting of sweet potato and yam bean in 2:1 ratio gave least weevil emergence (21.9 %) which was on par with alternate rows of sweet potato and marigold (22.3 %)/yam bean (24.1 %) in Andhra Pradesh. Planting of sweet potato and marigold in alternative rows gave highest marketable tuber yield (13.37 t/ha) which was on par with alternate rows of sweet potato and yam bean (12.33 t/ha), border row of marigold on all sides (12.33 t/ha), paired rows of sweet potato and one row of yam bean (12.28 t/ha), and border row of yam bean on all sides (12.26 t/ha). However, foliar spraying of dimethoate at 0.05 % gave least weevil infestation (14.68 %), weevil emergence (16.50 %), and highest marketable tuber yield (15.88 t/ha) (Table 3.5).

Barrier cropping of sweet potato with marigold at 2:1 ratio was found effective in reducing the weevil damage and weevil emergence/kg tuber (23.12 % and 12.41) which was on par with alternate rows of sweet potato and marigold (25.27 % and 14.66) in Maharashtra (Table 3.6).

Table 3.5 Effect of barrier cropping for the management of sweet potato weevil in Andhra Pradesh

Treatment	Tuber yield (t/ha)*	Marketable tuber yield (t/ha)*	% weevil infestation*	% weevil emergence*
Border row of yam bean on all sides	15.7^{b}	12.26^{bc}	$23.17\ (28.47)^{c}$	$28.7\ (31.81)^{c}$
Border row of marigold on all sides	15.3^{bc}	12.33^{bc}	$21.28\ (27.81)^{c}$	$26.9\ (31.36)^{c}$
Alternate rows of sweet potato and yam bean (1:1)	15.1^{bc}	12.33^{bc}	$19.31\ (26.72)^{bc}$	$24.1\ (29.44)^{bc}$
Paired rows of sweet potato and one row of yam bean (2:1)	15.5^{bc}	12.28^{bc}	$20.27\ (26.22)^{bc}$	$21.9\ (28.73)^{bc}$
Alternate rows of sweet potato and marigold (1:1)	15.2^{bc}	13.37^{b}	$16.13\ (24.80)^{b}$	$22.3\ (29.47)^{bc}$
Paired rows of sweet potato and one row of marigold (2:1)	14.9^{bc}	11.97^{c}	$17.99\ (26.76)^{bc}$	$26.9\ (30.12)^{c}$
Sole crop of sweet potato (control)	13.9^{c}	9.83^{d}	$32.36\ (34.77)^{d}$	$41.0\ (37.43)^{d}$
Foliar spray of dimethoate (0.05 %) at monthly intervals	17.9^{a}	15.88^{a}	$14.68\ (22.10)^{a}$	$16.5\ (25.04)^{a}$
CD (P=0.05)	**1.683**	**1.284**	**2.636**	**4.453**

*In a column, means followed by a common letter are not significantly different

Table 3.6 Effect of barrier cropping for the management of sweet potato weevil in Maharashtra

Treatment	Tuber damage at harvest (%)	Weevil emergence after harvest/kg tuber
Border row of yam bean on all sides	38.79 (38.31)	41.66 (7.42)
Border row of marigold on all sides	30.48 (33.26)	24.16 (5.89)
Alternate rows of sweet potato and yam bean (1:1)	33.78 (35.31)	28.92 (6.35)
Paired rows of sweet potato and one row of yam bean (2:1)	38.56 (38.24)	32.50 (6.57)
Alternate rows of sweet potato and marigold (1:1)	25.27 (29.99)	14.66 (4.86)
Paired rows of sweet potato and one row of marigold (2:1)	23.12 (28.50)	12.41 (4.47)
Sole crop of sweet potato (control)	47.65 (28.50)	47.25 (8.02)
Foliar spray of dimethoate (0.05 %) at monthly intervals	38.21 (43.60)	43.00 (7.28)
CD (P=0.05)	**3.90**	**0.45**

In Dharwad, Karnataka, border row of marigold on all sides resulted in significantly higher tuber yield and least weevil infestation and weevil emergence per kg tuber after harvest (Table 3.7).

(c) *Intercropping*: In Dholi (Bihar), intercropping of sweet potato with coriander at 1:1 ratio recorded lowest tuber infestation (9.8 %) of sweet potato weevil which was on par with foliar application of dimethoate (9.1 %) followed by sweet potato+coriander (11.3 %) and sweet potato+garlic (11.8 %) at 2:1 and 1:1 ratio, respectively. Growing of chilies with sweet potato recorded higher tuber infestation (14.9 and 16.2 %) at 1:1 and 2:1 ratio which was significantly lower than unprotected sole crop of sweet potato (19.3 %).

The highest marketable tuber yield (17.3 t/ha) was recorded with foliar spray of dimethoate (0.05 %) which was on par with sweet potato+coriander (16.6 and 15.4 t/ha) at 1:1 and 2:1 ratio at Dapoli, Maharashtra. The total productivity in terms of sweet potato equivalent yield (Table 3.8) revealed the combination of sweet potato with coriander (19.5 t/ha) at 1:1 ratio followed by sweet potato with garlic (21.4 and 18.0 t/ha) at 1:1 and 2:1 ratio, respectively. Maximum gross return (Rs. 171,200/ha) was recorded when sweet potato was intercropped with garlic (1:1 ratio) followed by sweet potato+coriander (Rs. 156,000/ha) and sweet potato+garlic (Rs. 144,000/ha) at 2:1 ratio. The maximum net profit (Rs. 32,800/ha) was derived when sweet potato was intercropped with garlic (1:1 ratio) followed by sweet potato+coriander (Rs. 17,600/ha) at 1:1 ratio.

In Rajendra Nagar (Andhra Pradesh), intercropping of garlic and coriander with sweet potato at 1:1 ratio recorded lowest tuber infestation of sweet potato weevil (17.5 and 20.3 %, respectively, compared to 39.6 % in control) and gave higher marketable tuber yield (16.94 and 16.99 t/ha, respectively, compared to 9.43 t/ha in control). Benefit-cost ratio was highest in the intercropping of coriander with sweet potato at 1:1 ratio (2.12 compared to 1.71 in control) which was on par with garlic and sweet potato at 1:1 ratio (2.11 at 1:1 ratio) (Table 3.9).

In Kalyani (West Bengal), intercropping of sweet potato with garlic at 1:1 and 2:1 ratio gave very good effect in reducing the infestation of the weevil (1.53 and 2.06, respectively) as evidenced by very low mean score of damaged tubers. The marketable tuber yield was improved with the intercrop of garlic (17.43 and 16.70 t/ha in 2:1 and 1:1 ratio, respectively). The percentage of weevil infestation was also significantly low when intercropped with garlic (8.88 and 9.10 % in 1:1 and 2:1 ratio, respectively) compared to sole crop (26.66 %). The total number of weevils emerged from one kg of tuber was least (22.11 and 25.33) with sweet potato+garlic in 2:1 and 1:1 ratio, respectively, compared to sole crop (104.11) (Table 3.10).

(d) *Botanicals*: Mulching with leaves of *Chromolaena odorata*/*Clerodendron infortunatum* at 3 t/ha at 30 DAP was found effective.

Table 3.7 Effect of barrier cropping for the management of sweet potato weevil in Dharwad, Karnataka

Treatment	Weevil-infected tubers (%)	Weevil emergence/ kg tuber	Marketable tuber yield (t/ha)
Border row of yam bean on all sides	26.64	39.67	29.69
Border row of marigold on all sides	17.31	25.33	37.00
Alternate rows of sweet potato and yam bean (1:1)	37.16	30.67	26.38
Paired rows of sweet potato and one row of yam bean (2:1)	24.18	38.67	30.80
Alternate rows of sweet potato and marigold (1:1)	33.14	30.00	24.52
Paired rows of sweet potato and one row of marigold (2:1)	35.37	33.33	23.92
Sole crop of sweet potato (control)	37.50	37.33	21.55
Foliar spray of dimethoate (0.05 %) at monthly intervals	27.33	21.67	22.90
CD (P=0.05)	**5.15**	**4.53**	**2.42**

Table 3.8 Yield equivalents and economics of sweet potato intercropping system (pooled mean of 2011–2012 and 2012–2013) at Dapoli, Maharashtra

	Tuber yield (t/ha)				
Treatment	Intercrops	Sweet potato (SP)	SP equivalent yield	Gross returns (Rs/ha)	Net profit/loss over chemical control
SP+coriander (1:1)	0.8	16.6	19.5	156,000	17,600
SP+coriander (2:1)	0.5	15.4	17.2	137,600	(–)1,800
SP+chilies (1:1)	0.8	13.1	16.7	133,600	(–)4,800
SP+chilies (2:1)	0.6	11.9	14.4	115,200	(–)23,200
SP+garlic (1:1)	1.3	14.9	21.4	171,200	32,800
SP+garlic (2:1)	0.8	13.8	18.0	144,000	5,600
Dimethoate – 0.05 %	–	17.3	–	138,400	–
SP sole crop	–	12.1	–	–	–

Prices of coriander, chilies, garlic, and sweet potato are Rs. 30,000, 35,000, 40,000, and 8,000/t, respectively

Table 3.9 Effect of intercropping for the management of weevil and tuber yield in Rajendra Nagar (Andhra Pradesh)

Intercropping treatment	Tuber yield (t/ha)	Marketable tuber yield (t/ha)	% weevil infestation*	Benefit: cost ratio
Sweet potato+coriander (1:1)	21.31	16.99	20.3 (26.74)[a]	2.12
Sweet potato+coriander (2:1)	18.55	13.64	26.5 (33.06)	1.70
Sweet potato+chilies (1:1)	19.08	14.83	22.3 (28.12)[ab]	1.85
Sweet potato+chilies (2:1)	16.44	12.33	25.0 (32.14)	1.54
Sweet potato+garlic (1:1)	20.53	16.94	17.5 (24.70)[a]	2.11
Sweet potato+garlic (2:1)	18.68	14.24	23.8 (29.14)	1.78
Dimethoate 0.05 %	20.09	14.53	27.7 (31.72)	1.81
Sole crop of sweet potato	16.61	9.43	39.6 (38.93)	1.71
CD (P=0.05)		**3.008**	**4.882**	

*In a column, means followed by a common letter are not significantly different

Table 3.10 Effect of intercropping for the management of weevil and tuber yield in Kalyani (West Bengal)

Treatment	Tuber yield (t/ha)		No. of weevils emerged/kg tuber	% tuber damage at harvest	Score of weevil-infested tubers
	Marketable	Nonmarketable			
SP+coriander (2:1)	15.03	2.06	31.44	13.77 (21.78)	2.40
SP+coriander (1:1)	16.50	2.20	44.1	12.88 (25.59)	3.13
SP+chilies (2:1)	13.05	2.18	49.33	14.66 (22.51)	3.53
SP+chilies (1:1)	12.90	2.66	51.33	18.66 (25.59)	3.84
SP+garlic (2:1)	17.43	1.86	22.11	9.10 (17.56)	1.53
SP+garlic (1:1)	16.70	1.98	25.33	8.88 (17.34)	2.06
Dimethoate at 0.05 %	17.05	1.90	22.77	9.16 (17.61)	1.93
SP sole crop	12.10	4.05	104.11	26.66 (31.09)	4.86
CD (P=0.05)	**1.73**	**1.10**	**6.39**	**2.78**	**1.33**

(e) *Pheromones*: Over 90 % of larvae are found in the upper 15 cm of the tubers and basal 10 cm of the vine. Early in the season larvae are found about equally in the vine and tuber, but later in the season most occur in the tubers. The distribution of sweet potato weevil in fields is aggregated. Pheromone traps show great promise for the monitoring of adult population density. Weevils respond to low concentrations of pheromone and apparently will move up to 280 m to a pheromone source. The sex pheromone also shows great potential for mating disruption and mass trapping.

Fig. 3.5 Pheromone trap for sweet potato weevil

Field evaluation of the indigenous pheromone technology developed by Bhabha Atomic Research Center (BARC), Mumbai, with pheromone-impregnated 4 mm ID rubber tube septa on SPW management indicated that sex pheromone-incorporated IPM showed only 10 % damage, whereas it was about 71 % in the control (Fig. 3.5).

A very effective IPM package with synthetic sex pheromone as the main component has been demonstrated successfully in nine states for the control of sweet potato weevil. In addition a kairomone, boehmeryl acetate, present in the periderm of sweet potato tubers is promising by attracting both male and female weevils.

Sweet potato weevil sex pheromone is produced commercially in several countries. It is produced in a laboratory and applied to small rubber capsules that are placed in traps in the field. The rubber capsules should be placed above the foliage and covered to protect them from rain and sunlight. A container of soapy water is usually placed under the capsule. Male adults that are attracted by the sex pheromone fall into the pail of water and can easily be collected and removed from the field. These traps are useful for indicating how large the weevil population is. In some countries research has shown that mass trapping using sex pheromone traps are an effective means to control the weevil. In Cuba the sex pheromone is often used together with an application of the fungus *Beauveria bassiana*. The fungus is applied on the soil surface beneath the sex pheromone trap or sprayed on the foliage around the trap. Weevils attracted to the sex pheromone will be infected by the fungus and killed after several days. Sex pheromones, however, are not yet widely available at the farm level.

Headspace volatile extracted were collected from sweet potato varieties Kishan (highly preferred by male and female SPW), Howrah, BX-86, and C-14 (least preferred by SPW). GC-EAD analysis revealed that varieties Howrah, BX-86, and C-14 release α-humulene (2, 6, 6-9-tetramethyl-1, 4-8-cycloundecatriene) which is also known as α-caryophyllene. α-Humulene is a monocyclic sesquiterpene. Another volatile compound, α-gurjunene, also got emitted from flowers of C-14. Both these compounds have shown 70–75 % repellency to female SPW. In addition to this, Howrah genotype emitted a compound from the storage root periderm [9, 12-(Z, Z)-octadecadienoic acid in ester form]. These compounds were absent from or expressed in very low quantities in SPW susceptible genotype Kishan. These compounds presumably play a role in SPW resistance in Howrah, BX-86, and C-14 genotypes by producing repellent compounds.

2. *Chemical Methods*: Planting time applications of insecticides are commonly made to the soil to prevent injury to the slips or cuttings. Either granular or liquid formulations are used, and systemic insecticides are preferred. Post-plant applications are sometimes made to the foliage for adult control, especially if fields are likely to be invaded from adjacent areas, but if systemic insecticide is applied, some suppression of larvae developing in the vine may also occur. Due to the long duration of the plant growth period, it is not uncommon for preplant or planting time applications to be followed by one or more insecticide applications to the plant or soil at mid-season. Insecticides are also applied to tubers being placed into storage to prevent reinfestation and inoculation of nearby fields.

Some of the chemicals recommended for the management of weevils are presented in Table. 3.11

Table 3.11 Recommendations for chemical control of weevils

Insecticide	Dose/Conc.	Application method
Carbaryl	0.1 %	3 times at 4 weekly intervals.
Carbofuran	1.4 kg/ha	Soil application around vine cutting at planting
Carbophenothion	0.035 %	–
Chlorfenvinphos dust	600–900 g/ha	30-day waiting period before harvest
Chlorfenvinphos EC	240 g/ha	–
Diazinon	0.3 %	Dip for planting material or foliar spray
Dieldrin	1 kg/ha	4 times at weekly intervals beginning 40 days after planting. Five weeks waiting period
Disulfoton	1.8 kg/ha	4 weeks after planting
Endosulfan EC	2.25 kg/ha	Application at planting time
Endosulfan granules	2.25 kg/ha	Soil application at planting
Fenitrothion	0.05 %	–
Fenthion	0.1 %	3 times at 4-week intervals
Malathion EC	28.5 ml	Dip for cuttings or foliar spray
Mephosfolan	1.4 kg/ha	Soil application around cutting at planting
Methomyl	1 kg/ha	Four times at 2-week intervals beginning 40 days after planting
Methyl demeton	0.05 %	Spray at the beginning of 2nd, 3rd, and 4th month of growth
Monocrotophos	0.05 %	4 times at monthly intervals
Phorate	1.4–2.0 kg/ha	Soil application around cutting at planting and 4 weeks after planting
Phosphamidon	0.05 %	Spray at beginning of 2nd, 3rd, and 4th month of growth

Foliar spray of dimethoate at 0.05 % recorded lowest tuber infestation (8.60 %) and highest marketable tuber yield (17.13 t/ha).

3. *Host Resistance*: Deep rooting varieties with long necks between the tubers and the stems being less susceptible because of adult weevil cannot burrow downwards more than 1 cm. Early maturing varieties escape serious damage.
4. *Biological Methods*
   (a) *Predators*: Among predators, ants (Hymenoptera: Formicidae) seem to be most important. A dipteran predator (*Drapetis exilis*) is common predator recorded on grubs.
   (b) *Parasitoids*: Wasps such as *Bracon mellitor*, *B. punctatus*, *Metapelma spectabile* (all Hymenoptera: Braconidae), and *Euderus purpureas* (Hymenoptera: Eulophidae) have been reared from sweet potato weevil larvae in the Southeastern United States. There have been no studies of parasitoid effectiveness, but these species seem to be infrequent.

      The parasitoids, *Rhaconotus menippus* and *Bracon* spp., parasitize up to 40 % of grubs in aerial parts of vines. Parasitization of sweet potato weevil by *Rhaconotus menippus* in the field ranged from 21 to 23 % and was comparable to that of the control field.
   (c) *Pathogens*: Diseases, especially the fungus *Beauveria bassiana*, have been observed to inflict high levels of mortality under conditions of high humidity and high insect density, but field conditions are rarely conducive for disease epizootics (Fig. 3.6).

The green muscardine fungus, *Metarhizium anisopliae*, is common in soil-inhabiting adults. *Beauveria bassiana* and *M. anisopliae* were found effective against *C. formicarius* on sweet potato (Khader Khan et al. 1990). Soil application of *M. anisopliae* at 11.25 kg/ha recorded lower weevil infestation (5.7 % compared to 24.2 % in control) and gave higher marketable yield (18 t/ha) in Bihar.

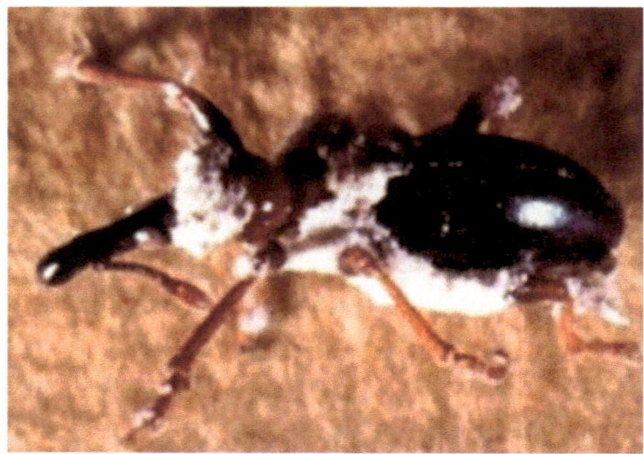

Fig. 3.6 *Beauveria bassiana*-infected sweet potato weevil

In Andhra Pradesh, soil application of *B. bassiana* at 6.75 kg/ha recorded lower weevil infestation (19.6 % compared to 30.7 % in control) and higher marketable tuber yield (15.3 t/ha compared to 11.6 t/ha in control) which was on par with *M. anisopliae* at 11.25 kg/ha (weevil infestation, 20.9 % compared to 30.7 % in control; yield. 15.1 t/ha compared to 11.6 t/ha in control) (Table 3.12).

In Maharashtra, soil application of *B. bassiana* at 6.75 kg/ha gave least tuber damage (20.48 %) and weevil emergence from tuber (7.86/kg tuber) after harvest which was on par with neem cake at 10 t/ha (23.94 % weevil damage and 10.20 weevil emergence/kg tuber) (Table 3.13). However, soil application of neem cake at 10 t/ha is not economical.

The weevil was susceptible to *Bacillus thuringiensis* (60–64 %) which is also infective to adults.

Entomopathogenic nematodes seem to be the organisms with the greatest potential for practical biological suppression of sweet potato weevil. Several strains of *Steinernema carpocapsae* (Nematoda: Steinernematidae) and *Heterorhabditis bacteriophora* (Nematoda: Heterorhabditidae) penetrate the soil and tubers, killing weevil larvae. At least in the soils of southern Florida, the infective nematodes are persistent, remaining active for up to 4 months. In some cases nematodes are more effective than insecticides at reducing damage. *Heterorhabditis bacteriophora* is an effective biocontrol agent.

Table 3.12 Effect of bioagents and botanicals for the management of sweet potato weevil in Andhra Pradesh

Treatment/dose	Tuber yield (t/ha)*	Marketable tuber yield (t/ha)*	% weevil infestation*	% weevil emergence*
Beauveria bassiana (bio-power 1.5 % WP)—6.75 kg/ha	17.9[bc]	15.3[b]	19.6 (26.27)[b]	20.8 (26.84)[b]
Metarhizium anisopliae (bio-magic 1.5 % WP)—11.25 kg/ha	18.4[ab]	15.1[b]	20.9 (27.16)	21.2 (27.57)[b]
Castor cake—10 t/ha	17.2	13.8	23.6 (28.95)	24.9 (30.04)[c]
Neem cake—10 t/ha	18.6[ab]	15.6[ab]	20.3 (26.70)[b]	20.0 (26.77)[b]
Yam bean seed extract—5 %	16.6	12.8	23.8 (29.08)	25.4 (30.44)[c]
Cassava rind extract—5 %	15.9[d]	11.6[d]	26.5 (30.86)[c]	28.0 (31.88)[c]
Garlic extract—5 %	16.3[cd]	12.3	26.5 (30.86)[c]	27.8 (32.07)[c]
Control	15.8[d]	11.2[d]	30.7 (33.46)[d]	32.9 (35.52)[d]
Foliar spray of dimethoate 0.05 % at monthly intervals	20.0[a]	17.5[a]	15.4 (23.03)[a]	16.3 (23.47)[a]
CD (P=0.05)	**1.706**	**2.050**	**1.988**	**2.060**

*In a column, means followed by a common letter are not significantly different

Table 3.13 Effect of bioagents and botanicals for the management of sweet potato weevil in Maharashtra

Treatment/	Dose	Tuber damage (%)	Weevil emergence/kg tuber
Beauveria bassiana (bio-power 1.5 % WP)	6.75 kg/ha	20.48 (26.85)	7.86 (3.75)
Metarhizium anisopliae (bio-magic 1.5 % WP)	11.25 kg/ha	27.09 (31.21)	13.33 (4.60)
Castor cake	10 t/ha	36.90 (37.25)	19.00 (5.27)
Neem cake	10 t/ha	23.94 (29.18)	10.20 (4.14)
Yam bean seed extract	5 %	38.01 (37.95)	21.93 (5.59)
Cassava rind extract	5 %	39.69 (38.96)	23.93 (5.72)
Garlic extract	5 %	32.96 (34.91)	15.25 (4.85)
Control	–	45.59 (42.40)	27.46 (6.16)
Foliar spray of dimethoate at monthly intervals	0.05 %	35.62 (36.40)	20.00 (5.24)
CD (P=0.05)		**2.41**	**0.51**

5. *Integrated Methods*
   (a) *Cultural and Chemical Methods*
   Preplanting
   - Removal and destruction of alternate host and residues of harvested crop.
   - Selection of pest-free planting materials.
   - Disinfest the vines by dipping in Monocrotophos (0.05 %).

   Post-planting
   - Re-ridging the crop at 30 and 60 days after planting (DAP).
   - Installation of pheromone traps, one trap per 100 m^2 (Rs 1,000 ha^{-1}).
   - Harvest the crop at 110 days after planting.

   (b) *Cultural and Biological Methods*: The solitary ectoparasitoid, *Rhaconotus menippus* (at 10 pairs/5 m^2), and the green muscardine fungus, *Metarhizium anisopliae* (at 3×10^9/ml), in combination with re-ridging (putting an additional layer of soil around the plant) at 65 days after planting is highly effective in reducing the weevil damage. This method is equally effective as that of chemical method. The fungus and parasitoid are recovered from the treated plots.
   (c) *Cultural, Biological, and Chemical Methods*: The following IPM technology was developed at Central Tuber Crops Research Institute, Thiruvananthapuram, Kerala (Palaniswami 2002):

  - Removal and destruction of *Ipomoea* weeds.
  - Selection of weevil-free planting material.
  - Dipping the vines in 0.05 % fenthion or fenitrothion or Monocrotophos for 10 min.
- Installation of traps with septa (sex pheromone impregnated in 4 mm rubber tube uniformly in such a way that one cm bits of such tube contains 1 mg sex pheromone) at 100 traps/ha. The traps are placed with the commencement of planting and continued a fortnight after the harvest. The male weevils trapped inside the bin trap (bottom of the trap contains water with a pinch of detergent) are removed on alternate days.
- Re-ridging (putting an additional layer of soil around the plant) the crop at 30 and 50 days after planting.
- Conserving the naturally occurring braconid solitary ectoparasitoids of weevil by not spraying insecticides especially on the vines.
- Harvesting the crop at 100–110 days maturity.
- Removal and destruction of crop residues like infested tubers and vines.
- Practicing crop rotation.

The above IPM technology was assessed, refined, and validated at National level in ten different centers under All India Coordinated Research Project on Tuber Crops. The farmers from different sweet potato-growing areas were convinced about the benefits of this technology in enhancing tuber production and reducing weevil infestation.

3.2.2 Leaf Folder: *Brachmia* spp.

3.2.2.1 Distribution

Brachmia convolvuli is distributed in Canary Island, Mauritius, Comoro Island, Southeast Asia, and Jawa. *Brachmia macroscopa* has been recorded in India.

3.2.2.2 Symptoms

Cream and black larvae spin webs close to the main veins of the leaves, on the underside, while older larger larvae spin webs across the upper surface of the leaf and draw the two sides together (Fig. 3.7). The leaves eventually dry up, die, and drop off.

3.2.2.3 Biology

Brachmia convolvuli eggs are laid on the lower surface of leaves and hatch after 3 days. There are five larval instars. The mature larva is 11–12 mm long and cream colored with dorsal black bands on segments 8–10 of the abdomen. The head, prothoracic shield, and all of abdominal segment 5 are black. The larvae pupate for 7 days in the folded leaf. Adults of *B. convolvuli* are 4 mm long with grayish-brown forewings, marked with white hind wings fringed with gray. The entire life cycle, egg to adult, takes 24 days.

Fig. 3.7 *Left* – fold produced by green leaf folder with web visible through feeding hole. *Right* – browning and holes inside the folds produced by brown leaf folder

3.3 Management

Sprays of 0.2 % carbaryl or 0.025 % fenitrothion are recommended in India for the control of *B. convolvuli*.

3.3.1 Hornworm: *Agrius convolvuli*

3.3.1.1 Distribution

Hornworm is distributed in most of Africa, Bangladesh, Burma, South China, Southern Europe, India, Indonesia, Iran, Malaysia, New Zealand, Pacific Islands, and Papua New Guinea.

3.3.1.2 Symptoms

Caterpillars feed on leaves, causing irregular holes (Fig. 3.8). They may eat an entire leaf, leaving only petiole. Insect frass can often be found near the infested plant part. One large caterpillar can defoliate a plant on its own. When older caterpillars are present in large numbers, they can defoliate a field overnight. Yield losses can occur if heavy defoliation takes place when the crop is young.

Fig. 3.8 Hornworm feeding on leaf

3.3.1.3 Biology

Spherical eggs (1 mm in diameter) are laid singly on the host plant, usually on the upper leaf surface. The larvae, which concentrate on the young shoots, are variable in color from green to brown and distinctively patterned. There are five larval instars and larvae mature after 3–4 weeks and reach a length of about 95 mm. Pupation takes place in the soil and may take 3 weeks or several months depending upon the climate. The adult hawk moths (wingspan 80–120 mm) are gray with black lines on the wings; the abdomen is marked laterally with pink bands. They feed upon nectar in flowers.

3.3.1.4 Management

1. *Cultural Methods*: Plowing the land between crops to expose the pupae reduces infestation. Handpicking of the larvae may be quite effective in small areas. Turning the soil over between crops exposes the pupae to predators and to the sun's heat.
2. *Chemical Methods*: A 0.2 % solution of carbaryl in a high volume spray, 0.05 % endosulfan or trichlorfon (7 days waiting period before harvest) at 100 g a.i. in 100 l of water, should be applied when the larvae are first seen. Early instars can be controlled more easily than fifth instars. Particular care should be taken to spray the young leaves.
3. *Biological Methods*: The eggs of *A. convolvuli* are parasitized by four species *Trichogramma australicum, T. Achaea, T. agriae*, and *Telenomus* sp.

3.3.2 Bihar Hairy Caterpillar: *Diacrisia obliqua*

3.3.2.1 Distribution

Bihar hairy caterpillar is distributed in South and East Asia.

3.3.2.2 Symptoms

Numerous larvae covered with yellow hairs (Fig. 3.9) skeletonize the leaves. The entire vines may be defoliated.

Fig. 3.9 Bihar hairy caterpillar

Fig. 3.10 Fields severely damaged by leaf miner

3.3.2.3 Biology

Eggs are laid on the leaves in rows or clumps and after 4–9 days the larvae emerge. Mature larvae are 30–40 mm long and have 7 orange bands across the body on which tufts of yellow hairs grow (Fig. 3.9). At first, the larvae feed in groups but when they are about a week old, they disperse to new feeding sites. The larval period lasts 28–34 days. Pupation takes place in the soil in reddish-brown cocoons and may take 51 days.

3.3.2.4 Management

Endosulfan and fenitrothion both at 0.05 % are recommended for the management of the pest. Diazinon or dichlorvos or carbaryl at 0.2 % are also suggested for the control of larvae.

3.3.3 Leaf Miners: *Bedellia* spp.

3.3.3.1 Distribution

Bedellia ipomoeae from Fiji, *B. somnulentella* from Papua New Guinea and Queensland, and a number of *Bedellia* spp. from Africa have been reported attacking sweet potato.

3.3.3.2 Symptoms

Larvae feed on the green tissue inside the leaf, leaving the transparent upper and lower membranes (epidermis) intact. Young larvae enter the leaf and form serpentine mines (narrow gray or silvery tracks), while mature larvae form broader "blotch" mines. Serpentine or blotched mine later become holes as the mined tissues are destroyed (Fig. 3.10). The lower surface of the infested leaves become dirty with blackish frass and show silken webbings where the insect pupates. With high infestation, the leaves become brown. Heavy infestation may occur during or after a prolonged dry period.

3.3.3.3 Biology

Eggs are laid on the leaf. The emerging larva is distinctly segmented with a rather pointed head and abdomen. Larvae may be red or yellowish gray in color. After making several mines, they pupate on a web of silken threads among the leaves. When moving outside the mines, the larvae travel by a looping movement. The life cycle can be completed within 4 weeks. Adult *Bedellia* spp are small (3 mm long with a wingspan of 8 mm) tan to gray moths.

3.3.3.4 Management

Generally *Bedellia* spp. are controlled by predators, but these may be killed by foliar sprays of insecticide. A serious breakdown can cut down the effective leaf surface and lead to reduced yields of tubers. Carbaryl at 1 kg a.i./ha, chlorfenvinphos at 0.6–0.9 kg/ha, diazinon at 55 ml/ha, dimethoate at 14 ml/ha, endosulfan at 70 ml/ha, malathion at 14 g/ha, or trichlorfon at 100 g/ha have been recommended for the management of the pest.

3.3.4 Tortoise Beetles: *Aspidomorpha* spp. and *Conchyloctaenia punctata* (Chrysomelidae)

3.3.4.1 Distribution

Conchyloctaenia punctata from South Africa; *Aspidomorpha adheares, A. miliaris, A. quadriradiata*, and *A. socia* from Papua New Guinea; *A. miliaris* from India; and *A. dissentanea* from Angola, Cameroon, and Guinea have been reported attacking sweet potato.

3.3.4.2 Symptoms

Adults and larvae feed on the leaves making large round holes on leaves (Fig. 3.11). Attacks are sometimes sufficiently severe to completely skeletonize leaves and peel stems.

3.3.4.3 Biology

The eggs are laid on the underside of the leaves in batches cemented to the leaf. The larvae are flattened and spiny and usually carry excreta and the skins of earlier instars on anal projections. Adult tortoise beetles (subfamily: Cassidinae) are broadly oval in shape and hence the common name. They are bright, sometimes patterned, or metallic in color.

3.3.4.4 Management

1. *Cultural Methods*: Weeds of the family Convolvulaceae should be controlled as tortoise beetles can breed on them.
2. *Chemical Methods*: Carbaryl (0.2 %) at 1,000 l/ha is recommended in India and Papua New Guinea, while endosulfan at 500–750 g a.i./ha elsewhere.
3. *Biological Methods*: The tortoise beetles were often parasitized by *Cassidocida aspidomorphae* and *Tetrastichus colemani*.

3.3.5 Striped Weevil: *Alcidodes* spp.

3.3.5.1 Distribution

Alcidodes dentipes from tropical Africa, *A. erroneous* from Kenya, and *A. fabricii* from India have been recorded attacking sweet potato.

3.3.5.2 Symptoms

Striped weevils (*Alcidodes dentipes* and *A. erroneous*) (Fig. 3.12) attack both stems and tubers. Adults girdle the stems causing wilting, while larvae bore into stems and tubers making tunnels slightly larger than those of *Cylas* larvae and causing galls to form on the stems. *A. fabricii* has been recorded breeding inside the seed capsules of sweet potato.

3.3.5.3 Biology

The life cycle of the weevils is similar to that of *Cylas*. *Alcidodes dentipes* is about 14 mm long and brown with 4 white streaks along its elytra. *A. erroneous* is similar in size but brown black in color with an irregular white patch on each elytrum. *A. fabricii* is reddish brown with four longitudinal white stripes on the elytra. The eggs

Fig. 3.11 Tortoise beetle damage on sweet potato leaves

Fig. 3.12 Adult striped beetle

Fig. 3.13 Flea beetles on sweet potato leaf

are laid at the base of the fruit; the larvae bore into the fruit and feed on the seeds. Pupation takes place within the fruit.

3.3.5.4 Management

Measures like crop rotation, use of clean planting material, sanitation and timely harvesting are effective in reducing incidence of this pest. Frequent earthing up around the plant base reduces the incidence of this pest.

The methods outlined for *Cylas* also should control *Alcidodes* spp.

3.3.6 Flea Beetles: *Chaetocnema confinis*

3.3.6.1 Damage

The flea beetles (Fig. 3.13) feeds on seedlings. They usually feed on the undersides of leaves, leaving numerous small round or irregularly shaped holes, although not generally all the way through the leaf. Because the beetle is small and active, it usually does not feed much in one spot.

The larvae are root feeders. They trim the root hairs and make circular pits in tap roots. Its feeding damage is external on the root which can reduce the marketability of tubers. The adults feed on the leaves and stems of emerging seedlings, on green pods, and heads. They chew small holes or pits, usually less than 3 mm in diameter, giving the leaves a characteristic "shot hole" appearance.

3.3.6.2 Description

The eggs are ovate and white when freshly laid and gradually become yellowish gray in color. The larva is white with a brown head and three pairs of brown legs near its head. The pupa is white, is shaped roughly like adult, and pupates in the soil. The adult is small, hard, elongated oval shaped, with enlarged black hind legs and slightly hairy wing covers. It is about 2 mm in size.

3.3.6.3 Management

The following management methods are recommended:

- Row covers keep flea beetles away. Put row cover immediately after sowing.
- White or yellow sticky traps should be placed every 5–10 m on the rows.
- Thick mulch in isolated planting interferes with the larva's feeding activities.

3.3.7 Cucumber Beetles: *Diabrotica balteata* and *D. undecimpunctata*

Two species of cucumber beetles damage sweet potatoes. Adults and larvae of the banded cucumber beetle, *Diabrotica balteata*, and the spotted cucumber beetle, *D. undecimpunctata*, feed on sweet potato (Fig. 3.14). Both species are highly mobile and also will feed on several other host plants including various vegetable plant species, soybeans, and corn.

Fig. 3.14 Banded (*Diabrotica balteata*) and the spotted cucumber beetle (*Diabrotica undecimpunctata*)

Fig. 3.15 Sweet potato roots injured by cucumber beetles

3.3.7.1 Damage

Adult beetles feed on sweet potato foliage, creating irregular holes in the leaves. Beetles usually do not consume enough foliage to negatively affect plant productivity; however, the presence of adult beetles is cause for concern. Adult beetles lay eggs in the soil, and larvae developing in the soil feed on developing sweet potato roots. Feeding on the roots can occur throughout the production season, but damage from these insects is increased in late season. Feeding injury results in unsightly blemishes on the roots at harvest (Fig. 3.15).

3.3.7.2 Biology

Adults lay eggs in the soil where larvae emerge in 1–2 weeks, depending on temperature. Larvae of these two species are very difficult to tell apart. The larval stage may last 8–30 days, depending on food supply and temperature. Pupae are found in cells just below the soil surface, and adults emerge in about 1 week. There are several generations per year.

3.3.7.3 Management

Because this species has several generations per year, a wide host range and is very mobile as adults, it is one of the most difficult soil insect pests to control. Repeated cultivation of the fields in the spring will help remove alternate food sources and expose the larvae to birds and other predators. A preplant soil insecticide (Mocap, Capture/Brigade) will kill cucumber beetles in the soil and provide residual control for 4–6 weeks after application. Adult beetles moving into the field will lay eggs, and resulting cucumber beetles will injure the roots once these soil insecticides break down. At this point, if adults exceed threshold levels, resort to foliar applications to prevent cucumber beetle injury. Depending on environmental conditions, the insecticide used and the coverage obtained with the spray application, a foliar-applied insecticide (Thiodan, Penncap, Imidan, Sevin, Capture/Brigade), may kill beetles for anywhere from 1 day to 1 week. Adult beetles can be monitored with sweep nets. Two banded cucumber beetles per 100 sweeps indicate the need to spray on a

7-day interval. With higher numbers of the beetle, reduce the spray interval to 5 days. These beetles are highly mobile. They may move into a field in large numbers and make management difficult.

3.3.8 Scarbee Weevil/West Indian Sweet Potato Weevil: *Euscepes postfasciatus* (Coleoptera: Curculionidae)

3.3.8.1 Distribution

The pest is distributed in Central America, Bermuda, Brazil, Cook Islands, Fiji, Futuna Islands, French Polynesia, Guam, Guyana, New Caledonia, Ryukyu Islands, Tonga, Southern United States, and West Indies.

3.3.8.2 Damage

Larvae are seen feeding in the tubers. The weevil is particularly important in dry localities.

3.3.8.3 Biology

Eggs are laid in the underground parts of the plant where they hatch within 7 days. Larvae are about 5 mm long, legless, white with a yellowish head, and reddish-brown mandibles when full grown. Development to pupation takes 25–28 days; pupation takes place in the tunnel made by the larva. The adult emerges from the pupal skin after 1 week but remains in the pupal cell until the integument hardens. Adults are grayish-brown weevils with white marks near the apex of the body.

3.3.8.4 Management

1. *Cultural Methods*: Crop rotation, use of insect-free cuttings, earthing up the plants, and cleaning up crop trash after harvest will reduce the risk of infestation.
2. *Chemical Methods*: Dipping cuttings in diazinon or malathion EC made up in water enables the farmer to use infested cuttings for propagation, although they should preferably be avoided. Chlorfenvinfos, diazinon, dieldrin and disulfoton are recommended in the West Indies and Hawaii for the control of both *Cylas* and *Euscepes*.

3.3.9 Pyralid Moth/Stem Borer: *Megastes grandalis* (Lepidoptera: Pyralidae)

3.3.9.1 Damage

Plants are stunted and may shed their leaves. Frass can be seen on the soil around the crown of the plant. Cuttings may die due to larvae feeding inside them. Harvested tubers, when cut open, are found to be riddled with clean tunnels. Even a slight infestation can be important because it may precipitate rot.

3.3.9.2 Biology

Green, plump oval eggs 1.2–1.5 mm in diameter are laid in groups beside the veins on the underside of the leaf or in the leaf axils of the petioles. As they mature, they turn purple.

The larvae are white with brown head and markings when they emerge, but after the second instar, the body is pinkish. The larvae begin to feed near the eggs as soon as they hatch. and a short time later, they move closer to the soil and bore into the stems and into the tuber. The frass produced by tunneling is usually pushed out through the first entry hole. When it has fed in the tuber for 5–7 weeks, the full-grown larva spins a cocoon not from the hole aboveground through which the adult will emerge. The pupal period lasts 13–16 days. The moths mate and begin to lay eggs the night after they emerge. Sweet potato is the only known host of this insect.

3.3.9.3 Management

This pest can be managed by careful selection of the planting material, application of systemic insecticides as per recommendation, and/or use of insect traps. Crop rotation and field sanitation can also help in the management of this pest.

1. *Chemical Methods*: The chemicals recommended for the management of *M. grandalis* in the West Indies are presented in Table 3.14.

Table 3.14 Recommendations for chemical control of *Megastes grandalis*

Insecticide	Dose (a.i.)	Application
Carbaryl	950 g/ha	In 900 L water (high volume) or 225 L (low volume) at 3 weekly intervals
Chlorfenvinfos EC	300 ml/ha	In 900 or 225 L water at 3 weekly intervals. 30-day minimum period before harvest
Dieldrin	0.5 %	Dip cuttings before planting
Malathion	1.4 kg/ha	
Methomyl	500–950 g/ha	In 900 or 225 L water at 3 weekly intervals. 7-day minimum period before harvest if feeding foliage to stock

Fig. 3.16 Clearwing moth

3.3.10 Clearwing Moth: *Synanthedon dasysceles* and *S. leptosceles* (Lepidoptera: Sesiidae)

3.3.10.1 Damage

The main stem develops galls at ground level, the plant is unhealthy, and the yield is reduced. Tubers are tunneled by white larvae with three pairs of thoracic legs and five abdominal prologs (Fig. 3.16).

3.3.10.2 Biology

Adults are day flying and lay batches of 70–100 eggs, probably on the vines. The larvae burrow into the vines on hatching and tunnel downwards, frequently reaching the tubers. Usually pupation takes place in the main stem of the plant just above ground level. When the adults have emerged, the brown pupal cocoon and pupal case may be seen protruding from the stem.

3.3.10.3 Management

The recommendations for *Cylas* control will reduce the incidence of this pest.

3.3.11 Stem or Vine Borer: *Omphisa anastomosalis* (Lepidoptera: Pyraustidae)

3.3.11.1 Distribution and Economic Importance

Generally the pest is widespread in Southeast Asia (China, India, Japan, and Sri Lanka) and Hawaii.

In Hawaii, reports of serious damage by this pest result in death of the sweet potato plant. In Malaysia, this pest reduces yield by about 30 %. In Taiwan, over 89 % of the plants sustain damage and the yield reduction is about 50 %. Infestation during the establishment phase of the crop can result in yield losses of 30–50 % or more.

3.3.11.2 Damage

The larva bores into the main stem and sometimes penetrates the storage roots (Fig. 3.17). Larval feeding produces large tunnels causing hollow cavities in the stem resulting in wilting and eventual death of infested plants. Attack at early stages of plant growth prevents storage root formation since translocation of food and its storage in the roots become severely affected. Further damage occurs when the larva tunnels into the storage roots making them nonmarketable. Infestation during the establishment phase of the crop can result in yield losses of 30–50 %

Fig. 3.17 Stem borer damage on sweet potato

or more. Infested plants usually have a pile of frass that can be found close to the attacked stem.

3.3.11.3 Biology

The flattened yellow, oval eggs are laid in groups of two or three, usually on the stem and occasionally on the leaves. The larvae emerge after an incubation of 5–8 days and eventually reach 30 mm in length. They tunnel in the stem for 28–50 days and pupate in a thin cocoon within the tunnel, close to the tubers. Adults emerge after about 14 days. The moth is 15 mm long with a wingspan of 33 mm, with a reddish-brown body and reddish-brown markings on the wings, which are white.

Egg is ovate with a flat base and greenish in color. It is about 0.5 mm in diameter. It is laid singly either on the upper or lower surface of the leaf lamina and on the leaf petiole. An egg hatches in about 7 days.

Larva is light purple and/or yellowish white in color. Its head is brown and has brown spots that form prominent stripes along its sides. A fully grown larva can measure up to 2.5–3.0 cm long. It undergoes 6 larval instars that usually vary from 21 to 90 days, depending upon the climatic conditions.

A pupa is light brown in color and is about 1.6 cm long. It pupates on the vine where the larva has been feeding but also in the soil near the roots.

The adult is white in color and has a brownish-yellow pattern on its wings. It has dark-brown forewings and lightly pigmented hind wings. The female has a broader abdomen than the male, with the latter having clasper-like structures on the last abdominal segment which is not found on the former. Both are immediately fertile upon emergence. A female lives for about 10 days.

3.3.11.4 Spread

Using planting material infested with stem borer eggs or planting in a new field next to an infested one are often the main means of disseminating this pest.

3.3.11.5 Management

Crop hygiene and crop rotation reduce infestation. Frequent earthing up around the plant base reduces the incidence of this pest. Treatment of planting material and crop rotation are valuable means of control. Hilling up, often practiced to reduce damage from sweet potato weevil, also contributes to the containment of a stem borer infestation. Hilling up is effective when the holes, made to provide the adults with a means of exiting the stems, are covered with soil. Earwigs and ants may attack the larvae developing within sweet potato vines. Sources of resistance have been identified by the Asian Vegetable Research and Development Center, Taiwan.

The following control measures are also recommended:

- After a heavy pest infestation, refrain from planting sweet potato for at least two cropping seasons. Always keep your area free of *Ipomea* weeds to eliminate all the possible food sources for the pests.
- Properly select your planting material. Never use cuttings that are taken from infested plants.
- Use a row cover when you want to grow sweet potato as your source of planting material. This will protect plants from the pests.

3.3.12 Armyworm: *Spodoptera* spp.

3.3.12.1 Damage

The larvae (Fig. 3.18) can eat the entire leaves of field crops and grasses. When feeding, they chew from the leaf edges until only the midrib is left. They feed on various crops and grasses during their migration and often bare crops of tender leaves after passing through. They travel from field to field in great numbers, hence the name "armyworm."

1. *S. exigua*: They are initially gregarious, becoming solitary at a later stage. The damage is characteristic: during the early stages, only the upper epidermis of the leaves is chewed and then the whole leaf is devoured apart from the midrib. Large numbers of these caterpillars are gregarious, moving around in groups.
2. *S. frugiperda*: The pest causes non-characteristic damage by way of defoliation.
3. *S. littoralis:* On hatching, the caterpillars remain in clusters on the lower surface of the leaves and eat the parenchyma. They subsequently disperse and are strictly nocturnal sheltering in cracks in the soil during the day. Being highly polyphagous, this is potentially one of the most hazardous pests.

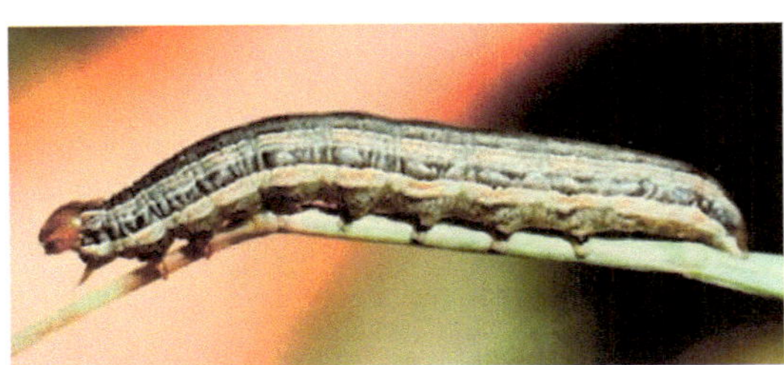

Fig. 3.18 Armyworm larva

Leaves are eaten by caterpillars with possible reduction in number of tubers or leaves per plant and size/weight of the tubers or leaves, if heavy infestation occurs on leaves at early stages of the crop. Damage to the leaves may delay harvest, increasing the likelihood of attack by the sweet potato weevil.

3.3.12.2 Biology

Eggs of armyworms are white and are found in clusters. One egg mass contains 200–300 eggs. These turn yellowish when mature. Hatching occurs in about 2–5 days. The larva varies its colors according to the species. Sizes vary from 2.0 to 4.5 cm long. The larva has six small legs just behind the head and ten fleshy legs at the rear. The larva eats voraciously as it grows. It digs 3–6 cm below the soil surface to pupate.

The pupa is brownish red in color and about 2–3 cm long. Pupation takes about 7–16 days depending on the climatic conditions.

The moth has 3–4 cm wingspan and 1.5 cm long. Colors vary from light green, light pink, and tan to dark brown in color. The tiny white dot in the center of each forewing is an important feature to identify the armyworm moth. The moth is seldom seen because it is active only at nighttime, either mating or searching for suitable egg-laying sites. The female moths lay eggs on the lower leaves of growing grasses or small grains.

3.3.12.3 Management

The following management practices are recommended:

- Remove weeds.
- Light traps can be hung over basins of water in the field to trap the adults at night.

- Manually remove the egg masses.
- Spray a contact insecticide as a matter of priority on gregarious caterpillars in the early stages of development.

1. *Physical Methods*: Plow a deep ditch. Keep it filled with water. This method is helpful, when larvae are found to be moving toward your field from the adjacent fields. Another method is to dig a deep ditch with vertical sides to trap the larvae and prevent them from crawling out. Dig a hole, a diameter of a fence post, in every 10 m within the ditch. Larvae are lured to congregate in the holes. Collect and properly dispose the trapped larvae.
2. *Cultural Methods*
   (a) *Chili and Neem Leaves Extract*: Pound 10–20 pieces of chili and 2.0–2.5 kg of neem leaves. Soak into 1 l of water overnight. Strain and add 20 l of water and 2 tbsp of powdered soap. Stir well before application.

3.3.13 Cluster Caterpillar: *Acraea acerata*

3.3.13.1 Damage

In clusters, they devour the upper surface of the leaves. Once solitary, the leaf is destroyed with the exception of the main ribs (Fig. 3.19).

Fig. 3.19 Caterpillars causing heavy defoliation

3.3.13.2 Management

The following management practices are recommended:

- Intercropping sweet potato with onion/or the silver leaf (*Desmodium uncinatum*) might reduce the number of eggs laid by the females.
- Monitor the crop on a regular basis and manually remove the nests of young insects.
- Spray a contact insecticide as a matter of priority on gregarious caterpillars in the early stages of development.
- Caterpillars are attacked by predatory ants, ladybird beetles, lacewings, and dragonflies. The fungus *Beauveria bassiana* has been observed on caterpillars in the field during the rainy season.

3.3.14 Lesser Grain Borer: *Rhyzopertha dominica* (Bostrichidae)

3.3.14.1 Damage

The pest attacks dry tubers. The quality of the dried tubers is reduced.

3.3.14.2 Management

Fresh sweet potato roots have to be stored in good condition in pit or clamp stores for 3–4 months during the dry season. It is very important that only roots with absolutely no evidence of damage from wounding during harvest, transport, or insect pests are selected for fresh storage in order to reduce losses due to rotting or insect damage during storage.

Indigenous knowledge (IK)-based control methods were used by farmers in the form of general ash, specific plant ash, and whole or dry plant powder. They used neem, *Chenopodium opulifolium* (Chenopodiaceae), *Ocimum suave* (Labiatae), *Senna siamea* (Fabaceae or Caesalpinioideae), tobacco and *Eucalyptus* spp. (Myrtaceae), and plant by-products (rice husks, ash from rice husks and red maize cobs, and general ash) to control storage pests. Most of these products were used together with one or two others in different formulation mixtures.

Fig. 3.20 Grub damage on sweet potato tubers

3.3.15 White Grub: *Phyllophaga ephilida* (Scarabaeidae)

3.3.15.1 Damage

White grubs feed on the main roots of the plants. They cut the main stems and roots of plants below the surface of the soil. They tunnel into tubers of root vegetables and root crops. These feeding habits affect the underground portion of the stems and roots causing the plant to wilt and eventually die.

The white grub adult lays eggs in the ground at a depth of 2.5–20.0 cm. After 3 weeks the eggs hatch and the larvae emerge and feed on the outer skin of the sweet potato (Fig. 3.20). This is when most damage is done. The tunnels formed render the tubers unmarketable. The larvae cause galleries on tubers. The quality of tubers is reduced.

3.3.15.2 Biology

The white grubs are the larvae of Japanese and May/June beetles. They are white or yellowish in color, with tan or brown heads and with six spiny legs.

They have fleshy, wrinkled C-shaped bodies (Fig. 3.21). Upon hatching they are quite small but can measure up to 2–4 cm when they mature.

3.3.15.3 Management

1. *Cultural Methods*
   - Ensure proper drainage. Grubs love moist soil, especially with decaying organic matter. Female beetles prefer to lay eggs on moist-decaying organic matter.

Fig. 3.21 The grubs in soil

   - If heavy infestation occurs, avoid planting the same crops the whole year. Practice crop rotation always.
   - Soybean is found to be resistant to white grubs.
2. *Chemical Methods*: Successful control has been achieved with Actara and Pronto.
3. *Biological Methods*: *Bacillus popilliae* and *B. lentimorbus* are microbial insecticide products that kill grubs and adult Japanese beetles. Infected grubs are filled with a milky white liquid full of bacterial spores. They inoculate the soil as they move. The advantages of using commercial preparations of *B. popilliae* are they are host specific, meaning only effective against Japanese beetles, and their persistence is longer lasting.

Natural enemy that control these white grubs include parasitic wasp, *Pelecinus polyturator* (Fig. 3.22) (Hymenoptera: Pelecinidae). Also, *Cordyceps* fungus infects the grubs (Fig. 3.23).

Fig. 3.22 A pelecinid wasp, *Pelecinus polyturator*, a natural enemy of white grubs, *Phyllophaga* spp.

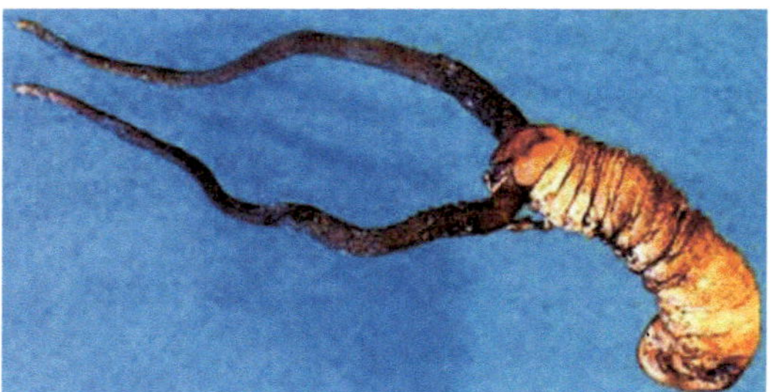

Fig. 3.23 White grub, *Phyllophaga* spp., infected with *Cordyceps* fungus

Fig. 3.24 Holes in a leaf caused by adult

Fig. 3.25 Thrips, *Frankliniella schultzei*

3.3.16 *Strobiderus* Beetle: *Strobiderus aequatorialis*

3.3.16.1 Damage

This is a small, yellowish beetle, 5–7 mm long occurring in East Africa. The adults perforate the leaves and cause damage similar to that of tortoiseshell beetles (Fig. 3.24).

3.3.16.2 Management

Spray a contact or systemic insecticide. Removal of nearby alternative host plants may reduce the populations or, alternatively, planting far away from alternative host plants.

3.3.17 Thrips: *Frankliniella schultzei*

3.3.17.1 Damage

Frankliniella schultzei (Fig. 3.25) can cause both direct and indirect damages to crop. Both adults and nymphs feed on pollen and floral tissue, leading to flower abortion. Severe infestations can cause discoloration and stunted growth of the plant (Amin and Palmer 1985). However, indirect damage by *F. schultzei* is due to the virus transmission.

The larvae are contained within the leaves, which curl up as a result of the biting action. The thrips grinds the epidermis of the leaves, even if they are in the form of buds, thus triggering distortion and patches where the leaf appears to be translucent.

3.3.17.2 Host Range

Frankliniella schultzei is a polyphagous pest feeding on various ornamental and vegetable hosts in different parts of the world (Milne et al. 1996). It has been recorded from 83 species of plants among 35 families (Palmer 1990). The major hosts of *F. schultzei* are cotton, groundnut, beans, and pigeon pea. However, due to its polyphagous feeding behavior, *F. schultzei* also attacks tomato, coffee, sorghum, chilies, onion, and sunflower (Hill 1975).

3.3.17.3 Management

1. *Cultural Methods*: Male thrips were most attracted to yellow sticky traps, while female

thrips were more attracted to pink sticky traps (Yaku et al. 2007).

2. *Biological Methods*: The efficiency of two predatory mites, *Amblyseius cucumeris* and *A. swirskii*, as potential biocontrol agents controlling *F. schultzei* populations.

3.3.18 Whitefly: *Bemisia tabaci* (Homoptera: Aleyrodidae)

3.3.18.1 Damage

Both the larvae and adults pierce and suck the sap of the leaves (Fig. 3.26). This causes the weakening and early wilting of the plant resulting in reduced plant growth. Their feeding may also cause yellowing, drying, and premature dropping of leaves that result in plant death. Whiteflies produce honeydew that serve as the substrate for the growth of black sooty molds on leaves and fruit. Ants are also attracted to the honeydew. The mold reduces photosynthesis causing the poor growth of the plant. They are the most important carriers of plant viruses that cause diseases of fiber crops, vegetables, fruit trees, and ornamentals.

3.3.18.2 Biology

The eggs are tiny, oval shaped, about 0.25 mm in diameter, and stand vertically on the leaf surface.

Newly laid eggs are white then turn brownish. They are deposited on the underside of leaves, sometimes in circle or oval-shaped patterns.

The larvae are transparent, ovate, and about 0.3–0.7 mm in size, and they move around on the plants looking for a feeding site upon hatching.

The pupae are dirty white and surrounded by wax and honeydew. During this stage, the red eyes of the emerging adults are visible.

The adults are about 1 mm long with two pairs of white wings and light yellow bodies. Their bodies are covered with waxy powdery materials. They are found feeding on top of the plants. A female can produce as many as 200 eggs in her lifetime and mating is not necessary. It takes about 40 days to develop from egg to adult.

3.3.18.3 Management

1. *Cultural Methods*: Do not plant a new crop next to one which is mature: The common practice of having mature crops adjacent to newly planted ones makes management of the pest very difficult since the cycle of the pest is never broken. An integrated control strategy is necessary for the effective management of this pest. Good farm sanitation, including removal of weeds around the cultivation, is necessary since weeds serve as hosts for the pests.

   (a) *Botanical Pesticides*

   Garlic Oil Spray: Chop 100 g of garlic finely. Soak the chopped garlic in mineral oil for a day. Add 0.5 l of mineral oil and 10 ml of soap. Dilute filtrate with 10 l of water. Constantly shake the

Fig. 3.26 Whiteflies on sweet potato leaf

container or stir the extract while in the process of application to prevent oil from separating.

Madre De Cacao and Neem: Shred 1 kg of Madre de cacao leaves and 1 kg of neem leaves. Soak leaves in 5 l of water for 3 days. Strain and add water to make up 20 l of filtrate. Spraying interval is 4–5 days.

Neem Oil Extract: Add 15 ml of neem oil into 1 l of soapy water. Constantly shake the container or stir the extract while in the process of application to prevent oil from separating. Spray solution.

Soap Spray: Mix 2.5 tablespoons of liquid soap to a gallon of water and spray.

Potato Starch Spray: Mix 2–4 tbsp potato flour to 1 l of water and add 2–3 drops of liquid soap. Shake the mixture and stir thoroughly before spraying.

2. *Chemical Methods*: Several new-generation insecticides are now available for the effective control of whiteflies. Targeting both nymphs and adults with *soap-based products* should be applied very early in the morning or late in the evening. Other chemicals which may be used include Admire, Pegasus and/or Basudin, and Vydate L at 10 ml to 4.5 l of water.

3.3.19 Leafhopper: *Empoasca* sp.

3.3.19.1 Damage

The saliva produced by the larvae and adults is toxic, which triggers color changes and distortion, chiefly in the tips of the leaves.

3.3.19.2 Management of Thrips, Whitefly, and Leafhopper

The following management strategies are recommended:

- Weed the field perimeters.
- Do not plant near another susceptible crop that has reached the end of its cycle.
- The presence of wind breakers slows down wind-borne infestations.
- Maintain the water and mineral balance of the crop.
- Spray a systemic insecticide or an insect repellent.
- Parasitic wasps and predators such as predatory mites, ladybird beetles, and lacewings are important in natural control of whiteflies.

3.3.20 Cutworm: *Agrotis* spp.

3.3.20.1 Damage

Seedlings are often cut off at ground level (Fig. 3.27). Cutworm larvae can be found in the soil (up to a depth of about 5 cm) near the plant host. They always curl up when disturbed. Cutworms feed only at night. Generally, they are not found on plants or on the soil surface during the day.

Young caterpillars eat the soft leaves of the plant. The fully grown caterpillars are capable of eating the entire plant. The newly hatched larvae feed from the base toward the tip of the leaf. At this stage, they first feed on the epidermis and may discolor the entire leaf surface.

3.3.20.2 Biology

Eggs are tiny pearl white and round and have a ridged surface. The newly hatched larvae are greenish and about 1 mm long. The full-grown larva has a cylindrical body, brown or brownish

Fig. 3.27 Cutworm damage on sweet potato

black with a tinge of orange. The thoracic segments have one to two dark spots near the base of the legs. The abdominal segments generally have two light-brownish lateral lines on each side, one above and one below the spiracles. Above the top lines are broken lines composed of velvet semi-crescent patches that vary in color among cutworms.

The pupa is black or brown in color and measures about 22.5 mm long and 9.2 mm wide.

The adult has dark-brown forewings with distinctive black spots and white and yellow wavy stripes. The hind wings are whitish with grayish margins. The total developmental period from egg to adult is about 35–40 days.

3.3.20.3 Management

1. *Physical Methods*: Place the protective collars made of plastic or paper cups, plastic drink bottles with ripped-out bottom, sturdy card board, and milk cartons around the young plant and push into the soil to prevent the cutworm from attacking the stem.

Sticky substances such as molasses, sawdust, or crushed egg shells should be placed around the base of each plant. When the cutworm emerges to feed, it will come in contact with the trap, get stuck, harden, and die.

2. *Cultural Methods*: Removal of weeds in and around fields will reduce egg-laying sites and will help in the prevention of cutworm infestation. This should be done at least 2–3 weeks before planting to reduce the incidence of cutworm larvae transferring to newly planted crops.

Plowing and harrowing of fields carried out properly before planting, will destroy eggs and expose larvae to chicken, ants, birds, and other predators.

Interplanting the main crops with onion, garlic, peppermint, coriander, or garlic every 10–20 rows helps to repel cutworms. Sunflowers and cosmos can also be planted as a trap crop in or around fields.

(a) *Neem Seed Powder Extract*: Add 20 g of finely ground neem kernels into 1 l of water, mix by continuously stirring, and steep for at least 6 h. Before application, add a drop of liquid soap and stir it again. This is also good for armyworm control.
(b) Spanish Needle (*Bidens pilosa*) Extract: Pound one whole plant. Soak in 2 l of water overnight. Squeeze out the plant sap. Strain and add a drop or two of liquid soap. Stir well and spray.

3.3.21 Variegated Grasshopper: *Zonocerus variegatus*

3.3.21.1 Damage

The feeding damage includes leaf notching and stripping, but as they mature they can consume an entire plant (Fig. 3.28). They can fly and move

Fig. 3.28 Variegated grasshopper damage to sweet potato leaf and adult

around easily in search of a wide variety of food sources.

3.3.21.2 Biology

Eggs are white, yellow green, tan, or various shades of brown in color and are elongated, curved, and pod-like. They are laid beneath the soil surface (also in non-crop areas including ditches, fence rows, and weedy areas) in compact clusters of 35–100 eggs that are covered by a frothy secretion. The froth bubbles, which protect the eggs from drying, are either white or brown, depending on the species. The eggs will hatch in about 1–3 months, depending on the species and environmental conditions.

The newly hatched nymph is white in color. After several hours of exposure to sunlight, it assumes its distinctive color and the markings of becoming an adult. Most nymphs immediately look for feeding sites and continue to feed on the same plants until they become adults. The nymphs have wing pads but they cannot fly. This is the reason why most of the damaged plants are concentrated and are found at the edges of the field. They will undergo 4–9 nymphal stages and become adults in about 2–7 months depending on the species and environmental conditions.

3.3.21.3 Management

The following control measures are recommended:

- Regularly scout the field and collect and destroy the grasshoppers. Catch the grasshoppers early in the morning when they are less active, using a butterfly net.
- An effective but laborious way to control the hoppers is to dig out the egg pods. Adults migrate toward oviposition sites where they oviposit at high density leaving clusters of egg pods behind.
- Use chemical control only when there is a heavy infestation using insecticides like Decis and dimethoate. Insecticides can be used to spray all instars (larvae). It can be an advantage to spray early instars which are still aggregated around the oviposition sites instead of blanket spraying of larger areas against adults.
- Rainfall is one of the natural factors against the pest.
- Use floating row covers (fine mesh nets) to protect seedlings and valuable crops.
- Yellow color attracts grasshoppers. To lure and trap them, make long sticky yellow strips.
- Use salty rice bran bait—moisten rice bran with salty water. Grasshoppers feed on salty rice bran.

3.3.22 Stink Bug: *Nezara viridula*

3.3.22.1 Damage

Adults and nymphs suck plant sap from leaves, flowers, buds, and tubers (Fig. 3.29). Feeding on tubers causes scarring and dimpling known as cat facing.

3.3.22.2 Biology

The eggs are yellow and barrel shaped. These are laid on the lower surface of the leaves in clusters of 20–130 in 5–8 parallel rows.

The nymph has heterogeneous colors (green, tan, brown, or gray). It is oval shaped and wingless but looks similar to an adult counterpart. The first nymphal instars do not feed. The nymphs form clusters at the natal site. The second and third instars are also found in clusters, but they disperse when disturbed. The fifth nymphal instars are sensitive to day length, which also causes the adults to begin diapauses. The nymphal development lasts for about 8 weeks.

The adult is shield shaped and green, tan, brown, or gray in color. Most of the adults are shiny, but some species are spiny and rough

Fig. 3.29 Stink bug

textured. The female starts mating 1 week after emergence and lives for about 30 days.

Stink bug emits a foul odor when disturbed, hence the name.

3.3.22.3 Management

1. *Cultural Methods*: Remove and control weeds from fields and also in the adjacent fields. Weeds serve as the pests' alternate hosts. Plant small flowering plants to attract native parasitic wasps and flies. Plow under all plant debris after the harvest to destroy all possible breeding sites.
2. *Biological Methods*
   (a) *Parasitoids*: Native parasitic wasp, *Trichopoda pennipes*, is a tachinid fly that develops in the nymph and adult stink bug or squash bug. The female tachinid fly lays its eggs on the body of the bug, it completes its cycle inside the bug's body, and the bug dies upon the fly's emergence.

3.3.23 Wireworm: *Conoderus falli*

3.3.23.1 Damage

Wireworms (Fig. 3.30) attack the stems of newly set plant by boring into the stems near the soil surface. Some plants may wilt and die within a few days or may have stunted growth. Often, the wireworm is found near the damaged or missing plant. Damage is most likely to occur where host plants like cotton, corn, sorghum, and tobacco have been previously planted.

Fig. 3.30 Wireworm

3.3.23.2 Description

Wireworms are the larvae of different species of click beetles. Larvae are slender and hard and feel somewhat like wires. Their bodies are segmented and shiny and are usually cylindrical but flat on the lower sides. There are three pairs of legs close together near the head and no prolegs. Some of the most common species are white and yellowish brown to reddish brown, although other species may vary in color.

3.3.23.3 Management

1. *Cultural Methods*
   - Flood fields for at least a month, during sunny weather, to eliminate wireworms.
   - Control weeds in and around fields, as click beetles feed on weed seeds.
   - Grow flax (*Linum* species) as a cover crop to kill wireworms. Flax, a tropical herb, is also good for controlling root-knot nematodes. White mustard and buckwheat plants are also found to reduce the wireworm population.
2. *Physical Methods*
   (a) *Corn/Wheat Seed Mixture Bait*: Soak mixture in water for 24 h to facilitate germination. Place 1/2 cup (of a 1:1 corn/wheat seed mixture) in a hole that is 12 cm wide and 30 cm deep. Cover with soil. Cover the topsoil with plastic to warm the surface and to speed up germination. Cover the edges with soil to prevent wind from blowing away the plastic. Remove the plastic, the soil cover, and the bait. Destroy larvae trapped in the baits. This method is also used to monitor white grub population.
   (b) *Potato/Carrot Bait*: Cut potatoes or carrots into chunks. Remove the potato "eyes" to prevent from further growing. Make the pieces big enough and put in sticks. Bury "staked potatoes/carrots" at a depth of 3–6 cm in the ground. The stick serves as the handle to easily pull the baits out. Bury randomly in the field. Leave baits in the soil for 2–3 days. Wireworm will feed on the baits. Dispose the baits and the wireworm properly.

Fig. 3.31 Adult cricket

3.3.24 Mole Crickets: *Gryllotalpa* spp. (Orthoptera: Gryllotalpidae) and *Acheta* spp. (Orthoptera: Gryllidae)

3.3.24.1 Damage

Crickets eat tubers of sweet potatoes. Fully grown crickets are brown in color and are about 2.5–3.5 cm long (Fig. 3.31). The various species of these insects usually live either in the soil or in bushes and under decaying crop residues and vegetation.

3.3.24.2 Biology

Mole crickets, which have heavily sclerotized front legs that are adapted for digging, are usually common in sandy soils. All crickets are nocturnal, feeding at night and secluded by day, under the soil. They feed at or slightly below the soil surface and can cause considerable damage before being discovered. Crickets spend their entire life cycle below the soil, which may go through a period of approximately 28–35 days. They are termed soil insects.

3.3.24.3 Management

1. *Cultural Methods*
   - The areas where sweet potatoes are grown should receive full sunlight.
   - Proper land preparation serves to control weeds, diseases, and soil insects and also helps in the destruction of large soil clods, which act as hiding places for cricket.
   - Good field sanitation—rid the field of weeds and plant residues from previous crops and burn.

Fig. 3.32 Termites

2. *Chemical Methods*: Any approved soil insecticide at the recommended rate may be applied, such as Basudin 60 % EC (diazinon) or Vydate L 40%EC at the rate of 10 ml in 4.5 L of water to cultivated areas.

3.3.25 Termites: *Coptotermes formosanus*

3.3.25.1 Damage

Termites (Fig. 3.32) damage stems and roots late or in the dry season. They chew and eat stem cuttings resulting in poor growth, death, and rot. In older plants, termites chew and enter the stem which weakens them and cause breaking.

3.3.25.2 Management

Plant early with the ensuing rains. Avoid planting on very dry land or on termite mounds.

Preplanting soil application of neem cake at 2 t/ha gave maximum protection against termites in sweet potato (8.2 % compared to 28.4 % in control) and gave highest marketable tuber yield (21.5 t/ha compared to 12.5 t/ha in control) in Bihar.

3.3.26 Mites

3.3.26.1 Distribution

Erinose is present in East Africa and the United States. The problem is more pronounced at lower altitudes where the climate is hot and dry. Research suggests that yields may be reduced.

3.3.26.2 Damage

1. *Tetranychus* spp.: Adults and nymphs of spider mites suck the sap from the leaves, causing the area around the feeding punctures to become chlorotic and appear as conspicuous whitish to yellowish stippling on the upper surface of the leaf (Fig. 3.33). Under heavy infestation, photosynthesis is greatly reduced, and the chlorotic areas may coalesce forming mottled yellowish interveinal patches. The leaves eventually turn yellow and may become brown and scorched and drop prematurely.

Sweet potato spider mite is a polyphagous species (i.e., it feeds on a wide range of plant species). Its common hosts are *Cucurbita maxima, Passiflora edulis, Centrosema pubescens, Ipomea triloba, Merremia vitifolia, Acalypha wilkesiana, Ricinus communis,* and many more.

Fig. 3.33 Spider mite adult

Fig. 3.34 Mite, *Aceria* sp.

2. *Aceria* sp.: The mites are minute in size, rather worm-like, and possess only two pairs of anterior legs. Sweet potato vines and leaves become excessively hairy, beginning at the shoot tip. Young expanding leaves may become distorted or puckered (Fig. 3.34).

3.3.26.3 Management

Control is through selection of mite-free planting material and good field sanitation. This might not be effective enough, however, because mite populations can build up rapidly.

3.3.27 Rodents: *Rattus rattus* and *R. norvegicus*

3.3.27.1 Damage

Rats and mole rats occasionally feed on sweet potato storage roots either by digging through the ridges or accessing the exposed roots. They often spoil more roots than they actually eat (Fig. 3.35).

3.3.27.2 Management

The following control measures are recommended:

- Rats and mice breed in burrows. Destruction of these burrows can help to reduce the populations.
- Rodents like to hide in vegetation and rubbish, as they do not like crossing open spaces where they may be seen and exposed to predation. Keeping the field and surrounding areas clean should reduce their damage.

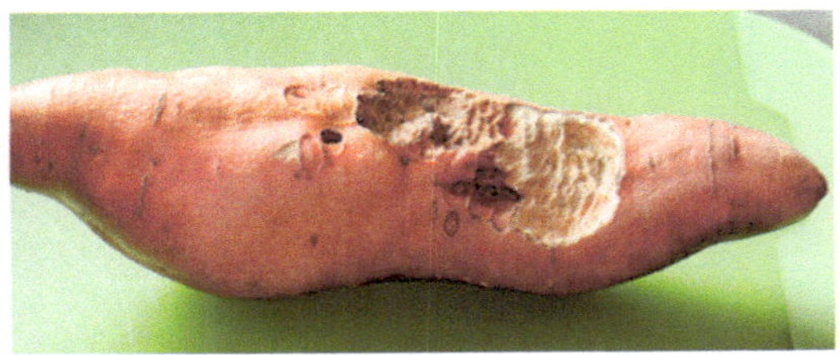

Fig. 3.35 Rat damage on sweet potato tuber

- Traps can be set but care must be taken to ensure they are placed in locations where livestock and children will not interfere or get hurt by them.

Some growers also use the following control methods:

- Some farmers dig a deep ditch around the perimeter of their field to deter rodents from digging tunnels straight into their fields.
- In Western Kenya a mixture of cow dung and pepper is made, placed in the burrows, and then burnt to smoke out the rodents.
- Experiments showed that mole rat damage could be reduced by planting on mounds rather than ridges and by planting the deep-rooted, poisonous shrub *Tephrosia vogelii* in the field.
- Farmers in Wangige, Kenya, chase away mole rats by pouring fermented cattle urine (1 week old) into their burrowing holes. They report this method as being very effective.

3.4 Beneficial Insects

The following are some of the beneficial insects that are indigenous on your fields and gardens or can be introduced for the control of pests (Fig. 3.36).

The following measures are recommended for the conservation of the beneficial insects:

- Provide hiding sites and alternative habitats such as mulches and other ground covers.

Ladybird beetles

Tachinid fly

Rove beetles

Fig. 3.36 Some of the beneficial insects indigenous in fields and gardens

- Plant small flowering plants on borders, hedges, and other perennial habitats as source of food and shelter.
- No indiscriminate use of synthetic pesticides.

3.5 Diseases

Sweet potatoes are affected by several different diseases in the field, some of which also affect the storage life of the tubers. Virus diseases are probably responsible for the greatest field losses, and some fungus diseases can also cause economic losses. The use of clean planting material is the most important disease control measure, and crop rotation aids control of many fungal diseases.

3.5.1 *Cercospora* Leaf Spot: *Cercospora ipomoeae*, *Pseudocercospora timorensis*, and *Phaeoisariopsis bataticola*

3.5.1.1 Distribution

Cercospora ipomoeae is very common throughout the tropics. *Pseudocercospora timorensis* is distributed in Brunei, Fiji, Formosa, Ghana, Guinea, Hong Kong, India, Malaysia, Mauritius, Nepal, New Guinea, New Hebrides, Sabah, St. Lucia, Sierra Leone, Solomon Islands, Sudan, Tanzania, and Uganda. *Phaeoisariopsis bataticola* is present in India, the United States, and Venezuela.

3.5.1.2 Symptoms

Leaf spots are small (about 6 mm in diameter) and angular with brownish coloration (Fig. 3.37). Leaves may drop from the lower parts of the stem in severe attacks. Damage to leaves is most severe during wet weather conditions.

3.5.1.3 Development and Spread

Conidia are produced abundantly on the leaf spots and are spread by wind and rain.

Fig. 3.37 *Cercospora* leaf spot on sweet potato

3.5.1.4 Management

Three sprays of 0.25 % Maneb (Dithane M-45) at fortnightly intervals was found effective in reducing the disease intensity (19.89 % as compared to 50.78 % in control) and in increasing the yield of tubers (21.76 t/ha as compared to 12.91 t/ha in control).

3.5.2 Wilt/Stem Rot: *Fusarium oxysporum* f. sp. *batatas*

3.5.2.1 Distribution and Economic Importance

This disease appears to occur in temperate regions and found in the United States, Japan, and Northern India. Stem rot is widespread in the United States and 10–50 % losses in susceptible varieties may occur. Losses up to 95 % have been recorded. In India, the disease is of common occurrence in North and Northeast India.

3.5.2.2 Symptoms

The fungus causes vascular wilt. The first symptom of the disease is yellowing of the youngest leaves. The vascular system becomes blackened causing stems to darken internally (Fig. 3.38). This darkening may extend several feet among stems. Infected plants wilt, collapse, and die.

Fig. 3.38 Browning of vascular tissue due to *Fusarium* wilt

Infected tubers have a blackened ring about 6 mm under the skin and develop a surface rot with shallow, sunken, circular lesions in storage. Plants produced from these tubers are likely to be diseased.

3.5.2.3 Development and Spread

The main sources of inoculum are infected tubers and plants. The fungus can live for several years on plant remains in the soil and can spread locally by water, wind, animals, machinery, etc. Diseased plants develop as a result of either diseased planting material or healthy plants becoming infected from the soil. Conidia, which are readily dispersed, are produced on the dead leaves and stems.

The incidence of the disease varies with the soil type, which may influence disease severity more than the inoculum density.

3.5.2.4 Management

The following management practices are recommended:

- Use resistant varieties, if available.
- Use disease-free planting stocks.
- Avoid fields with a history of wilt.
- To reduce infection near transplanting time is to hold transplants for 24 h at 29 °C to promote suberization of the injured surface, also yield loss can be reduced by planting more than one transplant per hill. Even though the percent of plants affected is not reduced, yield per unit land is maintained because there is less chance that all plants in a hill will be killed and also because productivity is similar for single-plant and multiple-plant hill.
- Whenever practical, remove and destroy infested plant materials after harvest. However, do not put these into your compost pit or pile. Compost from such materials will contain the fungi.
- Practice a proper crop rotation strategy. If soil is severely infested, planting Solanaceous and other susceptible crops is not advisable.

1. *Cultural Methods*: The most important means of controlling stem rot is to use healthy planting material and crop rotation. Crop rotation will not eradicate the fungus but it will help to reduce the losses. Tobacco should be avoided in the rotation as some varieties are attacked by the stem rot fungus. Sweet potato should

Table 3.15 Effect of pre-inoculation with nonpathogenic *Fusarium oxysporum* on the incidence of *Fusarium* wilt and on yield of sweet potato

	Field A			Field B		
Treatment	No. of plants treated	% of plants diseased	Yield (kg/100 plants)	No. of plants treated	% of plants diseased	Yield (kg/100 plants)
Nonpathogenic *Fusarium*	97	6.2	60.0	300	2.0	89.2
Benomyl at 0.1 %	97	5.2	60.8	500	1.5	95.6
Control	89	58.3	33.5	150	39.3	45.6

not be grown in the same field more often than once in 3–4 years.

2. *Chemical Methods*: The systemic fungicide thiabendazole has shown some promise as a dip treatment of tubers.
3. *Host Resistance*: Sweet potato cvs. 'Pusa Safed' and 'Kalmegh' have been reported resistant to wilt/stem rot disease.
4. *Biological Methods*
   (a) *Antagonistic Fungi*: Fusarium wilt of sweet potato has been reported to be successfully controlled by pre-inoculation of the cuttings with nonpathogenic strains of *Fusarium oxysporum* immediately before the cuttings are transplanted into the field (Ogawa and Komada 1986). It is believed that the nonpathogenic strains not only compete with the pathogenic ones in the rhizosphere and for infection sites, but they also enhance the resistance of the host toward the pathogenic strains. The level of control of *Fusarium* wilt of sweet potato using nonpathogenic strains of *Fusarium* was equivalent to that obtained with benomyl (Table 3.15). This method has been applied in commercial fields in Japan.

3.5.3 Chlorotic Leaf Distortion, *Fusarium denticulatum*

3.5.3.1 Distribution and Economic Importance

The disease has been reported from Brazil, Kenya, Peru, and the United States. However, the causal fungus was isolated from botanical seed from a number of additional countries. It appears that the disease is widely distributed, and it is likely that the disease occurs in many places from which it has not yet been reported.

The disease is found to occur in recent times and causes about 50 % yield reduction in susceptible variety (Celopia). The pathogen is responsible for reduction in length of petiole (24 %), leaf (13 %), and breadth of leaf (32 %).

3.5.3.2 Symptoms

Chlorotic leaf distortion (CLD) develops on young leaves near the vine terminals. The youngest 2–3 leaves at the tips of vines become generally chlorotic (Fig. 3.39), sometimes becoming bright yellow, almost bleached in appearance. As the leaves mature, they regain most of their normal color. On cultivars that normally have purple leaves, the leaves may become pink when affected with CLD. Following extended periods of conducive weather (sunny and humid), marginal necrosis may develop on affected leaves. Mycelia of the causal fungus may be seen growing out from between halves of leaves that have not yet opened or as scattered clumps on the upper surface of more mature leaves. Associated with the mycelia may be phialides that produce typical *Fusarium* macroconidia and microconidia. The latter clumps may appear similar to salt deposits. On some cultivars affected leaves may be twisted or distorted (Fig. 3.39).

3.5.3.3 Ecology

The relationship between the causal fungus and plant host is most unusual in that the fungus primarily colonizes the surfaces of the growing vine tip without invading the plant. Mycelia are found on apical meristems and between halves

Fig. 3.39 Symptoms of chlorotic leaf distortion disease on sweet potato

of developing leaves that have not yet opened. Once the leaves open and expose the fungal mycelia, the mycelia appear to stop growing. As a result, individual leaves appear to recover as they mature. Symptom development is strongly favored by warm, sunny, and humid weather. It is not known how the fungus overwinters in temperate areas.

3.5.3.4 Host Range

The disease is only known to occur on sweet potato, although other plants belonging to Convolvulaceae can be infected by artificial inoculation.

3.5.3.5 Management

It is difficult to eliminate the pathogen from planting material, but it has been successfully eliminated by using the meristem-tip culture techniques that are used for virus elimination. Spraying of 0.1 % carbendazim gives effective control of the disease.

3.5.4 Black Rot: *Ceratocystis fimbriata*

3.5.4.1 Distribution

The fungus occurs in North and Central America; parts of South America; West, Central, and Southern Africa; India; Southeast Asia; Japan; and the Pacific.

3.5.4.2 Symptoms

Black rot is an important field and storage disease of sweet potato. The fungus infects underground parts of the plant. Infection may extend a short way up the stems causing blackening above the surface of the soil giving rise to the common name "black shank" (Fig. 3.40). The first symptoms are small, slightly sunken black spots on the underground parts of the plant. The spots enlarge and may cover almost the whole tuber. Fungal spores are often produced in the center of these areas of infection. If the infected tubers are stored, the rot spreads rapidly causing heavy losses. Infected sweet potatoes have an unpleasant taste when cooked. Plants produced from infected tubers often have the disease.

3.5.4.3 Development and Spread

The fungus persists in the soil between sweet potato crops. The sources of inoculum are soil and infected tubers. Plants may become infected in the seedbed or in the field. The disease spreads particularly quickly in stored sweet potatoes.

3.5.4.4 Management

The following management practices are recommended:

- Crop rotation (4–5 years without sweet potato).
- Treat (fumigate) the soil.
- Use healthy planting material.

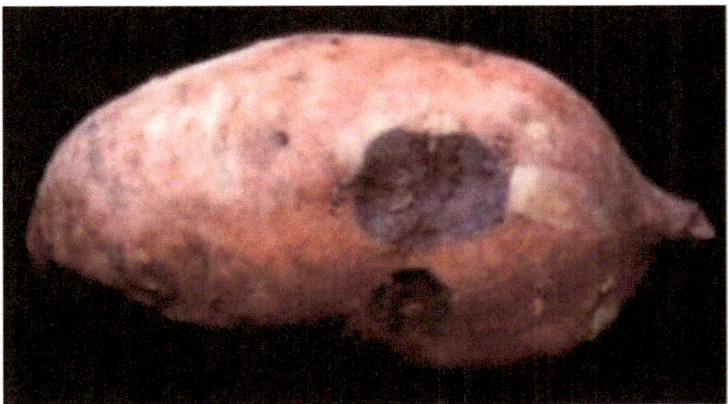
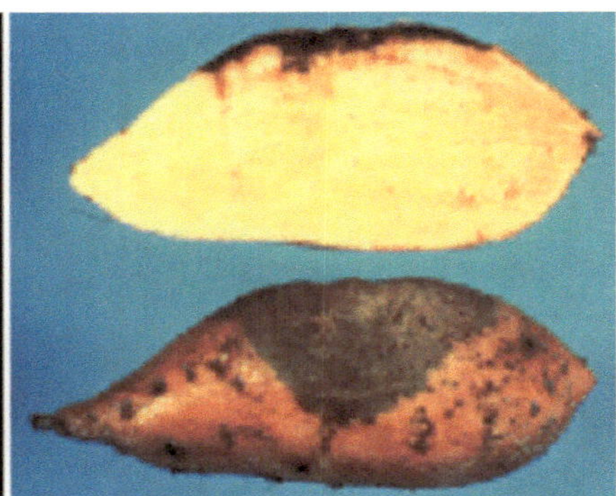

Fig. 3.40 Black rot of sweet potato tubers

- In places where it is difficult to find healthy mother plants, cuttings should be made 2 cm above the soil line to avoid infected portions of the plant.
- Use cuttings rather than parts of tubers for planting.
- Choose tolerant varieties.
- Allow the cuttings to take root in a disinfected soil before planting in the field.
- Apply a fungicide to the plant material.
- Avoid damaging the plant on harvesting.
- Cure wounds at the start of storage or prior to sale.
- Do not wash roots in water after harvest as contaminated water may spread the disease from infected roots to healthy.
- Hygiene in the storage facilities.

1. *Cultural Methods*: Proper selection of planting materials, curing the tubers after harvest, and placing the tubers in a clean and well-ventilated storage area.

The best method of control is to use disease-free planting stock. Healthy sweet potato seed can be grown from cuttings, which will be free from the disease as only underground parts are affected. Cuttings should be planted in the soil free from black rot pathogen to ensure the production of disease-free roots for field planting. As the fungus persists in soil, sweet potatoes should only be grown once in every 2–3 years.

Sweet potatoes should not be manured with dung from animals that have eaten black rot-infected tubers as it may contain spores of the fungus.

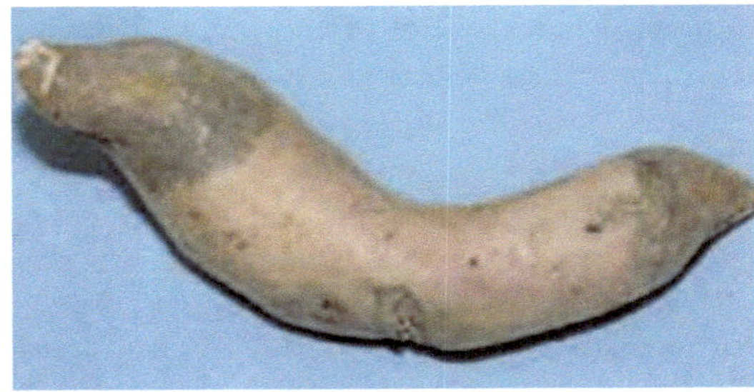

Fig. 3.41 Root rot at the tips of the sweet potatoes or in wounds

2. *Chemical Methods*: Tubers to be used for planting can be surface sterilized with fungicides. Thiabendazole has shown promise in trials. When inoculated sweet potatoes were dipped in a suspension of 3.32 g a.i./l, development of the disease was prevented. The control given by thiabendazole and benomyl dips was greatly improved by raising the temperature of the dips to 44–54 °C. Good control of black rot was also obtained with ferbam dips at 44–54 °C.

3.5.5 Root Rot: *Rhizopus stolonifer*

3.5.5.1 Symptoms

The fungus causes a soft, spongy moist decay in storage or transit (Fig. 3.41). An abundant growth of gray fuzzy mold is usually produced on the surface. When the rotting process is completed, or checked, the parts of the affected potato become shrunken, dry, and hard.

3.5.5.2 Spread

This disease is spread by infested soil, infected storage roots, and contaminated storage boxes, baskets, or tools.

3.5.5.3 Management

- Timely harvesting can reduce losses.
- The disease is spread by infested soil or airborne spores that enter wounds. Soft rot (*Rhizopus*) can destroy harvested roots in 48 h if they are left unprotected under sunlight.
- Avoid damaging the tubers during harvesting and storage.
- Sort the tubers prior to storage: remove tubers that are damaged, cracked, or starting to rot.
- Avoid washing storage roots since this practice is especially conducive to rot.
- The treatment of cuttings and the curing of the periderm during the postharvesting stage are the main methods used to control these organisms.
- Use dry, ventilated storage premises after the healing period or preferably store the tubers in a cold chamber at 13 °C with a relative humidity of 95 %.
- Some varieties rot faster than others because they are more susceptible.

3.5.6 Scab: *Elsinoe batatas*

Sweet potato scab of the leaf and stem is considered the most severe disease of sweet potato in the CNMI where it has been observed particularly on the Konabe variety, especially during the rainy season. The disease is common in Asia and on many Pacific Islands.

3.5.6.1 Symptoms

The first symptoms are small, brown lesions on the veins of the leaves. As the disease progresses, the lesions become corky in texture and result in distortion, causing the leaves to curl and fail to expand normally (Fig. 3.42). Stem lesions are slightly raised with rusty brown spots. A scab-like structure forms on the stems as the spots join.

Fig. 3.42 Sweet potato scab

3.5.6.2 Development

Infested planting materials and soil favor disease development.

3.5.6.3 Management

Little is known about controlling this disease. The use of resistant varieties is suggested where available. There are now sweet potato cultivars that are found to be resistant to the disease. Propagation material should be carefully inspected. Only disease-free cuttings should be used, and crop rotation with small grains and corn should be practiced. A new field should not be planted in the same place where an earlier field had much scab damage. Sanitation practices, removing and destroying infected leaves and stems as soon as possible, help keep the disease in check.

3.5.7 Leaf Spots: *Alternaria* sp. and *Phyllosticta* sp.

3.5.7.1 Symptoms

1. *Alternaria* sp.: This is mainly observed on old leaves, stems, and petioles. Small, round, blackish, clearly demarcated spots appear (Fig. 3.43). These spots then develop into lesions measuring several centimeters, which

Fig. 3.43 Anthracnose disease symptoms on stem

Fig. 3.44 Symptoms of *Phyllosticta* sp. leaf spot

subsequently crack and have a white center. These spots can form clusters triggering defoliation and causing the distal sections to dry out by choking the stem.

2. *Phyllosticta* sp.: Appearance of irregularly rounded leaf spots on both surfaces, measuring approximately 5 mm in diameter, can be seen. The pale brown to whitish center is surrounded by a reddish-brown border (Fig. 3.44). Fructification is often visible in the form of black dots in the center of the spots.

3.5.7.2 Management

- Choose less sensitive varieties.
- Apply fungicides.
- Use cuttings from disease-free plants.

3.5.8 Charcoal Rot: *Macrophomina phaseolina*

3.5.8.1 Symptoms

Charcoal rot is a storage disease. Symptoms in storage begin as a reddish brown to brown, firm, moist rot, initially restricted to the area just beneath the sweet potato skin. As the decay progresses, the pathogen moves toward the center of the sweet potato, causing further rot (Fig. 3.45). Two distinct zones become apparent within the infected tissue. The leading edge continues as a reddish-brown decay, and a zone of black develops behind the zone of active decay. Although the lesions are sometimes restricted, charcoal rot usually consumes the entire root, which eventually dries, becoming hard and mummified.

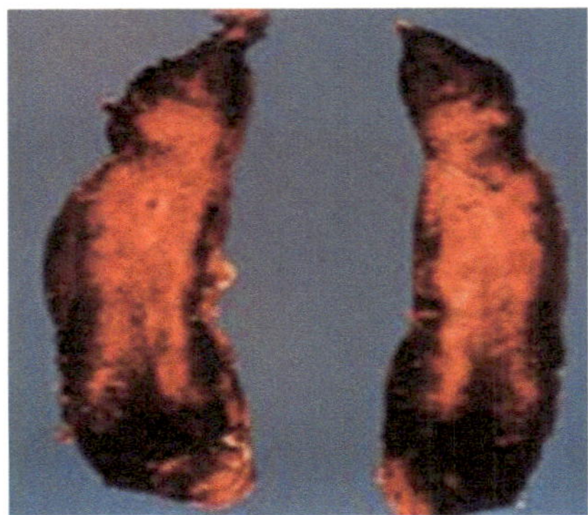

Fig. 3.45 Charcoal rot-infected sweet potato tubers

3.5.8.2 Host Range
The fungus has a wide host range which includes many crops grown in rotation with sweet potatoes, such as soybean, cotton, sorghum, and corn.

3.5.8.3 Survival and Spread
The fungus is soilborne and survives in plant debris or in the soil. There has been no evidence that the fungus can be carried in roots used for bedding. It requires a wound to enter the root.

High temperatures (84–88 °F) appear to favor disease development. Charcoal rot is more common in storage houses that are too warm or in crates located too close to heaters. Also, infection is greater if harvested roots are scalded by sunlight before being placed in storage.

3.5.8.4 Management

- Make sure that the tubers are not damaged during harvest. Damaged ones are to be discarded.
- Properly cure sweet potatoes immediately after harvest to reduce the incidence of charcoal rot.
- Store the cured tubers by placing them in a clean and well-ventilated storage area.

3.5.9 Scurf: *Monilochaetes infuscans*

3.5.9.1 Distribution
Scurf disease is distributed in Australia, Brazil, China, Hawaii, Japan, Rhodesia, Sierra Leone, and the United States.

Fig. 3.46 A hill of sweet potatoes showing scurf lesions that had spread from the stem of the mother plant onto the daughter roots

3.5.9.2 Symptoms
Infection with *M. infuscans* results in the production of black spots of various shapes and sizes all over the tubers (Fig. 3.46). The skin is usually not broken over these spots. The infection is only shallow and the spots can easily be scraped off. Usually, only a few scattered lesions are seen on the root, but in severe cases, almost the entire surface of the root can be affected. The presence of scurf does not markedly affect the nutritive value of the tuber or the total production; however, the market value is reduced considerably. Infected tubers become shriveled due to loss of moisture when stored.

3.5.9.3 Host Range
Monilochaetes infuscans has a very narrow host range, infecting only the sweet potato and closely related plants, including several of the morning glories that are common weeds in Louisiana.

3.5.9.4 Development and Spread
The fungus overseasons in stored tubers and in decayed plant material in the field. Infected seed tubers are the main source of inoculum. Stems produced from these carry the infection to the

field at transplanting. Soils which are heavy or rich in organic matter and moist conditions favor the disease. The use of animal manure may increase the incidence of scurf.

3.5.9.5 Management

The use of clean planting stock is the most important method for control of scurf. Cuttings grown on scurf-free soil are used to produce healthy tubers which are then planted in a clean nursery bed to produce sprouts for field planting. Visual sorting of clean roots is an inadequate method of control. Sweet potatoes should be grown in a 2–3-year rotation except on heavy soil and soils with high organic matter level, which should be avoided.

The benzimidazole fungicides benomyl (8.98 g a.i./L) and thiabendazole (1.08 g a.i./L) have been shown to be highly effective for the control of scurf when used as dip treatments of seed tubers or roots of transplants. Heated dips (49–50 °C) were found to be more effective than tap temperature ones (18 °C). Maneb also gives effective control when used as a dip.

Scurf is a disease that can be managed quite effectively through an integrated program of sanitation, crop rotation, and fungicide treatment of seed roots. As with most diseases, it is easier to prevent the disease rather than to try to eliminate it once it becomes a problem.

The following combinations of practices have proven effective in the management of scurf:

- Use only scurf-free sweet potatoes for seed. States in which sweet potatoes are grown for seed will certify this seed only if it meets standards for freedom from diseases, such as scurf, root-knot nematode, etc. If you cannot examine the seed before buying it, make certain that it is certified by the appropriate state agency.
- Treat seed roots with an effective fungicide, such as thiabendazole (Mertect) or dichloronitroaniline (Botran), being sure to cover the entire surface of the root as thoroughly as possible.
- Bed seed roots only in soil that is free of the pathogen. Do not bed in a field where sweet potatoes have been grown recently, and especially not in a field where scurf has occurred.
- Cut slips at least 1 in. above the soil line to avoid carrying the pathogen (and other pathogens and insects) to the field on infected stems.
- If scurf has become a problem in a field, it may be necessary to rotate out of sweet potatoes for 3–4 years. Shorter rotations may be possible if scurf has not become well established.

3.5.10 Foot Rot: *Plenodomus destruens*

3.5.10.1 Distribution and Economic Importance

The disease is distributed in Argentina, Hawaii, Tanzania, the United States, and Zaire. In Argentina losses in the field of 25–75 % have been reported in the past.

3.5.10.2 Symptoms

In the field, the first symptoms are small brown to black spots on the stem in the region of the soil line (Fig. 3.47). The infection develops slowly at first but eventually girdles the stem and extends a few inches upwards. Soon after the infection has reached this stage, the plant wilts and numerous black pycnidia (fruiting bodies) just visible to the naked eye appear in the infected region. Infected plants usually die, eventually. Plants infected by *P. destruens* often do not produce tubers despite good vine production. When tubers are produced, they may be affected by a brown, firm rot. Fruiting bodies appear on the surface of infected tubers giving them a pimply appearance. In storage, the infection can spread to healthy damaged tubers from the diseased ones.

3.5.10.3 Development and Spread

Foot rot is spread in the same way as stem rot and black rot but is not as common and does not usually cause such serious losses. However, local losses can be heavy. Initial infection usually

Fig. 3.47 Wilting, stunting, and yellowing caused by foot rot

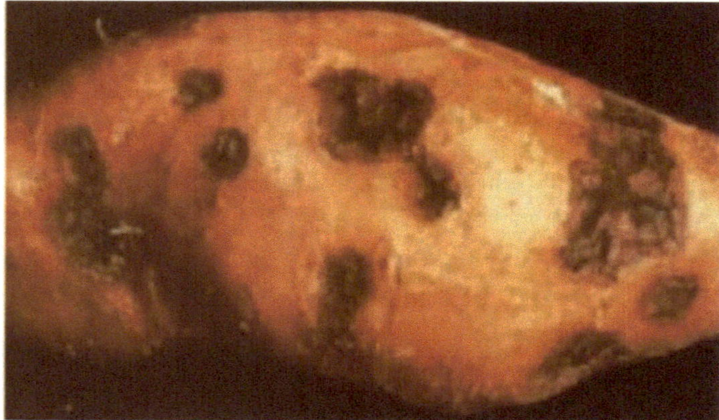

Fig. 3.48 Rootlet rot and scab lesions with irregular margins on tuber

occurs in the seedbed. The pycnidia produce vast numbers of spores which can be carried to healthy plants by insects and other vectors.

3.5.10.4 Management

The control measures are same as for stem rot and black rot, i.e., the use of healthy seed, clean seedbeds, seed treatment, and crop rotation.

1. *Cultural Methods*
   - Plant cuttings from healthy plants.
   - Two-year crop rotation is recommended.
2. *Host Resistance*: In Brazil the cultivar 'Princesa' has been found to show less disease than other popular cultivars.
3. *Chemical Methods*: Thiabendazole can be used for disinfecting planting materials.

3.5.11 Soil Rot: *Streptomyces ipomoeae*

3.5.11.1 Symptoms

A heavy infection results in dwarfed plants which only produce one or two short vines with small, pale green, and thin leaves. These symptoms result from damage to the underground parts of the plant. Lateral roots are attacked and many are destroyed, others are often distorted. The most obvious symptom is the development of pits on the tubers, the first signs of which are water-soaked dark spots.

As the spots on the tubers develop, the overlying skin breaks and the necrotic tissue underneath falls out, leaving the characteristic pits with jagged margins (Fig. 3.48). The pits range in diameter from 5 to 30 mm. The tubers may become girdled by lesions but continue to grow on either side of the constricted infected area.

3.5.11.2 Development and Spread

The organism which causes soil rot can survive in the soil for several years, and the disease is most severe in soils that are alkaline or only slightly acidic. The causal organism is spread through farm implements by people and animals and by transport of diseased tubers. Dry seasons and poor soils are conducive to severe losses.

3.5.11.3 Management

The most effective method to control this disease is to prevent the transfer of infected materials (soil and tubers) to uninfected fields. Planting material should be healthy. Fields known to be infested with the disease should be used for other crops for longest period possible. The addition of sulfur to the soil as an acidifying agent will reduce the disease incidence. Cultivating equipment should not be moved from infested fields to non-infested fields and/or should be disinfested before movement into a new area. Storage roots from affected fields should not be used for producing transplants.

The use of Banrot, Rizolex, captan, or Mankocide is recommended for effective control. These can also be used as a preplant soil drenches.

3.5.12 Feathery Mottle Virus

The most important disease in Kenya is the sweet potato virus disease (SPVD). Yield losses from the virus can be as high as 80 % according to Kenya Agricultural Research Institute (KARI). Kenya's average sweet potato yield stands at 6 t/ha, less than half the world's average of 14 t/ha. China has realized yields of 18 t/ha.

3.5.12.1 Symptoms

Sweet potato feathery mottle virus (SPFMV) occurs wherever the sweet potato is grown. Symptoms of SPFMV on the foliage are generally

Fig. 3.49 Symptoms of feathery mottle virus infection on sweet potato

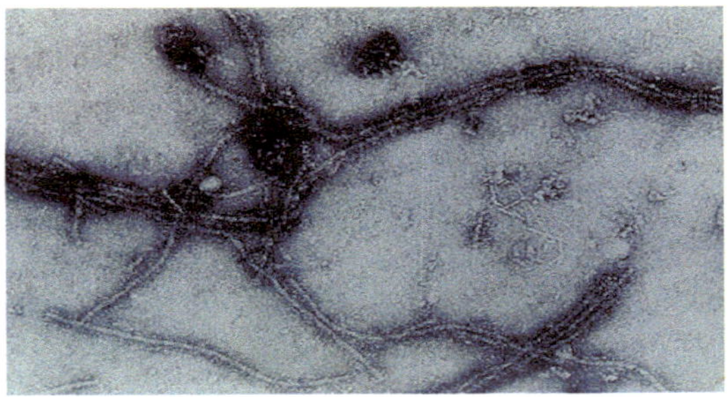

Fig. 3.50 Virus particles of sweet potato feathery mottle virus

slight or absent. The classic irregular chlorotic patterns (feathering) along leaf veins and faint-to-distinct chlorotic spots with or without purple margins occur in some cultivars (Fig. 3.49). The virus is capable of causing major crop losses of between 50 and 80 % especially when in combination with other sweet potato viruses such as the sweet potato chlorotic stunt virus.

3.5.12.2 Particle Morphology

Sweet potato feathery mottle viruses are flexuous rods in shape (Fig. 3.50).

3.5.12.3 Host Range

Sweet potato is the main natural host of SPFMV, although the virus occurs in wild *Ipomoea* species. The experimental host range of the virus is mainly restricted to the Convolvulaceae and Chenopodiaceae, but a few strains also infect species of the Solanaceae of which *Nicotiana benthamiana* is a good propagation host for purification of the virus. Several strains cause local lesions on *Chenopodium amaranticolor* and *C. quinoa*.

3.5.12.4 Transmission

SPFMV is sap transmissible and transmitted by a large number of aphid species such as *Aphis gossypii*, *A. craccivora*, and *Myzus persicae*, in a non-persistent manner. Vegetative propagation perpetuates the virus.

3.5.12.5 Management

The following management practices are recommended:

- Use cuttings from healthy plots.
- Use resistant varieties. "Namujuna" is considered resistant to the SPMMV virus.
- Crop rotation.
- Use clean or disinfected tools to take cuttings.
- Isolate new plantings from old plantings.
- Reduce the vector insect population.

1. *Virus-Free Tissue-Cultured Plants*

Technologies for eliminating viruses from plants by meristem-tip culture and for indexing sweet potatoes for viruses are not new (Fig. 3.51). However, most programs for providing farmers with propagating material that is relatively free of viruses were only deployed in the last 20 years. Farmers' responses to surveys and data from demonstration plots indicated yield increases of over 30 %. Virus-free plants have been obtained in many countries by meristem-tip culture and thermotherapy.

2. *Host Resistance*

(a) *GM Sweet Potato*

There is a need to develop sweet potato varieties resistant to SPFMV. Genetic engineering offers a means to introduce genes into sweet potato, which can confer tolerance to SPFMV. Approaches that have been used to produce transgenic sweet potato include expression of viral replicase genes, antisense RNAs, and viral coat protein genes. A Kenyan sweet potato variety KSP36 was subjected to somatic regeneration experiments and an *Agrobacterium-mediated* transformation.

GM sweet potato yields are double that of the regular plant and that the potatoes were bigger and richer in color, with more nutritional value. Monsanto isolated a viral coat protein responsible for virus resistance and donated it to KARI, royalty-free, to use in its sweet potato improvement program.

Background of the Case Study: Transgenic sweet potato (TSP) research project involved the development of varieties that are resistant to virus

Fig. 3.51 Tissue-cultured sweet potato plants free from viruses

diseases. The most common viral diseases affecting sweet potato are sweet potato feathery mottle virus (SPFMV), sweet potato chlorotic stunt virus (SPCSV), and sweet potato mild mottle virus (SPMMV). These diseases simultaneously cause the sweet potato virus disease (SPVD). Conventional approaches to breeding varieties resistant to SPVD have not been effective in combating the disease. It was realized that this could be complemented by nonconventional means involving the use of biotechnological methods.

Transgenic sweet potato is resistant to feathery mottle virus (FMV) and has the potential of increasing yields of sweet potato roots and foliage. With financial assistance from USAID/ABSP, a collaborative research project between KARI and Monsanto was launched in 1991 to develop a virus-resistant sweet potato. The initiative was an innovative undertaking based on public/private partnerships. The project is considered to be a groundbreaking initiative for the introduction of the first transgenic crop into Kenya.

Results of laboratory at Monsanto (USA) and initial field trials in Kenya reveal that transgenic sweet potato has good yield potential. Already selected lines of CPT560 achieved a minimum yield increase of 18 % of sweet potato from an annual average production of 6 t/ha. Apart from yield potential, the cost of transgenic sweet potato planting materials will be negligible. Hence, the technology will be more affordable to poor farmers. Furthermore, sweet potato is transplanted by cuttings, which are lighter to transport and do not require the use of new skills.

Collaborative Research and Training: The advent of modern biotechnology motivated researchers to conceive a collaborative program. The initiative to develop genetically modified sweet potato that is resistant to the virus is mainly composed of researchers from Monsanto and KARI—with some contribution from Central Research Institute for Food Crops (CRIFC) in Indonesia. Within the ABSP framework, the mandate of transgenic sweet potato is technology access/generation and technology transfer to developing countries such as Kenya. This includes germplasm collection; transformation and testing in the United States and developing countries; training scientists, administrators, and policy makers on the application of biosafety procedures; and intellectual property rights (IPRs) (Ives and Bedford 1998).

The development of virus-resistant sweet potato transformation began in 1991. Financial support for the project came from USAID and Monsanto. KARI and Monsanto scientists carried out research and technical aspects of transformation involving six Kenyan sweet potato varieties against SPFMV using a Monsanto-donated virus coat protein (*cp*) gene. The basic research components of the project such as the development of suitable biotransformation and plant regeneration protocols were conducted at Monsanto in St Louis, USA, in collaboration with KARI scientists (Gibbons 2000).

Initially, only one of the six original sweet potato varieties, CPT560, was successfully transformed using SPFMV *cp* gene (Fig. 3.52), but 195 lines of CPT560 had been transformed by the year 1997. Transgenic sweet potato is developed using genetic engineering techniques, while in other parts of the world (e.g. China), they are still using conventional plant breeding methods to induce disease resistance in sweet potato.

The actual transfer of the recombinant sweet potato technology from the Monsanto to KARI took place in April 2000. This process lasted up to 3 years because it coincided with global concern over transgenic crops especially with respect to IPRs and biosafety issues. Arrangements to introduce the transgenic sweet potato into the country took place at a time of establishing Kenya's Biosafety Guidelines and the National Biosafety Committee (NBC). As a result, the approval process for field evaluation was a slow and learning process given that this was the first case of a genetically modified organism (GMO) to be handled in the country. The NBC also acknowledged being constrained by human resources in terms of molecular scientists as well as the need to follow stringent and precautionary biosafety procedures and measures. Following 2 years of reviewing the application, NBC approved the introduction of transgenic sweet potato in the country. The Kenya Plant

Fig. 3.52 Sweet potato cv. Jonathan susceptible to feathery mottle virus (*left*), whereas the transgenic line resistant to the virus (*right*)

Health Inspectorate Services (KEPHIS), a regulatory authority responsible for enforcing biosafety regulations, issued the plant importation permit to KARI in December 1999.

At present, on-station trials are being conducted in at least five different agroecological zones in Kenya. The country is characterized by diverse agroecological conditions, which in turn influence farmers' preferences for particular sweet potato varieties or clones. The project is reported to be undertaking crop transformation of popular Kenyan varieties for disease resistance to develop a variety of clones that can satisfy the diverse varietal preferences of sweet potato producers and consumers.

Production: Transgenic sweet potato is promoted for its potential to reduce the incidence of disease infestation and increase yields. It has potential to increase yields of roots by 40 % for food and sale as well as fodder for livestock feed (Wambugu 2001). The technology is relatively cheaper because farmers can freely exchange planting materials. It is also easy to use because of its compatibility with existing production practices.

However, critics point that given that viral infection on sweet potatoes is complex of three viruses of which feathery mottle virus (FMV) is only one part, the extent to which CPT560 will control the complex virus remains uncertain. The complexity of the technology and biophysical conditions of farmers' fields will remain major constraints of generating and retaining satisfactory innovations for farmers. According to sources at CIP in Nairobi, the ideal situation is that there should be at least two sweet potato clones per each agroecological zone. Therefore, any research strategy ought to recognize the diversity of sweet potato varieties that farmers produce to meet different agronomic and nutritive qualities.

3.5.13 Witches' Broom

3.5.13.1 Distribution

The disease is distributed in Taiwan and the Pacific.

3.5.13.2 Symptoms

Plants infected with witches' broom have shortened internodes giving them a bushy appearance. Mycoplasma-like bodies of 200–2,500 nm have been observed in the phloem of infected plants.

Fig. 3.53 Proliferation of shoots and reduced leaf size in infected plant (*right*) compared with the adjacent uninfected plant (*left*)

Fig. 3.54 Reduced size of whole plant and of storage roots, in a plant infected (*right*) compared with the adjacent uninfected plant (*left*)

Some infected plants do not survive until harvest or they produce few harvestable roots.

The initial symptoms consist of a transient vein clearing followed by the development of new leaves that are distinctly smaller and more chlorotic (paler) than normal. Generally only scattered individual plants are infected, and they are conspicuously smaller than adjacent uninfected plants. Leaves may have a more rounded shape, often curling at the leaf margins. The growth habit tends to be more erect than in healthy plants, internodes are shortened, and there is a proliferation of axillary shoots which, together with a greatly reduced root system, result in weak plants with a compressed or bushy appearance (Fig. 3.53). The number and quality of storage roots are reduced, with few or no harvestable storage roots being produced on severely affected plants (Fig. 3.54). Production of latex in vines and roots is also reduced. Yield losses of 30–90 % have been reported.

Low annual rainfall and prolonged dry seasons favor the vector, and, under these conditions, the disease can reach epidemic proportions. Infected planting material is also important in the dissemination of the disease. As the disease has an exceptionally long incubation period in sweet potato (up to 283 days) following graft transmission, infected planting material can appear healthy.

Axillary buds are elongated with small leaves. Rosette, stunting, yellowing, mosaic, and necrosis symptoms have also been observed. Tubers are fewer and of poor quality.

It should be stressed that the field symptoms resulting from mycoplasma infection are so variable that they are very unreliable for diagnosis. The main tests for more accurate diagnosis are graft transmission to an indicator plant (usually *Ipomoea setosa*), insect and sap transmission, serology, and examination by electron microscopy.

3.5.13.3 Particle Morphology

Characteristic pleomorphic bodies ranging from 0.1 to 1.0 μm in diameter, with a well-defined unit membrane, are seen in phloem tissue.

3.5.13.4 Transmission

Witches' broom disease is caused by a mycoplasma-like organism which is transmitted by leafhoppers (*Orosius lotophagorum ryukyuensis*) and grafting. The high disease incidence was correlated with high populations of *O. lotophagorum ryukyuensis*. In the Solomon Islands, *Orosius argentatus* was shown to be a vector.

3.5.13.5 Host Range

Primary natural host is sweet potato. Wild hosts (*Ipomoea nil, Pharbitis purpurea, Ipomoea pes-caprae*) act as reservoirs.

Alternative experimental hosts of SPLL phytoplasma are *Ipomoea setosa*, *I. triloba*, *I. indica*, *I. ericolor*, and *Catharanthus roseus*.

3.5.13.6 Management

1. *Cultural Methods*: Sanitation (removal of all previous crop debris from the field and use of healthy planting material) has provided the best control to date. The removal and destruction of diseased plants and wild hosts reduces the spread of the disease.

In small, local cultivations, vines may be taken from more vigorous plants in order to establish a subsequent crop. While this practice may assist in selecting plants that are better able to tolerate infection by the phytoplasma, it probably contributes to the promulgation of the disease in the crop; this situation may be further exacerbated where crop plantings overlap or where the previous crop is not removed.

Phytoplasma-free vegetative material should be transferred as in vitro plantlets.

The following cultural control measures are important to reduce the disease:

- Remove plants with signs of the disease as soon as they appear, and burn them.
- Little leaf is spread mostly in infected planting material. If possible, take planting material from gardens where the disease has not been seen.
- If that is not possible, carefully select planting material from plants that are free of the disease.
- Look for symptoms after planting.

Meristem-tip culture and heat therapy have been used to produce pathogen-free plants. The use of healthy planting material and eradication of diseased material are the most promising practical control measures at present.

2. *Chemical Methods*: Chemical control of this disease is difficult as only tetracycline and related chemicals are effective.

Use synthetic pyrethroid insecticides, such as lambda cyhalothrin or permethrin. The choice of chemical is important: use those that are least persistent in the environment and have low toxicity to bees.

Derris (rotenone) may be effective against the leafhopper that spreads sweet potato little leaf. A local variety of Derris that has a high concentration of rotenone is being multiplied by MAL and the Kastom Gaden Association. The variety came originally from Papua New Guinea.

Some of the disease management practices recommended for sweet potato are presented in Table 3.16.

Table 3.16 Diseases of sweet potato and their management

Disease	Symptoms	Control
Black rot	Black depressions on sweet potato; black cankers on underground stem parts	Select disease-free potato seed. Rotate crops and planting beds. Use vine cuttings for propagation rather than slips
Internal cork	Dark brown to black, hard, corky lesions in flesh developing in storage at high temperatures. Yellow spots with purple borders on new growth of leaves	Select disease-free seed potatoes
Pox	Plants dwarfed; only one or two vines produced; leaves thin and pale green; soil rot pits on roots	Use disease-free stock and clean planting beds. Apply sulfur to lower soil pH to 5.2
Scurf	Brown to black discoloration of root; uniform rusting of root surface	Rotation of crops and beds. Use disease-free stock. Use vine cuttings rather than slips
Stem rot	Yellowing between veins; vines wilt; stems darken inside and may split	Select disease-free seed potatoes. Rotate fields and plant beds

3.6 Nematodes

The root-knot and the reniform nematodes are the two most widely occurring and important parasites of sweet potato throughout the world.

The burrowing nematode (*Radopholus similis*) causes a lesion on the plant's root that forms a canker, and the plant suffers from malnutrition, while the nematode completes its life cycle within the root. *Radopholus similis* caused significant reduction in root weight, tuber weight, and tuber length even at a low inoculum level of 100 nematodes/plant (Koshy and Jasy 1990).

3.6.1 Root-Knot Nematodes: *Meloidogyne* spp.

M. incognita is the major nematode species damaging sweet potato, but the crop is also attacked by *M. javanica, M. hapla*, and *M. arenaria*. In India, *M. incognita* and *M. javanica* are reported on sweet potato (Ray et al. 1990).

3.6.1.1 Distribution

The root-knot nematode, *Meloidogyne incognita,* is worldwide in distribution. It is widespread in Asia, particularly Southeast Asia, and usually occurs in warmer areas. In some countries, M. *javanica* is more dominant.

3.6.1.2 Economic Importance

The degree of damage depends upon the population density of the nematode, susceptibility of the crop, and environmental conditions, such as fertility, moisture, and the presence of other pathogenic organisms, which may interact with nematodes. In sweet potato an estimated annual yield loss of 10.2 % was reported. In susceptible varieties pathogenicity of *Meloidogyne incognita* showed a 50 % storage root reduction at a population density of 20,000/cm^3. Aside from yield loss, cracking can make storage roots unmarketable. Gaspin and Valdez (1979) reported tuber reduction to the extent of 47.7 and 50.6 % with an initial inoculum level of 20,000 eggs of *M. incognita* and *M. javanica*, respectively.

3.6.1.3 Symptoms

Nematode infection in primary roots cause swelling (knotting) of the entire root, and heavy infection can inhibit apical growth. The most obvious symptoms of damage on enlarged roots are longitudinal cracking and general rough appearance of the skin. In many cases lenticels are noticeably enlarged. Both the roots and tops of infected plants are reduced. In fibrous roots, the nematode infestation produces small swellings or knots which are visible in most cases to the naked eye (Fig. 3.55). Infected feeder roots are usually shorter and have fewer secondary roots and root hairs.

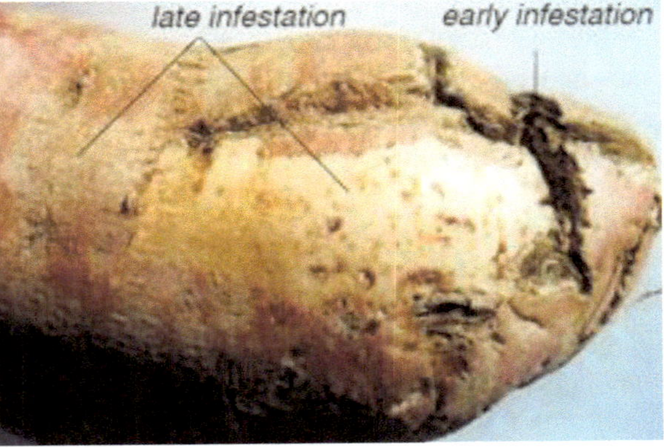

Fig. 3.55 Root-knot galls on sweet potato tubers (*left*), cracks in sweet potato tuber due to root-knot nematode infection (*right*)

Nonspecific aboveground symptoms include patchy stunted vine growth, discoloration and leaf chlorosis, excessive wilting during dry and hot conditions, stunting of whole plants, reduced yield and quality, and sometimes premature death. Root knot is generally more severe in sandy soils and under adverse environmental conditions. Abnormally abundant production of flowers could also be due to infestation by the nematode.

Presence can be diagnosed by the pearl-like swollen female nematodes in flesh of storage roots or in fibrous roots, within the galls or dark spots.

3.6.1.4 Biology

The nematode may complete several generations during the cultivation of this crop. Feeder and storage roots are attacked at the same rate. All races of *M. incognita* can attack sweet potatoes at varying degrees.

3.6.1.5 Survival and Spread

Meloidogyne juveniles and/or eggs survive in storage roots and can be disseminated in root, but not stem, and propagative material. Irrigation water and unclean farm tools and machinery can aid dissemination of the nematodes. Nematodes can survive on many alternate weed hosts.

3.6.1.6 Ecology

Meloidogyne species seem to do well in light, friable, sandy loam soil which happens to predominate and constitute the major portion of the world's sweet potato-growing areas. *M. incognita* requires warm temperature for completion of its life cycle. During a normal growing season, it can undergo 4–5 generations. Therefore, it is capable of increasing its population to a level of economic threshold in a short period.

3.6.1.7 Interaction with Other Pathogens

M. incognita interacts with *Fusarium* spp. and *Ralstonia solanacearum* causing severe wilting and premature death. Although there are several *Fusarium*-resistant cultivars, their resistance is broken in the presence of *M. incognita*.

3.6.1.8 Management

1. *Physical Methods*: Hot water treatment of tubers at 45 °C for 30 h, 46.7 °C for 65 min, or 50 °C for 3–5 min eliminates root-knot nematode infection without seriously impairing the viability of roots. Hot air treatment of roots, 5–8 cm in diameter, at 50 °C for 4–8 h eliminates most root-knot nematodes without significantly affecting their keeping qualities.
2. *Cultural Methods*: The clean cuttings of seed roots from the apex are less prone to root-knot damage when planted out than cuttings from the basal end.

Groundnut or maize in rotation with sweet potato decreases numbers of *M. incognita* to low levels.

Planting of trap crops and antagonistic crops such as *Tagetes erecta* and *Crotalaria spectabilis* in nematode-infested soil is effective against the root-knot nematode.

3. *Chemical Methods*: Application of carbofuran granules at 6.7 kg a.i./ha, 1–4 days before planting, and fensulfothion/ethoprophos granules each at 3.4 kg a.i./ha in a 30 cm band 2 weeks before planting has shown to control root-knot nematodes and increase the quality and yield of sweet potato.

Total yield and number of marketable roots can also be increased by treating nursery beds with ethoprophos and aldicarb granules at 1 g/30 cm^2.

Bare root-dip treatment of the propagating material in a solution of oxamyl or side dressing with nematicides at the time of planting will allow the establishment of the crop by providing early protection against nematodes.

4. *Host Resistance*: Sweet potato cvs. 'Heartgold,' 'Nemagold,' 'Redmar,' 'Jersey types,' and 'Kyushu No. 52' were reported to be resistant to root-knot nematodes (Giamalava et al. 1960). The cv. 'Jasper' is moderately resistant,

while cv. 'Painter' shows some tolerance. Mohandas and Palaniswami (1990a) reported the availability of high degree of resistance in high-yielding released varieties of sweet potato, viz., 'Sree Vardhini' and 'Sree Nandini.' Mohandas and Ramakrishnan (2002) reported that sweet potato cvs. 'Sree Vardhini,' 'Sree Nandini,' 'Sree Bhadra,' and 'Kanjangad' were resistant to *M. incognita*.

Sweet potato variety Sree Bhadra is a resistant trap crop of root-knot nematode. Sree Bhadra, a high-yielding variety of sweet potato was found to be resistant to root-knot nematodes. Experiments conducted at different locations recorded reduction in root-knot nematode population and consequent increase in yield in subsequent susceptible crops. RKN enter inside root of Sree Bhadra as if it is a susceptible crop. Cells around feeding site instead of becoming giant cells become necrotic (hypersensitive reaction). Nematodes die inside due to denial of food. Field population declines sharply to below detectable level during the crop period of 95 days. Subsequent susceptible crops escape nematode infestation and give better yield.

5. *Biological Methods*
   (a) *Antagonistic Fungi*: Galano et al. (1996) reported that application of different *Paecilomyces lilacinus* isolates grown on water hyacinth substrate, 50 days after inoculation of *M. incognita* eggs, controlled root-knot nematodes, and increased fresh vine and fibrous root weights by 32–26 % and 30–35 %, respectively, over untreated control in sweet potato.
6. *Integrated Methods*

   Control of root-knot damage to sweet potatoes involves integration of at least 3 methods:

- The selection of seed roots that is either free of nematodes or freed of nematodes by hot water treatment.
- Planting of the nematode-free seed roots into beds of sand or coarse-textured soil that is either nematode-free or, if infested, is preplant fumigated.
- Transplanting of the clean "slips" into nematode-free soil or preplant fumigated soil.

3.6.2 Reniform Nematode: *Rotylenchulus reniformis*

R. reniformis is another important nematode species reported from sweet potato from a large number of countries. Gaspin and Valdez (1979) reported 60.6 % yield reduction with an initial inoculum level of 5,000 larvae of *R. reniformis*.

3.6.2.1 Symptoms

The reniform nematode is associated with leaf yellowing and reduction in yield and size of swollen tubers (Fig. 3.56). The most serious injury occurs when nematode population levels are high at planting time, and there is a seasonal increase of the nematodes. Infestation by *R. reniformis* causes cracking of storage roots.

3.6.2.2 Biology

Juveniles of the reniform nematode are differentiated within the egg and undergo one molt before the second-stage juveniles hatch. Three additional molts occur without feeding while the juveniles are free in the soil. Adult stage and egg production occur 16 days after inoculation in susceptible cultivars. The life cycle from egg to egg is from 22 to 29 days in susceptible cultivars such as 'V20-436.'

3.6.2.3 Ecology

Male and female nematodes can survive in air-dried soil kept at 20–25 °C for 7 months. Local dissemination is through infested soil. Distribution is limited by low winter temperatures, and nematode and disease development are both greater at 29.5 °C than at 15, 21.5, or 36 °C.

3.6.2.4 Host Range

It has very wide host range including weeds. Major host are soybean, cowpea, cotton, pineapple, sweet potato, cassava, and other vegetable crops.

Fig. 3.56 Symptoms of reniform nematode infection on sweet potato

3.6.2.5 Interaction with Other Pathogens:

R. reniformis interact with *Fusarium* spp. in the development of disease complex.

3.6.2.6 Management

1. *Physical Methods*: Hot water treatment of sweet potatoes at 50 °C for 3–5 min eliminated the reniform nematode (Martin 1970).
2. *Chemical Methods*: Good control of *R. reniformis* and increase in sweet potato grade have been achieved by application of granules in 30 cm band of fensulfothion/aldicarb each at 5.6 kg a.i./ha or ethoprophos at 3.4–6.7 kg a.i./ha.
3. *Cultural Methods*: Planting of trap and antagonistic crops like *Tagetes erecta* and *Crotalaria spectabilis* in nematode-infested soil has been found effective against the nematode.
4. *Biological Methods*: *Paecilomyces lilacinus*, a fungal egg parasite, was found effective against the reniform nematode.
5. *Host Resistance*: The sweet potato variety 'Goldrush' is less susceptible to *R. reniformis* than other varieties but is highly susceptible to root-knot nematodes. Selection P-104 is reported resistant to cracking.

3.6.3 Lesion Nematode: *Pratylenchus* spp.

In Japan the principal species is *P. coffeae*, which caused significant losses, while in the United States, *P. brachyurus* is the most common.

3.6.3.1 Distribution

The lesion nematode is distributed in Asia, the Pacific, America, and other tropical and temperate areas where sweet potato is grown.

3.6.3.2 Economic Importance

Pratylenchus is considered second only to *Meloidogyne* in terms of plant species parasitized and extent of crop damage and loss. The thresholds vary somewhat with species, climate, soil type, and host crop, but a density of 1–2 nematodes g^{-1} soil at planting is a reasonable guide; thresholds range from 0.5 to 1.8 g^{-1} soil.

3.6.3.3 Symptoms

The nematode causes small, necrotic root lesions. Fibrous root necrosis may lead to some stunting of vines and a significant reduction in the quality of fleshy storage roots. Small, brown to black, necrotic lesions are also produced on storage roots, which make the roots unmarketable (Fig. 3.57).

Fig. 3.57 Sunken lesions induced by root-lesion nematode on sweet potato

3.6.3.4 Biology

This genus has the typical nematode life cycle with four juvenile stages, and the adults in some species are parthenogenic and in others amphimictic. The juveniles of *P. coffeae* mature and differentiate within the root, and adult females deposit eggs singly or in small groups within the tissues of the root. Commonly, the duration of the *Pratylenchus* life cycle is about 35–42 days, but this considerably varies with species, e.g., *P. penetrans* can range from 30 to 86 days (Corbett 1973), while *P. coffeae* in sweet potato is from 30 to 40 days at 25–30 °C to 50–60 days at 20 °C. The variability is due partly to the length of the life cycle being temperature dependent (Townshend and Anderson 1976).

3.6.3.5 Ecology

Lesion nematodes are migratory root endoparasites. They are also extremely diverse in their adaptive capacity. Individual species show special preferences for temperature. Some are less able than others to tolerate extremes, especially cold, or the extreme drought areas, where vegetable production is less common except under irrigation. The influence of moisture as rain is not as significant. However, *P. penetrans* moves best when soil water has drained so that 8–12 % of the soil volume is air-filled and survives a shorter time in wet than in dry soil. Preferred soil types include coarse-textured sandy loam, silt loam, or, occasionally, organic soil (muck). More *P. penetrans* penetrate roots in a sandy loam than in silt loam or loam. They are less common in clays and similar fine-textured soils; adults and juveniles of *P. penetrans* move farther in coarse than in fine-textured soil. *Pratylenchus* may be disseminated by transportation of plant root parts or soil and by surface or irrigation water.

3.6.3.6 Management

1. *Cultural Methods*: Crop rotation with peanut crops resulted in significantly lower population of *P. coffeae*. Paddy rice or increase fertilization with potassium reduced severity of infection.

   Addition of organic amendment such as chicken manure is very effective in reducing the nematode.
2. *Biological Methods*: *Paecilomyces lilacinus*, a fungal egg parasite, was evaluated and found effective against *Pratylenchus* sp.
3. *Chemical Methods*: Use of soil fumigants is effective.

3.6.4 Brown Ring Nematode: *Ditylenchus destructor* and *D. dipsaci*

3.6.4.1 Economic Importance

On some occasions, storage losses can be serious.

3.6.4.2 Symptoms

Fleshy roots, sometime after they are stored, show symptoms as depressed areas (Fig. 3.58). In cross sections, initial infections appear as necrotic isles of brown tissue scattered throughout the flesh. In advanced stages, the pulp becomes completely blackened, slightly soft, and corky (Fig. 3.58). These nematodes affect fleshy roots only during storage. No symptoms have been found in the field.

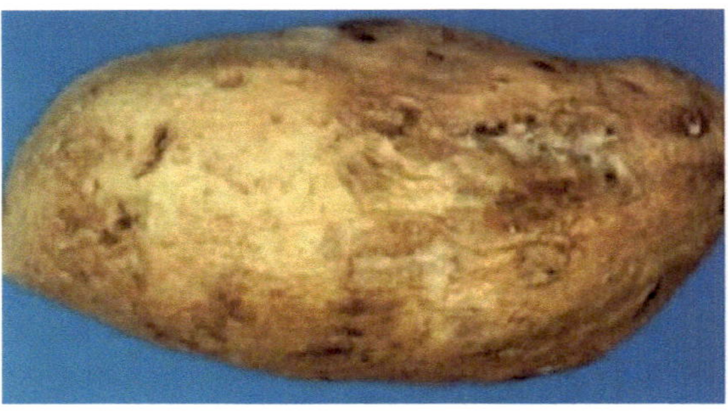
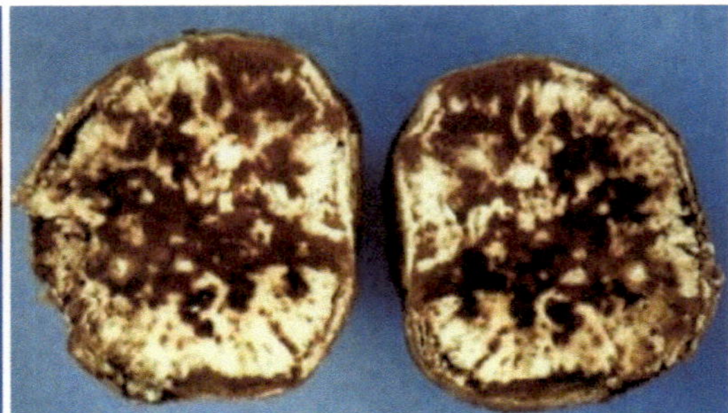

Fig. 3.58 Symptoms of brown ring nematode

3.6.4.3 Biology

The two species of *Ditylenchus* are distributed worldwide and have a vast host range. They are migratory endoparasites.

3.6.4.4 Management

The disease can be reduced by storing tubers at cooler temperatures (13–18 °C). Many weeds may harbor dense populations of this nematode; weed control is essential to reduce nematode populations. Soil fumigation and application of nematicides are also effective for managing the nematodes.

3.7 Future Thrusts

- More attractants or repellents for female sweet potato weevil (identifying attractants from the most susceptible genotypes or repellents/feeding arrestants from most resistant genotypes) should be identified, and suitable formulations for the application of sweet potato weevil (SPW) repellents under field conditions should be developed.
- Combi-trap for SPW with sex pheromone (male attractant) and female repellent [applying (Z)-3-dodecen-1-ol-(E)-2-buteonate + gurjunene + humulene] should be developed.
- The use of repellent genotypes as intercrops with high-yielding and susceptible varieties for reducing the SPW damage [intercropping 'Howrah' (resistant) + 'Kishan'/'Kalinga' (susceptible); 'BX-86' (resistant) + 'Kishan'/'Kalinga' (susceptible); yam bean + 'Kishan'/'Kalinga'] should be explored.
- Utilization of sterile insect technique (SIT) by irradiating SPW [In collaboration with Bhabha Atomic Research Centre (BARC), Mumbai, for using SIT and kairomones together] for controlling SPW should be explored.
- Multidisciplinary approach shall be adopted to tackle some of the problems limiting the productivity of sweet potato weevil.
- Development of disease diagnostic kit in sweet potato for SPFMV and SPLCV.
- Biologically based IPM for sweet potato weevil may be developed using biopesticides, biocontrol agents, and organic deterrents/repellants.
- Sweet potato weevil management through sex pheromone-based IPM has proved successful in different research centers. Commercialization of SPW pheromone technology should be carried out in Orissa, Bihar, and Uttar Pradesh where the crop is widely cultivated in order to make this effective IPM practice available to the poor and marginal farmers through participatory mode. Immediate action may be taken up in order to make the easy availability of pheromone traps to farmers.
- Sweet potato genotypes 'S-463,' 'Howrah,' and 'BX-12' showing resistant reaction under field conditions can be used for developing resistant cultivars. Breeding of SPW-resistant lines with popular cultivated varieties [crossing 'Howrah'/'BX-86'/'C-14' (resistant) with 'Kishan'/'Kalinga'/'ST-14' (susceptible)] needs to be undertaken.
- Identification of new pesticide formulations for the management of SPW (new formulations and combination products) as insecti-

cides recommended are already banned (fenthion, fenitrothion, monocrotophos).

- Detailed studies should be undertaken on sweet potato leaf curl in developing resistant varieties using biotechnological methods (pathogen-derived and elicitor-induced methods).
- Emerging pests and diseases due to climate change and introduction of invasive pests and pathogens are also main challenges. Appropriate technologies to contain such biotic stresses have to focus on refined eco-friendly methods. In vitro elimination of sweet potato feathery mottle virus has to be standardized using meristem and nodal culture combined with chemotherapy and thermotherapy. Strategy for mass multiplication of disease-free planting materials has to be improvised. The production of large quantities of polyclonal antibodies and developing cost-effective, easy, and sensitive diagnostic kits for indexing the planting materials are necessary.

References

Amin PW, Palmer JM (1985) Identification of groundnut Thysanoptera. Trop Pest Manag 31:268–291

Corbett DCM (1973) *Pratylenchus penetrans*, CIH Descriptions of plant parasitic nematodes. Set 2, No. 25. Commonwealth Institute of Helminthology, St Albans

FAOSTAT (2012) FAO statistical yearbook 2012. Food & Agriculture Organization of the United Nations, Rome

Galano CD, Gapasin RM, Lim JL (1996) Efficacy of *Paecilomyces lilacinus* isolates for the control of root-knot nematode [Meloidogyne incognita (Kofoid and White) Chitwood] in sweet potato. Ann Trop Res 18:12–23

Gaspin RM, Valdez RB (1979) Pathogenicity of *Meloidogyne* spp. and *Rotylenchulus reniformis* on sweet potato. Ann Trop Res 1:20–25

Giamalava MJ, Martin WJ, Hernandez TP (1960) Reaction of eight sweet potato selections to five species of root-knot nematodes. Phytopathology 50:575

Gibbons S (2000) Transgenic sweet potato research collaboration: a case study of ABSP involvement in Kenya. Linkages Newsletter Fourth Quarter Feature Article

Gopalakrishnan TR, Swadija OK, Pushpakumari R (2013) Tuber crops – climate resilient crops for food and nutritional security. In: International conference on tropical root and tubers for sustainable livelihood under changing agro-climate, Indian Society for Root Crops, CTCRI, Trivandrum, Kerala, India, pp 30–84

Hill DS (1975) Agricultural insect pest of the tropics and their control. Cambridge University Press, London

Ives CL, Bedford BM (eds) (1998) Agricultural biotechnology in international development. CABI Publishing, New York

Khader Khan H, Jayaraj S, Rabindra RJ (1990) Evaluation of mycopathogens against the sweet potato weevil. *Cylas formicorius* (F.). J Biol Control 4:109–111

Koshy PK, Jasy T (1990) Effect of the burrowing nematode, *Radopholus similis* on sweet potato (*Ipomea batatas* L) (Abstract). In: National symposium on recent advances in the production and utilization of tropical tuber crops, ISRC, Central Tuber Crops Research Institute, Thiruvanathapuram, India, p 47

Martin WJ (1970) Elimination of root-knot and reniform nematodes and scurf infection from rootlets of sweet potato plants by hot water treatment. Plant Dis Rep 54:1056–1058

Milne JR, Jhumlekhasing M, Walter GH (1996) Understanding host plant relationships of polyphagous flower thrips, a case study of *Frankliniella schultzei* (Trybom). In: Goodwin S, Gillespie P (eds) Proceedings of the 1995 Australia and New Zealand thrips workshop: methods, biology, ecology and management. NSW Agriculture, Gosford, pp 8–14

Mohandas C, Palaniswami MS (1990) Resistance in sweet potato (*Ipomea batatas* L) to *Meloidogyne incognita* (Kofoid and White) Chitwood in India. J Root Crops 16:148–149

Mohandas C, Ramakrishnan S (2002) Grow sweet potato and manage root-knot nematode problems of vegetables (Abstract). In: International conference on vegetables, Bangalore, p 256

Ogawa K, Komada H (1986) Induction of systemic resistance against Fusarium wilt of sweet potato with non-pathogenic *Fusarium oxysporum*. Ann Phytopathol Soc Jpn 50:1–9

Palaniswami MS (2002) Advances in integrated approaches for sweet potato weevil management. In: Proceedings of the international conference on vegetables, Bangalore, India, Dr. Prem Nath Agricultural Science Foundation, Bangalore, pp 304–308

Palmer JM (1990) Identification of the common thrips of tropical Africa (Thysanoptera: Insecta). Trop Pest Manag 36:27–49

Ray S, Sahoo NK, Mohanty K (1990) Plant parasitic nematodes associated with tuber crops in Orissa (Abstract). In: National symposium on recent advances in the production and utilization of tropical tuber crops, ISRC, Central Tuber Crops Research Institute, Thiruvanthapuram

Townshend JL, Anderson RV (1976) Pratylenchus neglectus. CIH descriptions of plant parasitic nematodes, Set 6, No 82, CABI Publishing House, 4 pp

Wambugu FM (2001) Modifying Africa: how biotechnology can benefit the poor and hungry, a case study from Kenya. Paperback – 2001

Yaku A, Walter GH, Najar-Rodriguez AJ (2007) Thrips see red – flower colour and the host relationships of a polyphagous anthophilic thrips. Ecol Entomol 32:527–535

4 Taro, *Colocasia esculenta*

Abstract

Distribution, economic importance, symptoms/damage, host range, biology, survival and spread, and management of insect and mite pests and fungal, bacterial, viral and nematode diseases of taro (*Colocasia esculenta*) using regulatory, physical, cultural, chemical, biological, host resistance, and integrated pest management methods are discussed.

Keywords

Taro • *Colocasia esculenta* • Pests • Diseases • Nematodes

4.1 Introduction

Taro, also known as dasheen or cocoyam or eddoe, is the most widely cultivated crop occupying 1 million hectares around the world and producing 6.61 million tons of tubers (Fig. 4.1). Taro is mostly cultivated in Asia, Africa, and the Pacific as well as Caribbean Islands. In the Pacific Islands, it is an important economic crop besides being a staple in countries like Fiji, Papua New Guinea, Western Samoa, Vanuatu, etc., in the South Pacific region (Table 4.1). In India, taro is cultivated in almost all the states, right from the foothills of the Himalayas to the coastal areas in the South (Table 4.2). Taro is believed to have originated in Southeast Asia including India (Watt 1989). From there, it probably spread to Egypt, Arabia, and the Pacific. The hypothesis that the Northeast India is the center of origin of taro was confirmed by many Indian workers. The edible aroid taro (*Colocasia esculenta*) is a member of the family Araceae and occurs wild in Southeast Asia. It is cultivated throughout the humid tropics and is of great importance in the Pacific Islands. It is mostly a staple food or subsistence crop grown commercially in some countries. It is also grown primarily in Hawaii being most well known in its cooked form as poi, the traditional staple food of native Hawaiians. It is well adapted to humid environment and flourishes under well-shaded conditions. Taro is the most shade tolerant of terrestrial food crops. In *Colocasia*, the richest diversity is noticed in the Northeast followed by Kerala and Western region. The taro germplasm includes cultivated and stoloniferous wild taro present in Western Ghats. *Moghania tuberosa*, a wild-type bearing succulent roots, has been located in coastal Konkan region of Maharashtra.

P.P. Reddy, *Plant Protection in Tropical Root and Tuber Crops*,
DOI 10.1007/978-81-322-2389-4_4

Fig. 4.1 Taro plants and tubers

Table 4.1 Estimated production of taro in 2007 from the 20 highest producers worldwide

Country	Production (tons)	Country	Production (tons)
Nigeria	4,996,000	Central African Republic	100,000
Ghana	1,662,000	Thailand	78,500
China	1,642,472	Ivory Coast	74,275
Cameroon	1,200,000	Democratic Republic of Congo	66,110
Papua New Guinea	260,000	Fiji	61,662
Madagascar	200,000	Burundi	58,125
Japan	195,800	Gabon	56,000
Rwanda	130,000	Solomon Islands	40,000
Philippines	113,954	Chad	39,000
Egypt	112,000	Guinea	31,000

Source: http://faostat.fao.org/

Table 4.2 Taro-growing states, regions, and production systems in India

State	Production system	Area/region
Andhra Pradesh	Plains irrigated	Prakasam and Guntur districts
West Bengal	Plains irrigated	Ghadabatha region
Odisha	Plains irrigated	Cuttack, Nayagarh, and Khordha districts
Uttar Pradesh	Plains irrigated	Jhansi region
Madhya Pradesh	Plains irrigated	Indore region
Tamil Nadu	Plains irrigated	Virudhunagar, Tirunelveli, Thoothukudi, North Arcot, and Thanjavur districts
Kerala	Rain-fed homestead farming	Entire state
NEH region	Rain-fed homestead farming	NEH states

Taro is a short-statured perennial plant, grown as annual. It stores starches in large corms at or below the sod surface. Vegetative propagules are taken from the top portion of the corms. Planting is labor intensive with plant populations ranging from 10,000 to 30,000 plants/ha. Harvest begins 8–12 months after planting, yielding from 7 to 30 t/ha of edible corms. Lifting devices similar to potato diggers are used as harvesting aids. Considerable amounts of hand labor are involved in this process. Many of the steps in both planting and harvesting could be mechanized further. In the Caribbean and Southeast Asia, young tender leaves are consumed as pot herb. Limited amounts of leaves are harvested and sold in US markets. The potential for expansion of these markets is considered to be very limited.

Corms which contain 25–35 % starch are plagued by the presence of an acrid factor, which causes itchiness and considerable inflammation of tissues. Cooking removes most, if not all, of this factor from domesticated clones. Shelf life of harvested corms varies considerably between taro and cocoyam and depends on the care taken during the harvesting and packaging process. Cocoyam has a considerably longer shelf life of several weeks. This can be extended further with curing and refrigerated storage.

Corms are usually peeled and boiled. Processing is limited to the production of deep-fried chips and poi from taro in Hawaii. The expansion of production for processing in the southern states of the US mainland is possible. The bulkiness of the propagules and lack of tolerance to freezing temperatures present problems. One area that deserves additional investigation concerns the digestibility of the starch. In Cuba, babies and people with mild ulcers are placed on diets of cocoyam.

4.1.1 Giant Taro, *Alocasia macrorrhiza* (Syn: *Alocasia indica*) (Family: Araceae)

A very large herb of the taro family, it has a stout erect trunk up to 4 m tall. This has upright leaves which are arrow shaped. The leaves have round lobes at the bottom. The leaves are leathery and are often wavy around the edge (Fig. 4.2). The secondary veins are not prominent. The leaf blade can be 1.0–1.2 m long. The leafy structure around the flower is yellow in the upper section. It forms a hood and drops off as the flower opens. The fruit are bright red berries. The corm is large, often curved and above the ground. It often has cormels, or small corms, at the side. Brown trailing fibers of the leaf bases often hang from the stem. The leaves and petioles, or leaf stalks, contain stinging crystals.

Fig. 4.2 Giant taro plant and tubers

Fig. 4.3 Giant swamp taro

4.1.1.1 Distribution

It is mostly grown in tropical regions in Asia and the Pacific. It is widely distributed in open wetlands and along streams and in some types of humid forest. The plant grows wild from sea level up to 2,600 m altitude in the tropics. Giant taro is a tropical plant and will not grow well below 10 °C. It requires a well-distributed rainfall and does not tolerate drought. Even though it grows along creek banks, it cannot tolerate wet soil. It is only used as food in a few coastal areas. Wild forms commonly seen growing are bitter and not used. It suits hardiness zones 11–12.

4.1.1.2 Cultivation

The top of the main corm is planted. The small round cormels can be planted but are slow to mature. A spacing of 1.2 m × 1.2 m is suitable.

4.1.1.3 Production

Because the giant taro takes more than a year to be ready to harvest, it often ends up left growing in an old garden site without much care or weeding until the owner wants to harvest it. Corms of 8.5–40 kg have been harvested from individual plants of unknown age. The time to maturity is about 12 months, but plants are often left for 2–3 years.

4.1.1.4 Use

Giant taro is an important crop in some areas of Solomon Islands and other Pacific Islands. The stems and corms are eaten after roasting or boiling. The main corm is cooked and eaten after being carefully peeled. The young leaves are edible. The leaf stalk is also eaten.

4.1.2 Giant Swamp Taro, *Colocasia esculenta* var. *stoloniferum* (Syn: *Cyrtosperma chamissonis*) (Family: Araceae)

Giant swamp taro is a "lowland" taro species and the largest of the taro family. It may reach heights of 4–5 meters (Fig. 4.3), with leaves and roots much larger than *Colocasia* taro. It is one of the few subsistence crops that grows well on atolls and within swampy areas of other islands, and its corm can reach weights of 80 kg or more! It may be stored in the field underground for very long periods—up to 30 years or more—and accordingly has traditionally been an important emergency crop in times of natural disaster and food scarcity. It is relatively resistant to disease and pests but is susceptible to taro beetle. It can take several years to mature but is commonly left for 15 years or more before harvest.

As indicated by its name, giant swamp taro has adapted to growth within freshwater and coastal swamps. It is also commonly grown in purpose-built swamp pits in low-lying coral atolls. In the case of constructed pits, various kinds of organic matter and fertilizer are usually added to the pit to improve soil fertility, physical properties, and water-holding capacity. Giant swamp taro is not suitable for growing in upland or rain-fed conditions. It exhibits some shade tolerance and is considered mildly tolerant of saline growing conditions compared to other taro species and can be grown in mildly brackish water.

4.1.2.1 Distribution

Swamp taro grows in Asia and most Pacific countries including Solomon Islands. It is a tropical plant and grows from 18°N–20°S. It grows in fresh or brackish swamps up to 150 m altitude. Water 0.6–0.9 m deep and rich in humus is the environment used. It can tolerate shade and withstand flooding. It is found in valleys in central and southern parts of the Philippines. It is grown in the Visayas and Camarines in the Philippines. It is an important plant in Bougainville. It grows in fresh or brackish water. It occurs on atoll islands.

4.1.2.2 Cultivation

Plants are normally put in swamps or ditches. Parts of the main corm, or suckers, can be used for planting. It can be intercropped with *Colocasia* taro. A spacing of 1.2 m × 1.2 m is suitable. It normally receives little management. It is known to respond to organic matter.

4.1.2.3 Production

Yields of 10–15 t/ha/year have been recorded. It takes 2–6 years to mature. Tubers become more fibrous with time, so 2 years is a suitable harvesting time. Individual corms can weigh 2–50 kg.

4.1.2.4 Use

The corms are peeled and boiled or roasted. They are also used to make flour. Leaves and young flowers can be eaten. Swamp taro is an important reserve food in damp areas of Solomon Islands.

4.2 Insect and Mite Pests

4.2.1 Aphid, *Aphis gossypii*

4.2.1.1 Distribution

The pest is cosmopolitan in distribution but absent from the cooler parts of Asia and Canada. Aphids have a wide distribution throughout the Pacific.

4.2.1.2 Damage

If aphids are present in high numbers (Fig. 4.4) and rainfall is low, leaves senesce faster than nor-

Fig. 4.4 Aphid infestation on *Colocasia*

mal. In severe cases, the plants wilt and may become stunted. Indirect damage is caused by the accumulation of honeydew produced by the aphids. Honeydew serves as a substrate for sooty molds, which blacken the leaves, reducing photosynthesis and plant vigor. Aphids are vectors of dasheen mosaic potyvirus.

4.2.1.3 Biology

Eggs: In temperate regions, aphids overwinter as eggs. In tropical regions, reproduction does not involve mating and egg laying.

Nymphs: Immature aphids are called nymphs—they look much like adults but are smaller and wingless. Females give birth to live female nymphs, which are oval and 0.1 mm in length. There are four nymphal stages, and development is complete in 4–10 days. The fourth-stage nymphs are approximately 1 mm in length.

Adults: Adult females give birth to approximately 50 nymphs. Adults range from 0.9 mm to 2.5 mm in length; they may be winged or wingless. Wingless adult aphids vary in color from pale green-yellow to dark green. Winged forms are generally darker: the head and thorax are black and the abdomen is yellowish-green, with dark-brown veins.

4.2.1.4 Management

1. *Cultural Methods*: Avoid planting *Colocasia* close to alternate hosts such as melon, cotton, cucumber, or other cucurbits.
   (a) *Botanicals*: In Rajendra Nagar, Hyderabad, Andhra Pradesh, spraying of yam bean seed extract at 5 and 2 % recorded maximum reduction in aphid population (67.33 % and 47.03 %, respectively) and increased cormel yield (24.28 and 22.4 t/ha, respectively) (Table 4.3).
2. *Chemical Methods*: The following chemicals have been recommended for the management of aphids (Table 4.4).
3. *Biological Methods*
   (a) *Predators*: Several aphidophagous predators observed on taro are coccinellids (*Menochilus sexmaculatus*, *Verania discolor*, *V. inops*, and *Pseudospidimerus circumflexus*) and syrphid (*Ischiodon scutellaris*). The predation is higher during larval stage, while it is less in adult stage. Among aphidophagous predators, *I. scutellaris* is a voracious feeder. The mean consumption per day ranges from 20 to 28 aphids. Coccinellid *Scymnus latemaculatus* feeds on aphids (Fig. 4.5).
   (b) *Parasitoids*: Four species of parasitoids *Aphelinus mali*, *Aphelinus* sp.,

Table 4.3 Effect of tuber-based biopesticides for the management of aphids on taro

Treatment	% reduction of aphid population*	Yield (t/ha)
Yam bean seed extract 2 %	47.03 (43.29)[b]	22.40[ab]
Yam bean seed extract 5 %	67.33 (55.19)[a]	24.28[a]
Cassava leaf powder 2 t/ha	24.40 (29.58)[de]	18.84[bc]
Cassava leaf powder 5 t/ha	30.77 (33.45)[cd]	22.13[ab]
Yam bean seed powder 2 t/ha	34.47 (35.86)[c]	19.22[bc]
Yam bean seed powder 5 t/ha	31.60 (33.99)[cd]	20.29[abc]
Cassava tuber rind extract 2 %	15.83 (23.26)[f]	18.08[bc]
Cassava tuber rind extract 5 %	20.57 (26.86)[ef]	19.78[bc]
Chemical treatment	53.24 (46.86)[b]	23.90[a]
Untreated control	3.37 (10.51)[g]	16.04[c]
CD (P=0.05)	**5.817**	**4.473**

Values with the same superscript (column wise) do not differ significantly

*Figures in brackets are $\sqrt{n+1}$ values

Table 4.4 Recommended chemicals for the control of aphids

Insecticide	Dose	Application method
Carbaryl WP	1 kg	In 450 l (high volume) or 112–168 l (low volume) of water
Diazinon EC	11 ml	In 13.5 l of water, 14 days minimum waiting period
Dimethoate EC	4.4 ml	In 13.5 l of water, 14 days minimum waiting period
Malathion EC	14.2 ml	In 13.6 l of water, 7 days minimum waiting period

Coccophagus cowperi (Aphelinidae), and *Aphidencyrtus aphidivorus* (Encyrtidae) are active on *Aphis gossypii* in fields of taro, tania, and elephant foot yam. The total parasitization in field ranges from 2 to 14 %.

4. *Positioning of Control Methods in Terms of the Development Cycle of the Plant* (Fig. 4.6)
   (a) *Choice and Preparation of the Field*
   - Provide hedges to limit movements of aphids from a crop to another and to encourage natural enemies.
   - Avoid nearby and under wind old taro field.
   - Uproot of host plants in the neighboring field.

   (b) *After Planting*
   - Sprinkler irrigation or sustained rain can reduce infestation.
   - Control ants in the field, as these will disrupt natural enemy activities.
   - Treatment with selective insecticides.

   (c) *After Harvesting*
   - Destruction of infested leaves after harvesting

Fig. 4.5 Coccinellid *Scymnus latemaculatus* feeding on aphids

4.2.2 Tobacco Whitefly, *Bemisia tabaci*

4.2.2.1 Distribution

The pest is recorded in most parts of the tropics and subtropics. Tobacco whitefly is widely distributed across the Pacific.

4.2.2.2 Damage

Adults, nymphs, and pupae are found on the lower leaf surface (Fig. 4.7). The adults and nymphs suck the sap from leaves—this can cause yellowing, wilting, and early death.

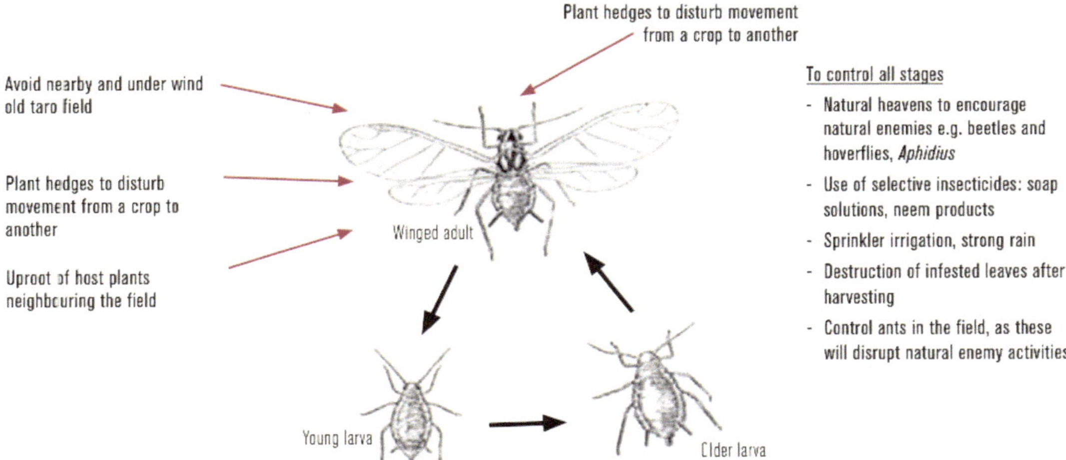

Fig. 4.6 Positioning of control methods in terms of the development cycle of the plant

Fig. 4.7 Tobacco whitefly infestation on taro leaves

However, symptoms this severe rarely occur on taro as numbers of whitefly are usually insufficient to cause direct damage, although they may increase wilting in times of drought. Indirect damage, that is, the buildup of sooty molds on honeydew deposits, is not frequently seen. The growth of the *Colocasia* may be stunted.

4.2.2.3 Biology

Yellow elliptical eggs, attached by filaments to the leaf, are laid on the undersurface of the leaves. The first instar nymphs emerge and feed. After the first molt, they are fixed to one place on the leaf as yellow elliptical scale-like insects. The third molt gives rise to the puparium which is less flat than the previous nymphal stages and the red eyes of the adult can be seen through the integument. Egg to adult development takes 2–3 weeks. The adult has yellow body, has white waxy broad-tipped wings, and is 1 mm long with a wingspan of about 2 mm.

4.2.2.4 Management

1. *Chemical Methods*: The following chemicals have been recommended for the control of whiteflies (Table 4.5).
2. *Positioning of Control Methods in Terms of the Development Cycle of the Plant* (Fig. 4.8)
   (a) *Choice and Preparation of the Field*

Table 4.5 Chemical control of whitefly on *Colocasia*

Insecticide	Dose (a.i./ 13.6 l of water)	Application
Carbaryl WP	34 g	7 days minimum waiting period before harvest
Malathion WP	14 g	7 days minimum waiting period before harvest
Trichlorfon WP	23 g	7 days minimum waiting period before harvest

   - Provide hedges to limit movements of whiteflies from a crop to another and to encourage natural enemies.
   - Avoid nearby and under wind old taro field.

   (b) *During the Production Cycle*
   - Sprinkler irrigation or strong rain will limit the spread of the pest.
   - Systemic insecticides, soap, and oils to control the pupae (puparium).
   - Selective insecticides (to limit the negative impact on natural enemies) used alternately (to limit the risks of resistance) to control adults.
   - Encourage natural enemies: *Encarsia formosa* and others.

   (c) *After Last Harvesting*
   - Destruction of harvest residues to prevent buildup and spreading

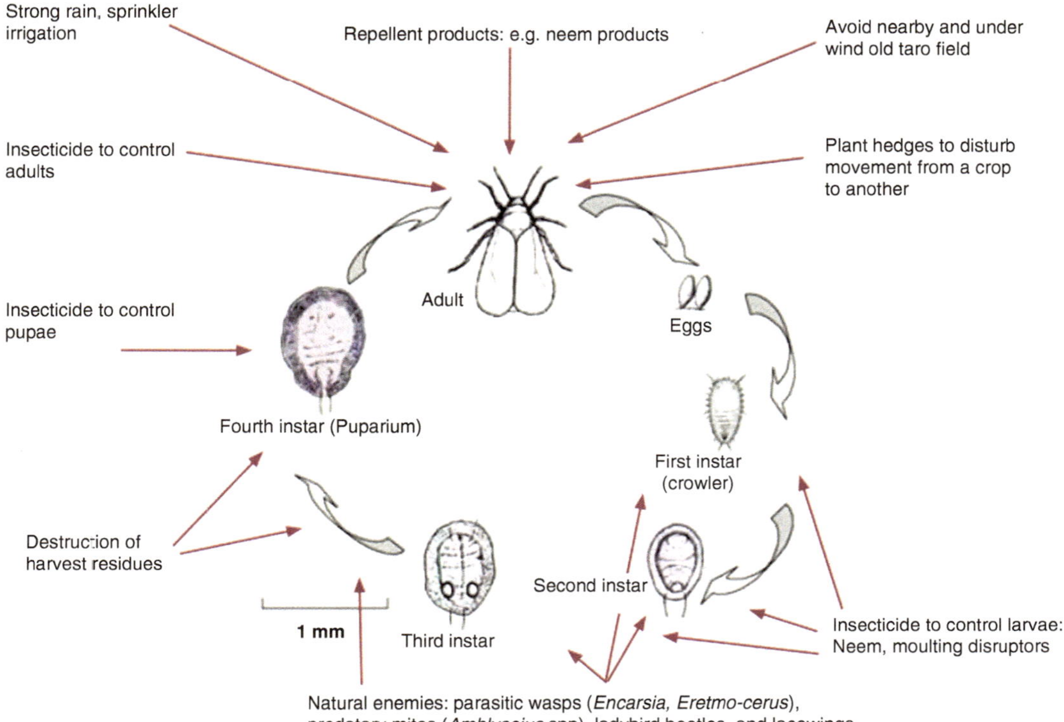

Fig. 4.8 Positioning of control methods in terms of the development cycle of the plant

4.2.3 Plant Hopper, *Tarophagus proserpina*

4.2.3.1 Distribution

The plant hopper is distributed in most of the South Pacific territories, the Hawaiian Islands, Malaysia, the Philippines, the Ryukyu Islands, and Queensland.

4.2.3.2 Symptoms

Plant hoppers are a sap-sucking insect and are mainly found on petioles and the undersides of leaves (Fig. 4.9). A reddish crust of sap exudate is produced when the leaf hoppers puncture the plant during feeding which is visible on the outside of the leaf stalks near the ground. Plants which are heavily infested may be stunted in growth and may wilt. Brown and white adult and white nymphal leaf hoppers can be seen.

Two serious virus diseases of taro, alomae and bobone, are transmitted by *T. proserpina* in a persistent manner, and once the virus is ingested, the plant hopper remains infectious for life (Jackson 1978).

4.2.3.3 Biology

The female cut slits in taro leaves and petioles with its ovipositor, usually laying two eggs in each slit. In Hawaii, the egg stage lasts for about 14 days, followed by five instar stages of 4, 3, 3, 4, and 5 days (Waterhouse and Norris 1987). These stages may be shorter in the warmer climate of American Samoa. There are two adult forms, short and long winged, the latter forming during periods of cool weather, plant senescence or overpopulation (Waterhouse and Norris 1987).

4.2.3.4 Spread

Long-distance spread is by the long-winged form and by planting tiapula with plant hopper eggs embedded in the petioles.

Fig. 4.9 Adults of plant hopper on the leaf

4.2.3.5 Management

1. *Cultural Methods*: Starting with plant hopper-free tiapula in uninfested fields could delay the onset of an attack. It is unlikely that farmers would adopt this strategy, however, given the planting methods used in taro production and the infrequency of damaging plant hopper infestations.

All petiole material and growing points of the corm must be removed. Topping of large corms of taro helps to remove the petiole bases that may carry eggs, nymphs and adults of *T. proserpina*.

By rotating crops grown in a field and moving taro to another area, insect populations that depend on taro to survive will not have the opportunity to build up to unmanageable numbers. In some areas, fields that have large pest populations are fallowed for a year or more to stop the populations from increasing.

In American Samoa, *Coleus blumei* or pate is frequently interplanted with the taro. Some believe that pate will repel plant hoppers and armyworms. Others contend that pate attracts the plant hoppers, which then feed on its juices and die from its toxic properties (Fatuesi et al. 1991).

Another method of controlling the plant hopper in American Samoa is through smoking or torching. Coconut leaves are bundled and then set afire and carried through the fields, causing the plant hoppers to jump off the plant. It is believed that the plant hoppers die as their wings get burned or that the smoke drives them away from the field in large numbers. This practice is done three times a week until all the plant hoppers are gone. The smoking method is sometimes thought to be effective against the armyworm as well (Fatuesi et al. 1991).

2. *Biological Methods*: In Hawaii, satisfactory control by reducing the injury of leaf hoppers was achieved by the introduction of the mirid, *Cyrtorhinus fulvus*, in 1938. *Cyrtorhinus fulvus* is an egg predator effective in controlling plant hopper in Hawaii, Guam, Ponape, and in many parts of the Pacific (Fig. 4.10). This mirid bug is present on taro in American Samoa and may be responsible for the low incidence and severity of plant hopper infestations. In a taro research plot at the ASCC Land Grant facility, however, a severe 2-month infestation by *T. proserpina* continued in the presence of a large mirid bug population. A study at ASCC Land Grant in 1990 demonstrated a typical biological control scenario between the plant hopper host and its mirid bug predator. An increase in the host population was followed by an increase in predators, followed by a decrease in the number of plant hoppers and a decrease in predators.

3. *Chemical Methods*: Insecticides are not commonly used against plant hoppers. Pesticides such as Imidacloprid (Confidor or Mustang) should only be used when there are no natural enemies present in the field. Other recommended pesticides are malathion, carbaryl, diazinon, and acephate (Crop Protection Reference 1995).

4.2.4 Spiraling Whitefly, *Aleurodicus dispersus*

4.2.4.1 Distribution

Spiraling whitefly is widely distributed across the Pacific.

Fig. 4.10 *Cyrtorhinus fulvus* (*arrowed*) eating *Tarophagus* spp. eggs on the underside of a taro leaf

4.2.4.2 Damage

Piercing leaf and sucking sap by whitefly lead to premature death of plant when infestations are high. Indirect damage is caused by the accumulation of honeydew and waxy, white, fluffy, woolly material produced by whiteflies (Fig. 4.11). Honeydew serves as a substrate for sooty molds, which blacken leaf, retarding photosynthesis, and reducing plant health.

4.2.4.3 Biology

1. *Eggs*: The eggs are smooth, yellow to tan, oval shaped, and 0.3 mm in length. The irregularly spiraling patterns of the deposited eggs are associated with trails of white wax.
2. *Larvae*: Upon hatching, crawlers (first instar larvae) move to find an appropriate leaf vein for feeding, where they settle. As the larvae mature, they develop waxy tufts. The first instars are mobile, while the last three instars are permanently attached to the leaf (sessile).
3. *Adults*: Adult males are 2.2 mm long and adult females are 1.7 mm. Adults are free moving and able to fly.

4.2.4.4 Management

There are two major control strategies:

- Effective quarantine measures to prevent international spread
- Biological control when quarantine incursions occur

Fig. 4.11 Spiraling whitefly damage on taro

Fig. 4.12 Corm borer damage on *Colocasia*

Several species of predators and parasitoids are natural enemies of whiteflies. The parasitic wasp, *Encarsia haitiensis*, has proved to be a very efficient means of controlling spiraling whitefly wherever it has been introduced. In some cases, it appears that the parasitoid has been introduced unwittingly, together with its host. In these cases, a balance has been established and further intervention is not required.

Chemical control is not an appropriate way of managing this pest, although dilute solutions of soaps and detergents can provide effective control in small plantings.

4.2.5 Corm Borer, *Aplosonyx chalybaeus*

Corm borer is a regular and endemic pest causing 20–30 % damage to the foliage and 80–90 % to the corms resulting in severe losses to the tribal farmers of the eastern Himalayan region of India. The pest is found up to 3,000 m MSL.

4.2.5.1 Damage

Adults are of bright metallic in color and emerge during June–July. They feed on leaves making circular trenches of 2.5 cm size (Fig. 4.12). They hide in the leaf sheath and each plant harbors 3–15 beetles. Crop losses are up to 50–60 % during severe infestation.

4.2.5.2 Biology

The adult female lays eggs in groups of 20–100 in shoot leaf sheath above the ground. Soon after hatching, the young ones bore into the shoot and consume the developing corm resulting in the death of the plant. A maximum of 75 grubs were found per plant.

4.2.5.3 Host Range

Wild relatives of taro are found in the eastern Himalayan region beginning from lower to higher altitudes. Successful establishment of the wild taro species cumulated the pest problem of corm borer.

4.2.5.4 Management

The adults are picked and destroyed by farmers and sometimes eaten by some of the tribes in the region. The corm borer incidence is reduced when taro is intercropped with ginger, sweet potato, yam, and maize than the monocrop.

Preliminary experiments suggested that entomopathogenic nematodes and fungi (*Beauveria bassiana*) cause significant increase in mortality of both adults and grubs (Rajasekhara Rao 2013). Earwigs are also found feeding on eggs of corm borer (Fig. 4.13).

4.2.6 Armyworm, *Spodoptera litura*

4.2.6.1 Distribution

Army worm, also called as cluster caterpillar, is found throughout Asia and the Pacific region. It has a wide host range but taro is the most important crop damaged.

4.2.6.2 Damage

The early larval stages remain together at first (Fig. 4.14), later radiating out from the egg mass, stripping the interveinal leaf surface, and skeletonizing the leaves as they advance. Later (solitary) stages eat all parts of the leaf, including the petioles. Older larvae are night feeders. Armyworms chew large areas of the leaf and, when numerous, can defoliate a crop. In such cases, the larvae migrate in large groups from one field to another in search of food.

Fig. 4.13 Earwigs feeding on the eggs of corm borer

4.2.6.3 Biology

Young caterpillars (2–10 mm) are pale green and become dark green to brown when fully grown. They have characteristic bright yellow stripes running lengthways on their back. The moth is nocturnal and its body is brownish green, measuring 15–20 mm with a wingspan of 30–40 mm. The larvae are gregarious in their early stages, with radical progression from the hatching site. Afterwards, the caterpillars become solitary and eat all parts of the leaf blade, sometimes cutting the leafstalks at ground level.

The female moth lays 200–300 eggs in a cluster on the underside of taro leaves (Waterhouse and Norris 1987). Eggs hatch in 4–8 days and the larvae feed side by side on the leaf. As larvae grow, they tend to move apart, their feeding producing large holes in the leaf; within a few days, only the skeleton-like leaf veins may remain. Unlike cutworms, larvae do not move into the soil during the day but remain on the leaf. The larval stage lasts from 13 to 30 days with five instars. Pupae form in the soil and the adult moth emerges in 7–18 days. A complete cycle is approximately 30 days, with up to eight generations per year (Waterhouse and Norris 1987).

Fig. 4.14 *Spodoptera litura* damage on *Colocasia*

4.2.6.4 Management

1. *Cultural Methods*: Most taro growers manually smash cluster caterpillars with their hands when infestations are light. Heavily infested leaves are removed and burned. Vargo (1990) reported that some American Samoan farmers used a 3-month fallow against *S. litura*, believing weed growth hid the taro and caused the pest to leave the field.
   (a) *Botanicals*: Two round spraying of 5 % aqueous yam bean seed extract starting from 50 % flowering stage at 15 days interval was found effective against yam bean spotted borer (*Maruca vitrata*) and *Spodoptera litura*. Rotenol is the active ingredient in yam bean seed extract which is responsible for the control of pest.

In Kalyani, West Bengal, foliar spray of yam bean seed extract at 5 % gave least damage (12.46 %) due to *S. litura* which was on par with yam bean seed extract at 2 % (15.17 %). The above treatments also gave maximum stolon yield (24.28 and 23.52 t/ha, respectively) (Table 4.6).

Table 4.6 Effect of plant products for the management of leaf eating caterpillar *Spodoptera litura* on taro

Treatment	% damage (leaf infestation)*	Stolon yield (t/ha) (% increase in yield over control)*
Yam bean seed extract 2 %	15.17 (22.920)	23.52 (19.08)
Yam bean seed extract 5 %	12.46 (20.67)	24.28 (22.93)
Cassava leaf powder 2 t/ha	24.42 (29.61)	20.53 (13.94)
Cassava leaf powder 5 t/ha	21.45 (27.59)	21.96 (19.75)
Yam bean seed powder 2 t/ha	16.61 (3.61)	22.25 (12.65)
Yam bean seed powder 5 t/ha	14.40 (22.30)	23.18 (17.36)
Cassava tuber rind extract 2 %	36.33 (37.07)	19.96 (11.06)
Cassava tuber rind extract 5 %	32.57 (34.80)	20.56 (13.79)
Untreated control	39.72 (39.06)	19.75
CD (P=0.05)	**3.54**	**1.84**

*Figures in brackets are $\sqrt{n+1}$ values

2. *Biological Methods*: On American Samoa, cluster caterpillar eggs are parasitized by *Telenomus* sp. and *Chelonus* sp. (Hymenoptera) and larvae by *Apanteles* sp., *Euplectrus* sp., and *Zele* sp. and by the fly *Palexorista* sp. (Waterhouse and Norris 1987). Due to the sporadic occurrence of pest outbreaks, large populations of predators are unlikely to develop unless they have other hosts. On the other hand, the infrequent, limited infestations of cluster caterpillar may be evidence of an effective predator or parasite population in American Samoa. Braune and Kan (cited in Waterhouse and Norris 1987) suggested the flowers of *Coleus blumei* may provide food for caterpillar parasites but this has not been documented. Several traditional methods of *Spodoptera* control are used in American Samoa; results are anecdotal. Some farmers believe *C. blumei* (pate) planted around or within the taro field emits an odor that repels the adult moth, and others believe that *S. litura* prefers *Coleus* to taro (Vargo 1990). Chickens are reported to pick caterpillar larvae from taro leaves (Fatuesi et al. 1991). Dipel (*Bacillus thuringiensis* subsp. *kurstaki*, Abbott Laboratories) has been used in the past against *S. litura*. Dipel is ASG EPA registered but there is no recent information regarding its use or effectiveness, including the 1999 PIAP survey.

Only products that are nontoxic to beneficial insects should be considered, such as biologically derived pesticides: Spinosad (derived from the soil Actinomycete *Saccharopolyspora spinosa*) and Bt.

3. *Chemical Methods*: Pesticides are seldom necessary. One application of malathion is recommended by the extension service to control heavy infestations.

4.2.7 Hornworm, *Hippotion celerio*

4.2.7.1 Distribution

Hippotion celerio is widespread throughout the Pacific region.

4.2.7.2 Damage

Infested plants have large areas of leaf missing and the leaf appears ragged. The caterpillars are voracious feeders (Fig. 4.15). Hornworms can defoliate taro when numbers are high. In the case of caterpillar infestation, defoliation may be severe. The larvae can be found on the leaves during the day, often on the underside.

4.2.7.3 Biology

1. *Eggs*: Eggs are laid singly on both the upper and lower surface of the leaves as well as on the petioles. Eggs are variable in size and shape, from nearly spherical (1 mm) to oval, and are clear to bluish green. Before emergence, they become greenish yellow.
2. *Larvae*: The first instar larvae are approximately 4 mm long with a reddish horn on the posterior end. As they age, they change from pale yellow to glossy green. In the second instar, two spots appear on the first and second abdominal segments, resembling eyes. In the third instar, a yellow, dorsolateral line appears running from thoracic segment 3 to the base of the horn and the eye spots assume their final coloration. Finally, larvae grow to 8–9 cm, becoming mid to dark brown, although a few remain green, before pupation.
3. *Adults*: Hawk moths have a wingspan of 4–9 cm. They are streamlined and robust in flight, with a conspicuous head and large eyes.

4.2.7.4 Management

1. *Physical Methods*: The larvae are large and relatively easily seen; they can be picked off the plants by hand. In small taro plantings, this is the best means of control.
2. *Chemical Methods*: Applications of chemical sprays may help control populations of hornworm. Present recommendations in the Pacific Island countries include: Indoxacarb (e.g., Steward), Spinosad (e.g., Success), Bt (e.g., Delfin, Thuricide, Dipel), and Imidacloprid (e.g., Confidor, Mustang).
3. *Biological Methods*: The caterpillars of hornworms are usually parasitized (6–10 %) by a Hymenopteran *Charops hersei* under field conditions. *Hippotion* spp. caterpillars are also preyed upon by a wasp *Polistes* sp. under field conditions.

4.2.8 Taro Beetles, *Papuana woodlarkiana*, *P. biroi*, *P. huebneri*, and *P. trinodosa* (Scarabaeidae: Coleoptera)

4.2.8.1 Distribution

Taro beetles are native and widespread throughout Papua New Guinea, and some species are also present in Vanuatu, Solomon Islands, Kiribati, and Fiji.

Fig. 4.15 Hornworm caterpillar damage on taro leaf and adult

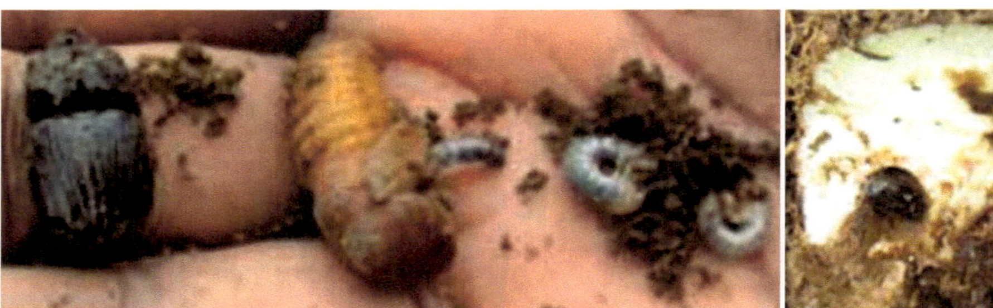
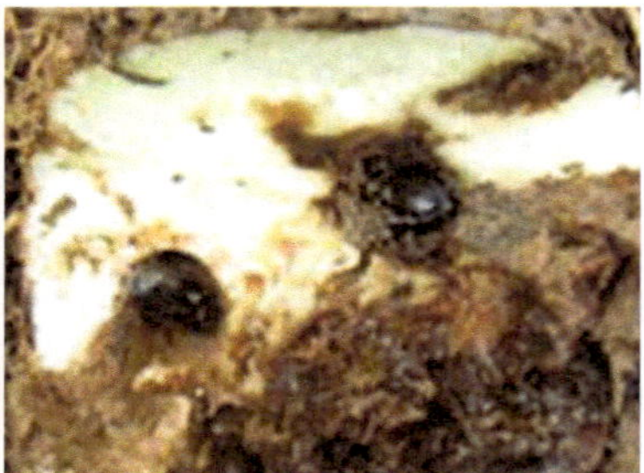

Fig. 4.16 Different stages of taro beetle and damage to corm

4.2.8.2 Damage

The beetles are major pests of taro in Papua New Guinea and Fiji. They measure 25 mm in length and are half as wide. Males have a horn on the head with a bulge at the base. Females sometimes have smaller versions of these attributes. The body is dark brown and very shiny during the first months.

The adult beetle feeds on underground taro corms, creating tunnels while feeding (Fig. 4.16). The impact of feeding is considerable as export markets do not tolerate any damage, and more than 15 % damage makes the crop unacceptable for local markets. Damage may be such that the corms cannot be used for home consumption or livestock feed. Aboveground, symptoms vary with the age of the plants: young plants may be killed as the beetle invades the shoot, while older plants grow more slowly and a few or all of the leaves will wilt.

4.2.8.3 Biology

The adult female can lay up to 300 eggs during its lifetime. Most (75 %) of these eggs are laid within the first 3 months. After mating, the female continues to lay fertile eggs for up to 3 months. There is a preoviposition period of 1 week after emergence; during this period the female feeds aggressively to obtain energy and then disperses to breeding sites for oviposition. Eggs are laid singly with an average of one per day. The adult beetles can live up to 22 months in captivity, although probably less under natural conditions. Males tend to remain in the taro gardens, while females disperse to breeding sites. The life cycle includes egg and first instar, second instar (approx. 18 days each), and third instar larvae (55 days). The prepupal and pupal stages are 7 and 30 days, respectively, giving a total life cycle of 146 days. The length of the life cycle is influenced by the availability of nutrients, soil moisture, and temperature.

4.2.8.4 Host Range

Papuana beetles have a wide host range. The host range includes tania, banana, potato, sugarcane, sweet potato, yams, coconut, areca nut, coffee, tea, cocoa, etc.

4.2.8.5 Favorable Conditions

The beetle prefers wet/moist soil, especially loams or silty loams with plenty of organic matter. Oviposition takes place in any suitable habitat that is moist and protected from natural enemies, such as under logs, along river banks, and in the fibrous roots of grasses. Adult beetles are common once the primary forest is disturbed and *Alocasia* and *Musa* spp. reestablish. Damage to taro in areas where logging has occurred is common.

4.2.8.6 Management

The lack of chemical pesticides registered for use on taro beetles means that control relies largely on populations of natural enemies and cultural control methods.

1. *Regulatory Methods*: Strict quarantine measures must be observed to prevent the spread of taro beetles into new areas within countries where they already occur and into countries where they do not occur.

Table 4.7 Natural enemies of taro beetle

Natural Enemy	Type of attack	Pest stage attacked	Countries where known to occur
Metarhizium anisopliae	Pathogen	Larvae, pupa, adult	Papua New Guinea, Kiribati, Solomon Islands, Vanuatu
Formosa sp. (Diptera: Tachinidae)	Parasitoid	Larvae	Papua New Guinea
Palpastoma sp. (Diptera: Tachinidae)	Parasitoid	Larvae	Papua New Guinea
Bufo marinus	Predator	Adult	Papua New Guinea

2. *Cultural Methods*: Taro gardens in Papua New Guinea are often flooded in an attempt to control the pest; farmers may also apply wood ash to the crop.

Planting material, soil, taro, and alternative hosts of the beetle must not be moved from infested areas to uninfested areas. A combination of measures should be used to manage populations of taro beetles.

Cultural control measures include—crop rotation, clean planting material (i.e., free from soil), and destruction of breeding sites near farms.

3. *Chemical Methods*: Spraying of Imidacloprid/Bifenthrin at recommended dose is suggested.
4. *Biological Methods*: The fungal pathogen, *Metarhizium anisopliae,* has been successfully isolated and grown and is being produced in large quantities and released in the field on an experimental basis. None of the natural enemies are effective in reducing damage to acceptable levels, except when combined with other methods, such as chemical and cultural control (Table 4.7).

4.2.9 Mealybugs

4.2.9.1 Distribution

Many species of mealybugs have been recorded from the Pacific Islands.

Fig. 4.17 Mealybugs on taro roots

4.2.9.2 Symptoms

On taro, mealybugs occur on the undersurface of the leaves, on and between the petioles, and on the roots and corms. On the roots, they occur as cotton-like masses, containing males and females, which are sometimes difficult to see clearly with the naked eye (Fig. 4.17).

Mealybugs have a long feeding tube that is used to pierce plant parts and suck the sap—in doing so; they cause a variety of symptoms. Direct feeding results in distorted foliage, yellowing, stunting, and wilting; indirectly, mealybugs cause a buildup of sooty mold fungi, which grows on the honeydew excreted as they feed. They also transmit viruses. In these ways, mealybugs are similar to aphids. On taro, mealybugs rarely harm the plants or promote sooty mold growth. However, some transmit taro badnavirus (TaBV), also known as taro bacilliform badnavirus.

4.2.9.3 Host Range

Pink hibiscus mealybugs have a very wide host range and may also be found on other crops such as cassava and mango as well as leguminous plants.

4.2.9.4 Management

The best course of action is to check if ants are present and take action against them, so that the activities of natural enemies are not curtailed. If this is not sufficient, then a spray of horticultural oil or soap should be considered.

Pesticides are often largely ineffective at controlling the insects as they hide in crevices on the plants and avoid contact with the chemical; in the Caribbean, predatory beetles have been very successful at controlling populations of mealybugs.

4.2.10 Dasheen Beetle, *Ligyrus ebenus* (Coleoptera: Dynastidae)

4.2.10.1 Distribution

The pest is distributed in Bolivia, Brazil, Colombia, Guyana, Mexico, Peru, Surinam, Trinidad, and Venezuela.

4.2.10.2 Damage

Corms become unmarketable when the beetle bores holes and tunnels through the corm causing secondary infection. The plants wilt due to both adults and larvae boring into the roots. Adults are dark brown to black, strongly convex dorsally, 25 mm long and 16 mm wide. If the attack is severe, death of the plants follows very quickly. The problem occurs mainly in the wet season. This pest appears to be attracted to plants growing in wet or irrigated areas during dry weather.

4.2.10.3 Biology

Eggs are usually laid singly in soil and organic material. After 13 days, the curved six-legged larvae hatch. After passing through four larval instars, pupation takes place in the soil and lasts 14 days; the adults emerge at night.

4.2.10.4 Management

1. *Cultural Methods*: In small gardens, wilting plants can be lifted and the insects killed by inserting wires into their tunnels. The taro can then be replanted.

   Cultural methods of control include the use of clean planting material, proper field sanitation, and crop rotation, followed by fallowing and plowing of harvested fields to expose grubs to sunlight.
2. *Chemical Methods*: A soil drench with Fastac (alphacypermethin) (a systemic soil insecticide) destroys the grubs (National Agriculture Marketing and Development Corporation 2003; Anselm 2004).
3. *Biological Methods*: The larvae may be attacked by a parasitic wasp, *Campsomeris dorsata*, and also by a muscardine fungus.

4.2.11 Root Aphid, *Patchiella reaumuri*

4.2.11.1 Distribution

Within the Pacific region, the aphid is known only from the islands of Hawaii and O'ahu in the State of Hawaii, in the United States. It is also recorded in Europe on *Arum* spp., where it occurs on the roots, and on *Tilia* spp. (lime or linden trees), where it causes a gall of the shoot tip.

4.2.11.2 Damage

The taro root aphid infests the roots and, when populations are high, the lower parts of the petioles of taro leaves. The taro root aphid feeds on the roots of taro, causing them to rot. Damage to the roots produces stunted plants with yellowish, small leaves. The effect of the aphid is greater on young plants than on mature plants and worse on all stages during periods of drought. Infestations appear as a white mold on the fibrous taro roots (Fig. 4.18).

4.2.11.3 Management

There is quarantine on transport of planting materials from the island of Hawaii to other Hawaiian Islands. It is recommended that where heavy

Fig. 4.18 *Left*: Signs of taro root aphid infestation appear as white mold on the fibrous taro roots. *Right*: When taro root aphid populations are high, colonies can be found both on roots and on the petioles, just above the corm

Fig. 4.19 Spider mite damage on taro and adult mite

infestations occur, the crop should be harvested and destroyed, and the land plowed and left fallow for a year or planted with a different crop. Insecticidal soaps (1 % active ingredient) have been recommended for the control of infestations of *Patchiella reaumuri* in the field and as a treatment for planting material before planting. Since the aphid is in the soil and has a waxy covering, soaps may not be effective for control in the field and are probably better used to rid planting material of aphids. Tests using hot water at 49 °C for 6 min, followed by immersion in cold water, have been effective in disinfesting planting material without damage. It is important that planting material is free from the aphid; otherwise yields will be unsatisfactory.

4.2.12 Red Spider Mite, *Tetranychus urticae*

4.2.12.1 Distribution

Tetranychus spp. probably occur in all Pacific Island countries, but there are few official records from taro or any other host.

4.2.12.2 Damage

Red spider mite is a menace in swamp taro (*Colocasia esculenta* var. *stolonifera*) cultivation. Their infestation causes typical burnt look of the leaves hampering photosynthetic activity of the plant (Fig. 4.19). Feeding injury caused by densities greater than 5 mites per leaf increases water stress and reduces photosynthesis.

Table 4.8 Relative field efficacy of different bio-rational acaricides against *Tetranychus urticae* infesting swamp taro

Treatment	Dose (g a.i.ha^{-1})	Pretreatment count of mites/leaflet	Reduction of mite population 7 days after spraying (%)
Abamectin 1.8 EC	5	16.36	80.46 (63.76)
	9	17.07	90.80 (72.34)
	12	16.66	98.55 (83.08)
Fenazaquin 10 EC	50	15.10	80.28 (63.63)
	75	16.21	90.45 (71.99)
	100	18.51	97.75 (81.37)
Diafenthiuron 50 SC	200	17.62	35.88 (36.79)
	250	20.33	49.74 (44.85)
	300	19.10	57.40 (49.25)
Propagate 57 EC	500	18.36	58.91 (50.13)
	750	16.92	68.08 (55.59)
	1,000	18.36	75.64 (60.42)
Dicofol 18.5 EC	225	17.32	69.83 (56.68)
Control	Water	20.21	0.00 (4.05)
CD (P=0.05)		**NS**	**5.38**

4.2.12.3 Host Range

Many crops are hosts to spider mites in the Pacific Island countries, among them okra, papaya, sweet potato, cassava, tomato, beans, and cucumber and other cucurbits are important.

4.2.12.4 Management

Spraying of Abamectin, a GABA (γ-amino butyric acid)-ergic pesticide, at both 9 and 12 g a.i. ha^{-1} was found to be the best among the treatments and can be recommended for IPM program as it is less toxic toward the natural enemies (Antonin et al. 1997). Fenazaquin, a quinazolin compound acting as mitochondrial electron transport inhibitor (METI), at higher dose (100 g a.i. ha^{-1}) was more persistent and also had quick knockdown effect and hence was a satisfactory performer (Sarkar and Tarafdar 2013) (Table 4.8).

4.2.13 Giant African Snail, *Lissachatina fulica*

Lissachatina fulica is a very large snail that eats numerous vegetable and ornamental species. It is a major quarantine threat to countries not yet infested. It is also the carrier of the rat lungworm, *Angiostrongylus cantonensis*.

4.2.13.1 Distribution

Lissachatina fulica has been recorded feeding on taro (*Colocasia esculenta*) in American Samoa and other host plants in Federated States of Micronesia, French Polynesia, Guam, Marshall Islands, New Caledonia, Palau, Papua New Guinea, Samoa, Vanuatu, and Wallis and Futuna.

4.2.13.2 Damage

Taro and other edible aroids (e.g., *Alocasia*, the giant taro) do not appear to be a preferred food plant for the giant African snail in the Pacific Island countries. Damage is related to population level: when a population is high, soon after the snail has been introduced, plants may be severely defoliated. Often the leaves are "skeletonized," with only the main veins remaining. When populations decline and there are other plants available for food, taro is rarely eaten—*Brassica* species are favored foods.

The giant African snail is a large and conspicuous pest (Fig. 4.20), which hides during the day and feeds at night. Damaged plants exhibit extensive rasping symptoms and defoliation. The weight of large numbers of individuals can break the stems of some host species.

Fig. 4.20 Giant African snail on taro

4.2.13.3 Biology

Lissachatina fulica is readily identified by its large size and its long, narrow, conical shell, which is light brown with alternating brown and cream stripes on the upper whorls of larger specimens, with the color becoming lighter toward the tip of the shell. Eggs are spherical to ellipsoidal in shape (4.5–5.5 mm in diameter) and are yellow to cream colored.

4.2.13.4 Management

1. *Regulatory Methods*: Inspection and treatment of commodities that are likely to harbor the snail are an important means of preventing further spread. The snails are easily carried from country to country on plant parts, vehicles, and other machinery, and in packages of all kinds.
2. *Physical Methods*: The giant African snail rarely moves onto bare ground, so a 1.5 m strip of bare earth around crops will provide some protection.
3. *Cultural Methods*: The collection of individuals and their destruction in salt water, or by crushing or burying, are effective as long as they are done frequently.
4. *Chemical Methods*: Metaldehyde and methiocarb can be used in pellet form as a poison.

4.3 Diseases

4.3.1 *Phytophthora* Leaf Blight, *Phytophthora colocasiae*

Leaf blight is a major and widespread disease common in wetland and is favored by humid, cloudy conditions and poor soil fertility.

4.3.1.1 Distribution and Economic Importance

This disease occurs widely in India, Indonesia, Malaysia, Sarawak, and the Pacific. It has also been recorded from parts of Africa and the Caribbean. Losses up to 50 % have been reported.

4.3.1.2 Symptoms

Leaf blight is the most destructive leaf disease of *Colocasia*. Lesions are initially small, dark, and round but rapidly enlarge to 2.5–5.0 cm in diameter and become purplish to brownish in color (Fig. 4.21). Drops of a clear liquid exude from the spots and turn yellow, orange, or purple when dry. There is usually chlorotic halo round the spots. As the disease progresses, the spots coalesce and have characteristic rings of yellow or brown color. Eventually the whole leaf may be affected and may die.

Lesions may form concentric pattern and develop white fuzz on both sides of leaves; holes

Fig. 4.21 Symptoms of leaf blight on taro foliage

may form in leaves where lesions have dried and dropped out; leaves collapse and die; disease may occur as a postharvest rot of corms and causes large areas of dark gray to blue rot with indistinct margins.

4.3.1.3 Survival and Spread

The pathogen survives in infected tubers and self-sown *Colocasia* plants. Spores are produced on the leaf spots and are readily spread by rain. Disease spread primarily by splashing rainwater.

4.3.1.4 Predisposing Factors

Frequent rainfall and high RH increases disease severity. Most favorable weather conditions are 20–22 °C and 100 % RH during night and 25–28 °C and 100 % RH at day with cloudy weather. Disease emergence is favored by rainy overcast weather with low night temperatures.

4.3.1.5 Management

1. *Cultural Methods*: Various cultural methods have been recommended for the control of *Colocasia* leaf blight. Removal of infected leaves has been effective during the early stages of disease development in a number of countries (Hunter et al. 2001). According to Jackson et al. (1980), regular rouging of diseased leaves in plots affected by a severe blight did not eradicate the pathogen. Disease increased rapidly after rouging ceased and corm yields were greatly decreased. Rouging may delay the start of epiphytotics (Ashok and Mehrotra 1987). Wide spacing of plants has been reported to reduce disease severity, but this appears to have a negligible effect when conditions favor disease development (Hunter et al. 2001).

Other cultural methods that have been recommended include delayed planting on the same land for a minimum of 3 weeks, avoiding plantings close to older infected ones, and preventing the carryover of corms or suckers which can harbor the pathogen from one crop to another (Jackson 1999).

Amosa and Wati (1997) reported that disease incidence and severity of leaf blight were lower in *Colocasia*/maize intercropping system than those grown in monoculture. The effect of planting density and relative time of planting on *Colocasia*/rice intercropping system yielded similar results (Agyekum 2004). Fertilizer treatment may help the plant cope with leaf blight (Tilialo et al. 1996).

Wide spacing minimizes spread of the disease due to rain splashing. Wide spacing together with interplanting and open sides to the *Colocasia* field helps to improve the air flow and thus keeps the humidity down. Farmers should avoid walking through the field when it is wet as this encourages the spread of spores. Where possible the *Colocasia* should be cultivated on higher land. Crop remains should be burnt after harvest.

In case of mild infection of young plants (up to 2–3 months), leaf rouging can be effective in controlling or delaying the buildup of the disease. Rouged leaves should be burnt.

Strict hygiene should be observed at planting, and infected leaves should be removed twice a week for 80–90 days after planting, or until a dense canopy is formed.

Crop sanitation and rouging, removing infected plant parts or whole plants from the field, are possible measures against TLB. Used early in the season or during mild infections, they can limit disease severity (Jatala and Bridge 1993). During moderate to severe infections, however, removing infected leaves becomes counterproductive, reducing corm yield even more effectively than leaves lost to the disease. Increased spacing between plants may slow disease spread by lowering relative humidity and wetness in the field and by decreasing spore transfer due to leaf contact. The movement of spores in a field during heavy rains and wind, however, quickly leads to the same level of disease. Total yield is reduced with wider spacing as fewer plants are grown in the same area (Jatala and Bridge 1993).

The planting of *Coleus blumei* (a plant whose nectar and pollen attract adult parasitoids) (Fig. 4.22) in the taro fields favors the development of natural enemies.

Fig. 4.22 *Coleus blumei* whose nectar and pollen attract adult parasitoids

2. *Host Resistance*: *Colocasia* cvs. 'Jhankri' and 'Muktakeshi' are tolerant to leaf blight disease. Taro cvs. P1, P5, P7, P10, and P20 (tiapula planting sets) were found resistant to taro blight and most popular among farmers in American Samoa.
3. *Chemical Methods*: The use of fungicides such as copper and copper metalaxyl-based compounds is the most reliable and popular with farmers because of their quick and effective action (Adejumo 1997). Jackson (1996) reported that blight disease can be controlled by spraying with Copper fungicides. Ashok and Mehrotra (1987) observed in field trials that excellent control of *Colocasia* leaf blight was obtained when plants were treated with chloroneb and captafol, good control with metalaxyl, fair control with copper oxychloride, and poor control with thiophanate methyl and zineb.

Field experiments conducted to study the effect of fungicides in controlling leaf blight caused by *P. colocasiae* in *C. esculenta* revealed that 0.2 % metalaxyl and mancozeb as Ridomil MZ-72 was the most effective treatment, followed by 0.2 % captafol, Bordeaux mixture (1 % copper sulfate and lime), and 0.25 % mancozeb (Ashok and Saikia 1996). A significant increase in yield was recorded for all treatments over the untreated control.

The frequency and time of spray application have been reported to affect the effectiveness of fungicides (Adegbola 1993). Bergquist (1974) confirmed the effect of fungicide rate, spray interval, timing of spray application, and precipitation in relation to control of leaf blight disease of *C. esculenta*. In an experiment conducted by Bergquist (1974), *C. esculenta* was sprayed with mancozeb at rates of 4.48, 2.24, or 1.12 kg/ha at intervals of 5, 7, 10, or 14 days at drier and wetter sites. Rate of fungicide had no effect in the drier sites, while at wetter sites, the highest rate of 4.48 kg/ha was the most effective. Spraying every 5 days was significantly more effective than

Table 4.9 Field efficacy of fungicides for the control of leaf blight disease and yield of *Colocasia*

Fungicide	Conc. (%)	Disease incidence (%)	Disease index (%)	Corm yield (t/ha)
Indofil M-45	0.25	21.88	21.25	19.028
Blitox-50	0.30	20.31	18.12	17.500
Captaf	0.20	18.75	17.75	19.514
Ridomil MZ	0.20	12.06	8.13	20.347
Bordeaux mixture	0.50	23.44	20.00	17.222
Control	---	46.88	32.50	14.790
CD (P=0.05)	**---**	**5.505**	**4.487**	**0.9465**

spraying every 14 days. Applications of fungicide at 7-day intervals gave substantial disease control.

Tribasic copper sulfate and copper oxychloride have been used successfully at intervals of 7–15 days. Zineb and mancozeb have been shown to give effective control. The best control was obtained with sprays of mancozeb at 2.24–4.48 kg/ha at intervals of 5–7 days. It has been suggested that copper oxychloride sprays and phytosanitary measures could be used together in a practical control program.

Four sprays of 0.25 % zineb at 15-day interval was found effective in reducing disease intensity (23.24 % as compared to 56.15 % in control) and increasing the tuber yield (20.91 t/ha as compared to 15.49 t/ha in control).

Two sprays of Ridomil MZ at 0.25 % at 20-day interval starting just after the appearance of the disease were found significantly superior to other treatments in terms of percent disease incidence (12.06 %) and percent disease index (8.13 %) yielding 20.3 t/ha of corms which was at par with Indofil M-45 and Captaf (Table 4.9).

4. *Biological Methods*: Several potential biocontrol agents have been reported on various plants. These include *Aspergillus niger*, *Penicillium* spp., and *Trichoderma viride* (Frais and Garcia 1981), *Bacillus* spp. (Odigie and Ikotun 1982), and *Anoplolepis longipes* (McGregor and Moxon 1985). Effect of soil application, seed treatment, and foliar spray of rhizobacterial cultures that were isolated from *C. esculenta* on *Phytophthora* blight reduced the disease incidence and severity and increased the yield, compared to untreated pathogen-inoculated control plants (Sriram et al. 2003). Biological control agents may be used judiciously as a complement to chemical application and cultural practices. In such a situation, compatibility with the synthetic fungicide would be desirable, as it is often possible to schedule both in control programs (Coffey 1991).

Prophylactic spray of *Pseudomonas* 1–2 % checked the incidence of leaf diseases like blight. *Trichoderma viride*, *T. harzianum*, and *Pseudomonas fluorescens* were found effective against *Colocasia* leaf blight.

Under polyhouse conditions, the antagonistic rhizobacterial cultures S1B3, S11B4, S13B5, and S23BS, when applied as seed tuber treatment, reduced the *Phytophthora* blight disease severity. In these treatments there was no disease incidence compared to control where the disease severity was 2.92 on a 0–5 disease rating scale. Soil application of rhizobacterial cultures S4BS, S13B5, and S23BS gave complete control of leaf blight disease compared to control where the disease severity was 2.83 on a 0–5 disease rating scale. Foliar application with S1B4 and S11B3 reduced the disease severity to 0–0.33 rating compared to 2.66 in control.

Under field conditions, tuber treatment with S1B3, soil application of S13B5, or foliar application with S1B4 and S11B3 reduced the disease severity and increased the yield compared to untreated pathogen-inoculated control plants. Seed treatment with S1B3 resulted in tuber yield of 255 g/plant compared to 95.42 g in control. Soil application with S13B5 resulted in 232.65 g/plant, while in foliar application with S1B4 or S11B3, yields were 274 g and 605 g per plant, respectively. These treatments promoted the plant growth also.

The field application of bacteria in combination (seed treatment, soil treatment, and foliar spray) helped in reducing the leaf area damaged due to blight by 41 % during the first peak of the disease spread and by 28 % during the second peak of the disease spread. Rhizobacteria treatment also helped in reducing the storage losses. The storage loss of tubers harvested from rhizobacteria-treated plots ranged from 4.14 to 21.24 % compared to 26.02 % in fungicide-treated plots, resulting in 18–36 % increased yield in the field trials (Sriram and Misra 2007).

Potent biocontrol agents of vermicompost origin against taro leaf blight were identified as *Bacillus subtilis*, *B. cereus*, *Providencia rettgeri*, and *Trichoderma asperellum*. The bacterial isolates expressed N_2 fixing, P and K solubilizing, and growth promotion properties. Disease suppression varied with the origin of vermicompost. Minimum concentration of 10 % was required for pathogen suppression. The application of vermicompost reduced taro leaf blight incidence by 10–30 % in pot culture. Vermin wash reduced the lesion development by *P. colocasiae* on detached taro leaves in vitro. The production of defense enzymes, chitinase, glucanase, and total phenols, increased with the application of vermicompost in taro plants.

5. *Integrated Methods*
   - Use of healthy planting material.
   - Growing resistant varieties—Muktakeshi and Jankhri.
   - Suitable adjustment in planting dates to escape disease/crucial stage.
   - Field sanitation and removal of infected leaves.
   - Intercropping with nonhost crops.
   - Removal of self-sown plants from the vicinity.
   - Preventive sprays with mancozeb (protective) and metalaxyl (systemic) as a last resort. Spraying of metalaxyl (0.05 %), folpet (0.2 %), and dithane M-45 (0.25 %) is also effective.
   - Early planting (in May) with early maturing variety like telia.
   - Use of biocontrol agents—*Trichoderma viride*.

   (a) *Biological and Chemical Methods*: Tuber treatment with *T. viride*+0.25 % mancozeb spray after the first appearance of the disease+0.2 % Ridomil spray 20 days after mancozeb spray was effective in reducing the percent leaf area damaged and increased the yield.

   (b) *Host Resistance, Cultural, and Biological Methods*: IDM package (selection of healthy tubers+cormel treatment with *Trichoderma viride* at 5 g/kg mixed with cow dung slurry before planting+application of *T. viride-enriched* compost before planting+mulching with paddy straw after planting+spray of CTCRI formulation at 60 and 90 DAP) recorded minimum incidence of taro leaf blight (12.9 % compared to 20.3 % in control) and intensity (7.8 % compared to 10.8 % in control) with highest sprouting (99.4 % compared to 94.74 % in control) and cormel yield (13.5 t/ha compared to 9.5 t/ha in control) in Bihar.

In Andhra Pradesh, seed tuber treatment (cv. Telia) with *Trichoderma viride* (5 g/kg cow dung slurry) and soil application of *T. viride*-enriched compost at 1 kg/pit gave maximum sprouting (97.93 %) and yield (16.5 t/ha) and least disease incidence (PDI—17.36) (Table 4.10).

6. *Positioning of Control Methods in Terms of the Development Cycle of the Disease* (Fig. 4.23)
   (a) *Choice of the Plot Before Planting*: Choose plots that are isolated from other taro fields, if possible surrounded by high vegetation (forest clearing). When the plots are close, synchronize the taro planting dates if possible, which prevents the young plants from being contaminated by affected plots that represent a significant stock of inoculum.
   (b) *At Planting*: Use cultivars that are resistant to the *Phytophthora* (Muktakeshi, Jankhri) obtained from varietal selections or creations carried out by agronomic research. Resistant material can be obtained in particular from the Regional Germplasm Centre of the Pacific Community based in Fiji.

Table 4.10 Effect of cultural methods, bioagents, botanicals, and host resistance for the management of leaf blight disease and cormel yield

Cultivar	Treatment	Sprouting (%)	% Disease incidence on leaves (PDI)	Cormel yield (t/ha)
Telia (susceptible)	Mulching with dry leaves (15 cm thickness)	95.83 (78.27)	24.09 (29.34)	12.7
	Mulching with black polythene	94.98 (77.19)	23.71 (29.08)	13.4
	Seed tuber treatment with *T. viride* (5 g/kg cow dung slurry) and soil appln. of *T. viride-enriched* compost (1 kg/pit)	97.93 (81.78)	17.36 (24.47)	16.5
	Foliar spray with CTCRI formulation[a] at 60 and 90 DAP	94.23 (76.10)	23.46 (28.97)	12.8
	Control untreated	94.98 (77.09)	30.55 (33.51)	11.0
Muktakeshi (resistant)	–	96.55 (77.32)	11.01 (18.60)	14.2
CD (P=0.05)		**2.075**	**5.544**	**2.998**

[a]CTCRI formulation: Water, 10 L; neem cake, 1 kg; and neem, bhel, custard apple, eucalyptus, and castor leaves, 21 each were taken in an earthen pot and mixed thoroughly, and the mouth of the pot is sealed with polythene sheet and allowed to decompose for 3–4 weeks. Finally dilute the mixture with 100 L of water and filter through two layers of muslin cloth and use for spraying

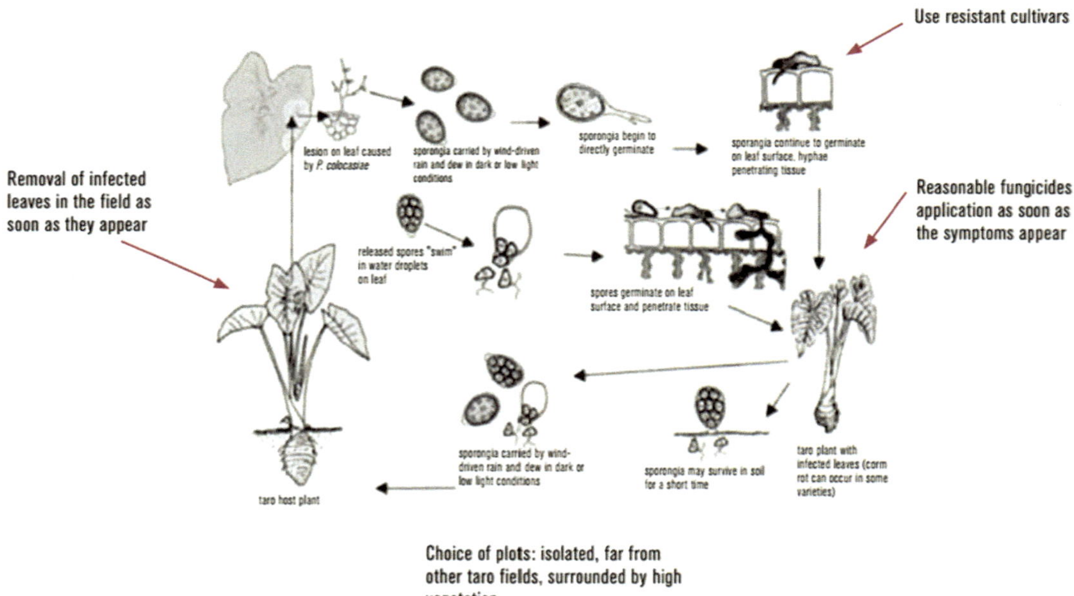

Fig. 4.23 Positioning of control methods in terms of the development cycle of the disease

(c) *Throughout the Plant Cycle*: Remove the diseased leaves as soon as the symptoms of wilting appear and burn them. To do so, it is necessary to regularly inspect the plots, in particular 3–4 days after strong rainfall or in the event of morning dew. Apply fungicides on the leaves by spraying as soon as the symptoms appear and following the removal of the attacked leaves.

Fig. 4.24 Symptoms of brown leaf spot on taro leaves

(d) *After the Harvest*: To avoid rotting due to the *Phytophtora* (and to the *Pythium*), the corms will be treated before being put on the market by soaking them in a solution with a 1 % bleach (sodium hypochlorite) content for 2 min. Dry the corms well before packing in a polyethylene bag.

4.3.2 Brown Leaf Spot, *Cladosporium colocasiae*

4.3.2.1 Distribution

The disease is distributed in Borneo, Ethiopia, Fiji, Formosa, Guinea, Hawaii, Hong Kong, India, Nepal, New Caledonia, New Guinea, Nigeria, Pakistan, Sarawak, Salmon Islands, and Tahiti.

4.3.2.2 Symptoms

Brown leaf spot is a fungal disease of older leaves. It is also called ghost spot because the lesions are often less evident on the opposite surface of the leaf. *Cladosporium colocasiae* causes brown leaf spot (or ghost spot) of older leaves—reddish-brown, circular or irregular, diffuse spots or blotches on either leaf surface, sometimes with dark, diffuse centers (Fig. 4.24). The spots are usually less evident on the opposite surface of the leaf. Sometimes the spots are surrounded by a yellow halo or have a dark-brown, diffuse border. Spots can be up to 15 mm in diameter but are usually much smaller when there are many spots on a single leaf.

4.3.2.3 Management

1. *Regulatory Methods*: Plant quarantine authorities might require certification that consignments of leaves are free from this pathogen when leaves are moved internationally. However, it is not considered to be a pest of "potential economic importance."
2. *Cultural Methods*: Removal and destruction by burning of infected leaves will reduce inoculum levels.
3. *Chemical Methods*: Fungicidal sprays applied for leaf blight control will also control brown leaf spot.

4.3.3 *Curvularia* Leaf Blight, *Curvularia* sp.

Blight disease of taro not managed early led to yield reduction of more than 50 % (Jackson 1999).

Fig. 4.25 Advancing *Colocasia* leaf blight symptoms

4.3.3.1 Symptoms

The disease is mainly a foliar disease. Initial symptoms of the disease are small brown water-soaked flecks on the leaf that enlarge to form dark-brown lesions, often with a yellow margin (Fig. 4.25). Secondary infections lead to rapid destruction of the leaf, which may occur in 10–20 days or less in very susceptible varieties (Hunter et al. 2001). The normal longevity of a healthy leaf is about 40 days (Ooka and Brennan 2000). The disease significantly reduces the number of functional leaves and can lead to yield losses (Jackson 1999).

4.3.3.2 Management

In the field trial, *C. esculenta* treated with Topsin M 70 WP (thiophanate methyl) and Sundomil 72 WP (metalaxyl 8 % and mancozeb 64 %) at a rate of 4 g/15 L fortnightly reduced disease symptoms and produced higher corm yields than the control treatment (Joshua 2010).

The yield of *Colocasia* ranged from 19.28 to 25.95 t/ha with the Sundomil-treated plants recording the highest yield. There were significant differences ($P<0.05$) between fungicide-treated plants and water-treated plants. Fungicide-treated plants also differed significantly from each other (Table 4.11).

Table 4.11 Effect of different fungicides for the management of taro *Curvularia* leaf blight

Treatment	Disease severity (%)	Corm yield (tons/ha)
Topsin M 70 WP (thiophanate methyl)	3.8	22.75
Sundomil 72 WP (metalaxyl 8 % and mancozeb 64 %)	1.1	25.95
Water	14.3	19.28
CD (P=0.05)	–	0.34

4.3.4 Corm Soft Rot, *Pythium* spp.

The root rot disease (RRD) is the most devastating disease in cocoyam at present (Tambong et al. 1998) and can cause a total loss of yield (Saborío et al. 2004a). Several different pathogens have been reported to be associated with the RRD including *Rhizoctonia* (Giacometti and León 1994), *Sclerotium rolfsii* (Bejarano-Mendoza et al. 1998), *Aspergillus niger*, *Fusarium solani*, *F. oxysporum*, *Botryodiplodia theobromae*, *Corticium rolfsii*, and *Geotrichum candida* (Ugwuanyi and Obeta 1996; Saborío et al. 2004a). However, the pathogen that seems to be the main causal agent is *Pythium myriotylum* (Nzietchueng 1984; Pacumbaba et al. 1992; Onwueme and Charles 1994; Tambong et al. 1999).

4.3.4.1 Distribution

The disease occurs in India and the Pacific area. Various *Pythium* species occur throughout the Pacific.

4.3.4.2 Symptoms

The symptoms are stunting, yellowing of the foliage, and a reduction or elimination of the

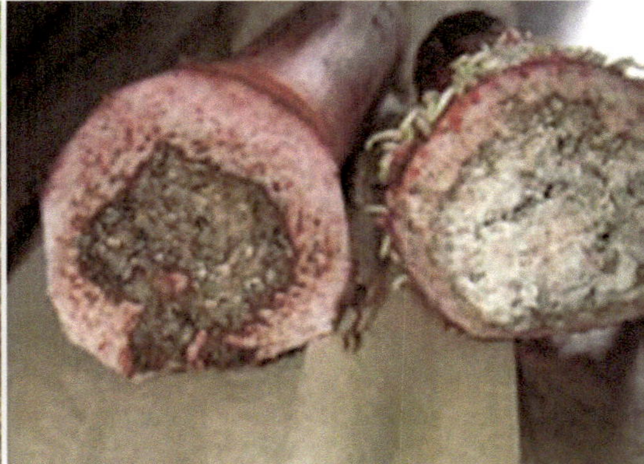

Fig. 4.26 Symptoms of corm rot on taro

root system. If young plants are attacked, they may be killed without producing leaves or they may become stunted. In plants over 3 months old, the first symptoms observed are wilting, chlorosis, and stunting which may be followed by the collapse of the whole plant. The rotting of the corms usually starts at the base and progresses upwards so that eventually the whole corm may be affected. The rotted tissue is firm and may be whitish yellow to gray and blue or dark purple (Fig. 4.26). Healthy and diseased tissues are clearly divided. In later stages, the diseased tissue crumbles like cheese and has an offensive smell.

Roots decay in association with the corms. In some cases there is little damage to the corm but many roots are killed leaving the plant loose in the ground. Rotting may also spread from the main corm to lateral suckers.

Root damage caused by *Pythium* interferes with water and nutrient uptake, and plants show slowed leaf production, poor color, and wilting of the outer leaves. Plants not killed before harvest are usually stunted and produce small corms. Root rot may enter the corm from the base or sides and turn the insides into a stinking, mushy mass of decayed tissue. If the corm rot moves up into the growing point, the plant is killed.

4.3.4.3 Survival and Spread

The disease is spread through soil and planting material (Nzietchueng 1984) and can persist in the soil for many years.

The fungi are encouraged by plentiful moisture, warm temperature, and poor soil conditions. Any physical damage to *Colocasia* plant predisposes them to infection. If *Colocasia* is left in the field after maturity, it causes a buildup of the disease. These fungi survive between crops in crop residues. Damage may be confined to a few corms or may be widespread and cause 100 % loss depending upon the conditions. Damage is particularly severe in wetland areas and poorly drained soil.

4.3.4.4 Management

1. *Cultural Methods*: Wide spacing, high mounds, regulation of the time of planting (Onwueme and Charles 1994), drainage improvement, use of disease-free planting material (Saborío et al. 2004b), planting on ridges, crop rotation (Giacometti and León 1994), and use of organic fertilizer (Torres-Portuguez 1996) have been suggested in order to control the disease.

The best control is exclusion of the pathogen by planting disease-free material in an uninfested field. Once *Pythium* spp. become established in the soil, they are difficult and expensive to control and almost impossible to eradicate. In wet, shallow, or poorly drained soils, the subsistence farmer may be forced to leave the field fallow for at least 5 years (Jackson et al. 1980). Gollifer et al. (1980) found that drying tiapula for 2–3 weeks before planting killed most fungal spores of *Phytophthora colocasiae*. Based on similarities between the two genera, this may be effective in fields infested with *Pythium* spp. Heavy mulching and incorporation of organic matter into the soil can moderate soil temperatures, improve drainage, and possibly reduce the severity of root and corm rot. Fertilizing to promote optimum plant health is an important component in disease control. In Hawaii, the addition of phosphorus appeared to be related to a decrease in root and corm rot (Jackson et al. 1980). Removal of diseased plants and soil when symptoms first appear may reduce the pathogen population during a cropping cycle; it will be of little benefit, however, in a continuous cropping system.

Colocasia should be harvested as soon as possible after it is mature and crop debris should be removed from the fields. Diseased plants should be rouged from the growing crop taking care to remove as much of the corm and decayed roots as possible. If a field is badly infected with the disease, *Colocasia* should not be cultivated for 2 or 3 seasons, and the soil should be allowed to dry out by deep cultivation. In irrigated wetland *Colocasia*, water should be kept flowing and not be allowed to stagnate. Dryland *Colocasia* should be planted in well-drained soil. Healthy planting material should always be used for planting.

Significantly lowest percent loss (2.7–6.0 %) due to rots was recorded when the seed cormels were stored in zero energy cool chamber (ZECC) as a heap or in gunny bags or in bamboo baskets than open storage conditions. Less damage of seed cormels in ZECC may be due to less temperature (5–8 °C) maintained when compared to open storage (Bhagavan et al. 2013).

In Solomon Islands, storage in leaf-lined, shallow soil pits has been shown to reduce damage.

2. *Chemical Methods*: Preplant and postharvest treatments of taro corms with Ridomil (metalaxyl) or Aliette (aluminum phosphonate) (Crop Protection Reference 1995) are neither sustainable nor cost-effective in subsistence agriculture.

Dipping of planting material in copper oxychloride or captan has been recommended. Soil drenches with these fungicides have also been found to be effective.

Preharvest application of Carbendazim (0.1 %) + Streptomycin (1,000 ppm) and postharvest application of copper oxychloride (0.3 %) + Streptomycin (100 ppm) on harvested cormels was found to be significantly most effective in reducing storage rots and percent damage of seed material up to 2 months from harvest (Bhagavan et al. 2013).

Lowest cormel rotting (7.8 %) was observed when treated with Carbendazim (0.05 %) and stored on concrete floor. The minimum weight loss (28.5 %) and lowest sprouting (48.7 %) were recorded when cormels were treated with Carbendazim (0.05 %) and stored in pits lined with polythene sheet containing moist sand in Bihar. In Andhra Pradesh, the minimum weight loss (21.05 %) was observed when cormels were treated with Carbendazim (0.05 %) and stored in pits. The cormel rotting (7.57 %) and sprouting (9.6 %) were lowest, and shelf life was highest (148.65 days) when cormels were treated with Carbendazim (0.05 %) and stored on concrete floor (Table 4.12).

3. *Biological Methods*: Seed tuber treatment with *Trichoderma viride* (0.5 %) and stored in pits lined with polythene sheet containing moist sand gave least rotting of seed tubers (18.90 %) followed by treatment with Carbendazim (0.05 %) and stored in pits lined with polythene sheet containing moist sand (20.74 %).
4. *Host Resistance*: Some varieties are more resistant to the disease than others (e.g., Hawaiian varieties Kai Kea and Kai Uliuli) and should be planted if disease is known to be a problem in the area.

No taro varieties are immune to *Pythium* rots, but several have demonstrated some resistance, including the Samoan varieties Tusi Tusi, Talo Vale, Pule Mu, and Pula Sama Sama (Jackson et al. 1980). These varieties are susceptible to taro leaf blight, however, and are not an option in American Samoa at this time.

Table 4.12 Effect of type of storage, chemicals, and bioagents for the management of cormel rot

Treatment	Cormel rotting (%)	Cormel weight loss (%)	Sprouting (%)	Shelf life (days)
Cormels treated with 0.05 % Carbendazim and stored on kachha floor	10.04	25.04	15.64	139.13
Cormels treated with 0.5 % *Trichoderma viride* and stored on kachha floor	10.53	26.53	15.72	135.53
Cormels stored on kachha floor	16.61	31.53	20.55	120.76
Cormels treated with 0.05 % Carbendazim and stored on concrete floor	7.57	28.29	9.60	148.65
Cormels treated with 0.5 % *Trichoderma viride* and stored on concrete floor	8.65	29.18	9.73	144.33
Cormels stored on concrete floor	14.55	31.46	16.15	129.20
Cormels treated with 0.05 % Carbendazim and stored in pits	10.04	21.05	17.83	137.11
Cormels treated with 0.5 % *Trichoderma viride* and stored in pits	12.55	21.75	20.13	133.89
Cormels stored in pits	17.41	28.02	22.71	118.67
Cormels treated with 0.05 % Carbendazim and stored in pits lined with polyethylene sheet containing with moist sand	11.08	23.10	17.61	134.90
Cormels treated with 0.5 % *Trichoderma viride* and stored in pits with polyethylene sheet containing with moist sand	11.22	25.63	17.04	138.1
Cormels stored in pits with polyethylene sheet containing with moist sand	17.66	26.06	21.25	117.67

4.3.5 Corm Rot, *Athelia rolfsii*

4.3.5.1 Distribution

The disease is widespread in distribution. Corm rot has been recorded in American Samoa, Cook Islands, Fiji, French Polynesia, Federated States of Micronesia, New Caledonia, Palau, Papua New Guinea, Samoa, Solomon Islands, Tonga, Vanuatu, and Wallis and Futuna.

4.3.5.2 Symptoms

Athelia rolfsii is a soilborne fungus that infects taro at the soil level, causing corms and roots to rot and leaves to wilt. Infection starts at the soil level, at the base of the petioles. Fans of white mycelia grow over the infected area and sometimes on nearby organic matter. Leaves often wilt. The disease is characterized by the presence of more dead leaves than normal (Fig. 4.27). Pale cream to reddish-brown sclerotia of 1–2 mm in diameter are usually present at the site of infection. The fungus also causes a postharvest pinkish corm rot, infecting corms through wounds made when suckers are detached. The disease may be more severe in *Colocasia* that is over mature.

4.3.5.3 Development and Spread

A. rolfsii is a common pathogen and attacks many crops; sources of inoculum are therefore readily available. The disease is more common in upland *Colocasia*.

4.3.5.4 Management

1. *Cultural Methods*: The following cultural methods are recommended:
   - Removing and destroying infected plants by burning.
   - Applying good cultural practices—for example, deep plowing and encouraging the growth of microorganisms (especially *Trichoderma* spp.) that inhibit fungal growth.
   - Using soil solarization—covering the soil with plastic and allowing the sun to heat the trapped air.

Fig. 4.27 Wilting and leaf death associated with infection by *Athelia rolfsii*

- Liming the soil.
- Crop rotation with nonhosts or tolerant hosts can greatly reduce numbers of infective propagules in a field, although the sclerotia can remain viable for many years. Cereal crops (Poaceae) are relatively resistant to the fungus.

2. *Chemical Methods*: Soil drenches with Quintozone or Dicloran have been recommended for control.

4.3.6 *Phyllosticta* Leaf Spot, *Phyllosticta colocasiophila*

4.3.6.1 Symptoms

Symptoms include oval or irregular beige to reddish-brown spots on leaves, dark-brown spots with chlorotic area around lesion, and holes in leaves where lesion centers have dried and dropped out.

4.3.6.2 Epidemiology

Disease emergence is favored by cloudy and rainy weather.

4.3.6.3 Management

No control is needed unless the disease is causing severe defoliation of the plants. Remove and destroy any diseased leaves.

4.3.7 Spongy Black Rot, *Lasiodiplodia theobromae*

4.3.7.1 Distribution

Lasiodiplodia theobromae has been recorded on taro in Guam, Papua New Guinea, Samoa, and Solomon Islands. It has been recorded on other host plants in American Samoa, Australia, Cook Islands, Fiji islands, French Polynesia, Federated States of Micronesia, New Caledonia, New Zealand, Niue, Palau, Papua New Guinea, Tonga, Vanuatu, and Wallis and Futuna.

4.3.7.2 Symptoms

Lasiodiplodia theobromae in taro corms causes a postharvest rot that is initially whitish cream, later becoming blue/black. *Lasiodiplodia theobromae* is frequently isolated in decayed corm tissues behind advancing rot caused by *Phytophthora colocasiae*

Fig. 4.28 *Lasiodiplodia theobromae* infection on taro corms, showing advanced decay

and *Pythium splendens*. Even in the absence of other fungi, it enters corms through wounds made during harvest and causes complete decay in 10–14 days (Fig. 4.28). *Lasiodiplodia theobromae* causes a spongy rot, which occasionally becomes dry and powdery, with an indistinct margin between healthy and diseased tissue. Spongy black rot can be detected by cutting the corm to reveal the black interior; it has a strong, sour smell and black spore masses form on the corm surface.

4.3.7.3 Management

1. *Cultural Methods*: The traditional practice of the Sikaiana Island people (of Polynesian descent) is to store taro for up to 4 weeks buried in pits situated in shaded, well-drained soil.
2. *Chemical Methods*: Dipping corms in bleach (1 % sodium hypochlorite) for 2 min before storing in polyethylene bags is effective in controlling this fungus.

4.3.8 White Spot, *Leptosphaerulina trifolii*

4.3.8.1 Distribution

Leptosphaerulina trifolii has been recorded on taro in American Samoa, Papua New Guinea, Samoa, Solomon Islands, and Tuvalu. It is found on other hosts in Fiji, Marshall Islands, Niue, Tonga, and Vanuatu.

4.3.8.2 Symptoms

Leptosphaerulina trifolii produces yellow spots on taro leaves. These later turn white. Spots sometimes merge and show "shot hole" symptoms as the centers fall out. Infections are initially visible as small, yellow-green spots on the upper leaf. As spots mature, they become edged by a thin (1 mm), reddish-brown border and surrounded by an intense yellow halo, 1–2 mm wide. Mature lesions are 2–5 mm in diameter with paper-white centers (Fig. 4.29). Small, black fruiting bodies can be seen on close observation against the white tissue of mature lesions. Centers often fall out, creating a "shot hole" appearance. In severe infections, spots may coalesce, and the leaves look tattered.

4.3.8.3 Management

Control measures are usually not necessary. The impact of this disease is very low. In American Samoa, only a few plants have been seen to be severely infected (25–50 % leaf area) and usually the disease is unremarkable.

Fig. 4.29 Leaves infected by white spot

Fig. 4.30 Symptoms of corm and leaf spot disease

4.3.9 Corm and Leaf Spot, *Marasmiellus stenophyllus*

4.3.9.1 Distribution

Marasmiellus stenophyllus has been recorded on taro in American Samoa, French Polynesia, and Wallis and Futuna and on other hosts in Fiji.

4.3.9.2 Symptoms

Marasmiellus stenophyllus infects taro at the base of the plant, destroying leaves, corms, and roots and commonly producing toadstools on the dying parts (Fig. 4.30). Corm and leaf spot lead to leaf collapse due to the development of large brown rots at the base of the plant associated with white fungal growth. The leaves are often stuck together by the fungal threads (mycelia). Toadstools form in large numbers on the withered leaves at soil level. The fungus grows over the roots and kills them, and soil particles become fastened to the roots in the process. Infection with *M. stenophyllus* can kill the plant, which appears desiccated or mummified. Corms become inedible and, even at an early stage of decay, may be unsightly with mycelium growth causing small "pocket" rots. However, the incidence of infection is low.

4.3.9.3 Management

1. *Cultural Methods*: The removal and destruction of infected plants by burning is helpful in controlling the fungus.

Fig. 4.31 Symptoms of orange leaf spot disease

4.3.10 Orange Leaf Spot, *Neojohnstonia colocasiae*

4.3.10.1 Distribution

Neojohnstonia colocasiae has been recorded on taro in American Samoa, Fiji, Federated States of Micronesia, Palau, Papua New Guinea, Samoa, Solomon Islands, Tuvalu, Vanuatu, and Wallis and Futuna.

4.3.10.2 Symptoms

Orange leaf spot is a fungal disease of older leaves causing symptoms very similar to those of *Cladosporium colocasiae* (brown leaf spot). *Neojohnstonia colocasiae* causes yellowish-brown, circular or irregular blotches on either leaf surface. These become darker with the onset of sporulation. Spots are sometimes surrounded by a yellow halo or have a brown border. They can be up to 15 mm in diameter, but tend to be smaller when there are many spots on a single leaf (Fig. 4.31).

4.3.10.3 Management

1. *Regulatory Methods*: Plant quarantine authorities might require certification that consignments of leaves are free from this pathogen when leaves are moved internationally. However, it is not considered to be a pest of "potential economic importance."
2. *Cultural Methods*: No control measures are recommended; however, removal and destruction by burning of infected leaves will reduce inoculum levels.

4.3.11 Shot Hole, *Phoma* spp.

4.3.11.1 Distribution

Phoma colocasiae has been recorded in Palau and Samoa. The Pacific taro fungus, *Phoma* sp., has been recorded in American Samoa, Cook Islands, Federated States of Micronesia, Fiji islands, French Polynesia, Marshall Islands, Niue, Palau, Papua New Guinea, Samoa, Solomon Islands, Tokelau, Tonga, and Vanuatu.

4.3.11.2 Symptoms

Phoma spp. (*Phoma* sp. and *Phoma colocasiae*) produce relatively large lesions on the leaf. As the spots age, their centers fall out, giving the "shot hole" effect. The first symptoms of a *Phoma* infestation are small, round, brown spots on the second or third leaves. As the spots enlarge to 2 cm in diameter, the brown centers fall out, resulting in the typical "shot hole" symptom. The holes have a narrow, brown margin, which is surrounded by an intense yellow halo (Fig. 4.32). The holes may merge, so that large areas of the leaf are destroyed. This leads to premature leaf death.

Fig. 4.32 Symptoms of shot hole disease

4.3.11.3 Management

There is no evidence that the disease warrants control. Fungicides may control the disease; however, they cannot be recommended at present, since recent studies have not established that the disease reduces corm yields.

4.3.12 Leaf Blotch, *Pseudocercospora colocasiae*

4.3.12.1 Distribution

Pseudocercospora colocasiae has been recorded in American Samoa, Fiji, French Polynesia, New Caledonia, Samoa, Solomon Islands, and Vanuatu.

4.3.12.2 Symptoms

Leaf blotch is a fungal disease, mostly affecting older leaves. The symptoms are similar to those caused by *Neojohnstonia colocasiae* (orange leaf spot) and *Cladosporium colocasiae* (brown leaf spot). This fungus has little impact of consequence in taro. It causes blotches with indistinct, circular, yellow-reddish to whitish-green discoloration on the upper surface of the leaf and black mold growth on the corresponding lower surface. The blotches can be up to 1.5 cm in diameter (Fig. 4.33).

4.3.12.3 Management

This disease is not considered to be of economic importance; therefore, no control measures are necessary. It is a disease of older leaves.

Fig. 4.33 Symptoms of leaf blotch disease

4.3.13 Bacterial Soft Rot, *Erwinia chrysanthemi*

4.3.13.1 Distribution

Erwinia chrysanthemi has been recorded on taro in Solomon Islands and on other host plants in Cook Islands and Papua New Guinea.

4.3.13.2 Symptoms

In the field, infection causes a foul-smelling, creamy-white corm soft rot, and plants wilt suddenly (Fig. 4.34). A similar rot occurs in harvested corms stored at high temperature and humidity. In Solomon Islands, soft rot is associated with plants infected by *Pythium myriotylum*, sometimes together with *P. splendens*.

A sudden collapse of the leaves of mature plants is often indicative of bacterial soft rot of the corm. Leaf collapse occurs in plants that have wilted due to root infection by *Pythium* spp. At this stage, corms are usually so decayed that plants can topple over in the wind. In storage, in soil pits or plastic bags, the bacterium can be detected by the presence of soft rot with a strong, unpleasant smell.

4.3.13.3 Management

The "tops"—the petiole base with corm piece—from corm-rot-affected plants should not be used

Fig. 4.34 Taro plants infected with soft rot; note collapse of plant on the *right*

as propagating material. Rots are more important in corms stored at high humidity, either in soil pits or in plastic bags. This type of storage would otherwise extend the shelf life by preventing infection from *Phytophthora colocasiae*, *Pythium splendens*, and *Lasiodiplodia theobromae*. A reduction in the incidence of these types of rots is possible if corms are pretreated with bleach (1 % sodium hypochlorite).

4.3.14 Dasheen Mosaic Virus

From economic point of view, this disease is not considered to be significant; however, the use of infected material over the years will reduce the yield significantly.

4.3.14.1 Distribution

The disease occurs in Egypt, Hawaii, India, Japan, Papua New Guinea, Puerto Rico, Trinidad, the United States, and Venezuela. Dasheen mosaic virus is widespread throughout the Pacific; however, only in French Polynesia have severe distortions been reported.

4.3.14.2 Symptoms

Plants show a variety of mosaic patterns: small, irregular, and scattered gray, green, yellow, and sometimes white patches along or between the major veins or brilliant white or yellow feather-like patterns along the veins and sometimes throughout the leaf blade (Fig. 4.35). Invariably, plants recover from the symptoms, producing leaves that are healthy in appearance. Plants in French Polynesia have been reported with small, stunted, and severely distorted leaves. Some leaves are reduced to strap-like structures without leaf blades. These plants often fail to recover.

The infection results in a feathery mosaic pattern adjacent to the veins. Yellow-green streaking of the petioles may also be observed. In severely affected plants, petioles are short and leaves are dwarfed and fail to unfurl properly (Fig. 4.35).

4.3.14.3 Particle Morphology

The virus has long flexuous rod-shaped particles measuring about 75 nm in length (Fig. 4.36).

4.3.14.4 Transmission

The virus is transmitted non-persistently by the aphids, *Myzus persicae*, *Aphis gossypii*, and *Pentalonia nigronervosa*. This virus is also sap transmissible.

4.3.14.5 Management

Although DMV is widespread, caution is still required when moving taro (or other aroids)

Fig. 4.35 Symptoms of dasheen mosaic virus

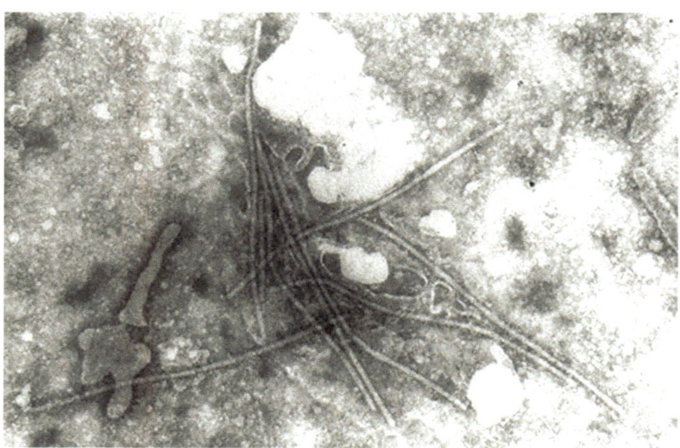

Fig. 4.36 Virus particles of dasheen mosaic virus

internationally, especially as there appear to be severe forms of the disease. Strict quarantine measures should be followed. Guidelines suggest that all transfers should be as sterile, pathogen-indexed plantlets growing in a tissue culture medium.

Only certified healthy planting material should be used for planting. The healthy planting material may be obtained by meristem tip culture.

4.3.15 Alomae

Three viruses are associated with alomae: *Colocasia* bobone rhabdovirus, taro vein chlorosis rhabdovirus, and taro badnavirus.

4.3.15.1 Distribution

Alomae virus has been recorded in Solomon Islands and Papua New Guinea.

4.3.15.2 Symptoms

Initial symptoms of alomae vary. Symptoms are similar to those in plants infected with *Colocasia* bobone disease ?rhabdovirus (stunted, thickened, twisted, dark-green leaves) or in plants that are stunted with leaf blades bent under at the tip. Symptoms of alomae include chlorosis of the veins, appearance of feather-like pattern, rolling of the youngest leaves, and other leaves that are short with tips that curl under (Fig. 4.37).

Fig. 4.37 Symptoms of alomae disease (rolling of the youngest leaves) on taro

4.3.15.3 Transmission and Spread

These diseases are perpetuated by infected planting material of taro huli, and possible transmission of the virus particles is by insect vectors from older plantings to new plantings. It is suspected that the large bacilliform virus particle is transmitted by the taro plant hopper, *Tarophagus proserpina* (Kirk); the smaller bacilliform particle could be transmitted by mealybug, *Planococcus citri* and *Pseudococcus longispinus*.

The plant hoppers and mealybugs suck up the viruses as they feed; the viruses multiply in the insects, which then spread them as they move and feed on healthy plants.

Apart from being spread by insects, the viruses are also spread in planting material. It is likely that all taros are infected with taro badnavirus, without showing symptoms, or only rarely do they show symptoms. The viruses spread from mother plants to suckers. Also, it is likely that all "female" taros are infected with the rhabdovirus. Bobone often occurs after planting, when plant hoppers are absent. Possibly, stress at planting causes the virus to multiply.

Alomae is spread from one garden to another in diseased "tops" and suckers used for planting.

4.3.15.4 Management

1. *Cultural Methods*: Remove any plants showing symptoms of disease and destroy by burning; removal of outer leaf sheaths may control the disease by reducing the number of leafhopper eggs.

Alomae is best controlled by cultural methods. These methods rely on farmers' understanding on how the disease spreads. It is very important that farmers understand the following points:

- Understand that insects spread alomae, as they fly between plants and gardens.
- Work together, each one applying the methods in their garden.
- A village group is formed, which meets regularly and exchanges ideas, and the members help each other.
- Plant resistant varieties: "Female" taros are resistant to alomae, and they can be used where the disease is severe, especially in the lowlands.
  Farmers should do the following:
- Make new gardens as far away from old ones as possible.
- Avoid taking planting material from diseased to new gardens.
- Pull out alomae plants carefully, making sure that any plant hoppers are trapped in the leaves or between the stalks—or put a rice bag over the plants before pulling them out and then burn them.
- Do not pull out plants and leave them in the garden or throw them into the bush—the insects will come back again.

2. *Biological Methods*: *Cyrtorhinus*, a bug that feeds on the eggs of *Tarophagus*, reduces the population of the plant hopper, but experience shows that it is not enough to stop the spread of alomae.
3. *Chemical Methods*: Regular spraying with pyrethroid insecticides (Permethrin or Lambda cyhalothrin) will kill *Tarophagus* and reduce alomae. It will not prevent bobone.

4.3.16 *Colocasia* Bobone Rhabdovirus

4.3.16.1 Distribution

Colocasia bobone disease ?rhabdovirus is widespread throughout Papua New Guinea and Solomon Islands.

4.3.16.2 Symptoms

Symptoms differ between taro varieties, whether they are "male" or "female"—a distinction made by farmers in the Solomon Islands based on the size of the plants and their reaction to alomae. In female varieties, *Colocasia* bobone disease ?rhabdovirus (CBDV) causes severe stunting with distorted, thickened, and brittle leaves, which sometimes fail to unfurl (Fig. 4.38). Galls may be present on the petioles and sometimes on the larger veins. Plants usually recover after three to five leaves have been affected. Sometimes symptoms recur after recovery. In Solomon Islands, plants with these symptoms are said to have bobone. Translated, this means "the plant grows small."

Male taro contains many varieties on which CBDV causes only mild symptoms: leaves show localized dark-green wrinkled patches and petioles are normal in appearance. This is a mild form of bobone from which the plants recover. There is no local name for this phenomenon.

4.3.16.3 Management

Growers do not normally apply control measures against this disease as the plants recover and produce symptomless leaves. However, the removal and subsequent burning or burial of infected plants is recommended, as these plants may serve as sources of infection for alomae. It is not yet clear which viruses interact to cause alomae, but *Colocasia* bobone disease ?rhabdovirus is suspected to be one of them.

Care must be taken to ensure that all insects on the infected plants are killed before the removal of the plants for destruction or that the plants are pulled up carefully so the insects are not spread. The virus is reportedly eliminated from taro plants if meristem tip culture is used. The restricted occurrence of this disease—confined to Papua New Guinea and Solomon Islands—means that strict quarantine measures should be followed when germplasm is moved internationally. Guidelines suggest that all transfers should be as sterile, pathogen-indexed plantlets growing in a tissue culture medium.

Fig. 4.38 Symptoms of *Colocasia* bobone disease (distorted, thickened, and brittle leaves) on taro

4.3.17 Taro Badnavirus (TaBV)

There is no widely accepted common name for taro badnavirus disease, although "vein clearing" has been suggested.

4.3.17.1 Distribution

Taro badnavirus is widespread in American Samoa, Australia, Cook Islands, Fiji, Papua New Guinea, Samoa, Solomon Islands, Tonga and Vanuatu and is present in Federated States of Micronesia, Marshall Islands, and New Caledonia.

4.3.17.2 Symptoms

Taro badnavirus (TaBV) is often latent. Where symptoms do occur, they appear as intermittent, indistinct areas of vein chlorosis, often near the leaf margin (Fig. 4.39). Frequently, the leaf blades are bent backwards, and sometimes the leaves are puckered. Rarely (but particularly on the Philippine variety PSB-G2 in Samoa), distinct vein chlorosis occurs throughout the leaf. Plants may be stunted, but eventually recover, producing apparently healthy leaves. Alone,

Fig. 4.39 Symptoms of taro badnavirus (distinct vein chlorosis)

TaBV appears to have a minimal effect on taro plant growth. However, in association with another virus, *Colocasia* bobone disease ?rhabdovirus (and possibly taro vein chlorosis virus), TaBV is thought to be responsible for causing alomae in the Solomon Islands and Papua New Guinea.

4.3.17.3 Management

No control measures are known to prevent the spread of taro badnavirus in the field. The virus is latent. Although the virus is widespread, caution is necessary when moving taro (and other aroids) internationally, as sequence variability has been observed. Strict quarantine measures should be followed. Guidelines suggest that all transfers should be as sterile, pathogen-indexed plantlets growing in a tissue culture medium.

4.3.18 Taro Vein Chlorosis Virus (TaVCV)

Taro vein chlorosis virus causes chlorosis (yellowing) of the veins, often near the leaf margin. The effect of the virus on corm yield is unknown, but there is a possibility that it causes alomae when it is associated with other viruses.

4.3.18.1 Distribution

Taro vein chlorosis virus has been recorded in Federated States of Micronesia, Fiji, New Caledonia, Palau, Papua New Guinea, Solomon Islands, Tuvalu, and Vanuatu.

4.3.18.2 Symptoms

Unlike other taro virus diseases, symptoms of taro vein chlorosis virus (TaVCV) are commonly seen on plants when they are at maximum growth rather than after planting or at maturity. Leaves show distinct vein chlorosis (Fig. 4.40), which is more pronounced than vein chlorosis sometimes associated with taro badnavirus. As the leaves age, chlorosis spreads between the veins, which then form a network. In some cases, the veins later become necrotic and the leaf margins have a tattered appearance. The number of leaves showing symptoms varies, but in most cases it is three or four. Leaves formed subsequently appear healthy. In contrast to *Colocasia* bobone disease virus, which is also a rhabdovirus, infection with TaVCV does not lead to galls on the leaf blades and petioles, and plants are generally not stunted.

4.3.18.3 Management

Growers do not normally apply control measures for this disease as the taro plants recover and produce symptomless leaves. There is no indication that infection affects corm yield. However, removal and subsequent burning or burial of infected plants are recommended in Papua New Guinea and Solomon Islands, as these plants may serve as sources of infection for alomae. It is not yet clear which viruses interact to cause alomae, but taro vein chlorosis virus (TaVCV) is a candidate.

Care must also be taken to ensure that either all insects are killed before removal of the infected plants for destruction or the plants are pulled up carefully so that the insects are not spread. The virus is reportedly eliminated from taro by using meristem tip culture. The restricted occurrence of this virus, to only nine countries, means that strict quarantine measures should be

Fig. 4.40 Symptoms of taro vein chlorosis virus (distinct vein chlorosis on the middle leaf and feathering symptoms on right side leaf)

followed when germplasm is moved internationally. It is possible that mixed infection of taro by TaVCV, *Colocasia* bobone disease ?rhabdovirus, and taro badnavirus causes the lethal alomae disease present in Papua New Guinea and Solomon Islands.

4.3.19 Leaf Blight and Mosaic

4.3.19.1 Integrated Management

The treatment IDM package gave increased (97.2 % compared to 91.0 % in control) and early sprouting and recorded least incidence of leaf blight (10.04 %) and mosaic (9.9 %) diseases and recorded higher tuber yield (21.3 t/ha) which was on par with chemical treatment (Table 4.13).

Table 4.13 Integrated disease management of leaf blight and mosaic in taro

Treatment	Sprouting (%)	Leaf blight (PDI)	Mosaic (PDI)	Tuber yield (t/ha)
IDM package[a]	97.47	10.4	9.91	21.3
Farmers' practice	91.17	17.4	14.32	16.9
Chemical control	94.37	11.8	9.07	21.6
CD (P = 0.05)	**NS**	**2.83**	**1.78**	**3.10**

[a]IDM package (selection of healthy tubers + corm treatment with *Trichoderma viride* at 5 g/kg mixed with cow dung slurry before planting + application of neem cake at 200 g/pit along with *T. viride*-enriched compost at 1 kg/plant + dry leaf mulching + spray of CTCRI formulation at 60 and 90 DAP)

4.4 Nematodes

Of five potentially parasitic nematodes isolated from roots and soil in American Samoa by Grandison (Grandison 1996), two were of concern: the root-knot nematode, *Meloidogyne incognita*, and the lesion nematode, *Pratylenchus coffeae*. Root-knot nematode was found in all taro fields surveyed but damage appeared minor; lesion nematodes were isolated in high numbers (Grandison 1996; Kirby 1987). Symptoms caused by plant parasitic nematodes are slow plant growth, pale-green or yellow-green leaves, and possibly wilting in warm weather. Parasitic nematodes may cause yield losses that range from barely noticeable to complete. The damage they cause to roots or corms may create entry points for bacteria and fungi, leading to root rot, corm rot, or plant death.

4.4.1 Root-Knot Nematode, *Meloidogyne incognita* and *M. javanica*

The root-knot nematode is an important limiting factor for successful production of *Colocasia* crop. *M. incognita* and *M. javanica* are widely reported on *Colocasia*. Nirula (1959) was the first to report root-knot nematode on *Colocasia* from India.

4.4.1.1 Economic Importance

M. javanica causes severe losses to *Colocasia* in India. *M. incognita* was responsible for 24 % loss in tuber yield of *Colocasia* (Anon 1990).

4.4.1.2 Symptoms

The foliage of nematode infested plants at first becomes yellow and then turns brown and ultimately dies back. No corms were found in plants attacked in the earlier stages, and late-stage attack resulted in deformed and galled corms (with blister-like swellings varying in size from 2 to 15 mm) of little market value. Rotting of corms associated with the nematode occurs during storage (Fig. 4.41). It causes reduction in yield and irregular-shaped smaller branched tubers.

4.4.1.3 Survival and Spread

Meloidogyne spp. can be carried over from one *Colocasia* crop to next in the wide range of other host crops and weeds. As the nematodes feed and reproduce in corm tissues, they can be spread in corms and cormels if infested material is used for propagation.

4.4.1.4 Management

1. *Physical Methods*: Root-knot can be controlled in taro by dipping the corms in hot water at 51 °C for 10 min (Martin and Hancock 1985).
2. *Cultural Control*: The use of nematode-free planting material will prevent spread into the field. Seed corms or cormels should be free of root-knot damage. It is best if the "tops" are selected from fields without any history of infection. There is no specific effort to use nematode-free planting material in American Samoa, though farmers inadvertently do so by planting corms with no roots and less than a centimeter of corm tissue.

Rotation with crops bred for resistance to *M. incognita* or nematode-resistant cover crops planted during periods of fallow can lower nematode populations and decrease root and corm damage in subsequent taro plantings. Cover crops such as marigolds produce chemicals that are toxic to nematodes. Cover crops have the added benefits of stabilizing topsoil and improving soil quality. Fields must be kept free of weeds and other plants that may act as hosts for this polyphagous parasite. The Secretariat for Pacific Communities (formerly the South Pacific

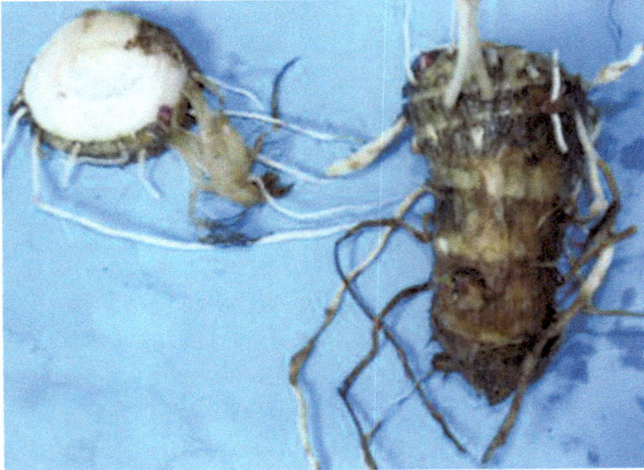

Fig. 4.41 Root-knot nematode symptoms on taro corms and roots

Commission or SPC) recommends *Panicum maximum* var. *trichoglume* or a mixture of *P. maximum* and *Macroptilium atropurpureum* as rotation crops. Most farmers, however, do not rotate crops but grow taro in a year-round cycle with no distinct planting or harvest times.

Meloidogyne populations could be suppressed when *Colocasia* is grown in very wet or flooded conditions.

3. *Chemical Methods*: Nematicides (e.g., dazomet, oxamyl, and fenamiphos) are commonly applied as pre-planting treatments to reduce nematode numbers. Srivastava et al. (1971) found that the application of DBCP at 3.6 l/ha as soil drench was more effective than Aldicarb (2.5 kg a.i./ha), Fensulfothion (3.0 kg a.i./ha), and Dichlofenthion (2.0 l a.i./ha) for the control of *M. javanica* in India.
4. *Biological Control*: The bacterium *Pasteuria penetrans* and the fungus *Paecilomyces lilacinus* have been used as biological control agents. *P. penetrans* can be cultured easily by planting root-knot-infected tomatoes where the bacterium is found. The roots can then be harvested, chopped and dried, and incorporated into the soil where there is a nematode problem.
5. *Host Resistance*: Mohandas and Palaniswami (1990b) reported resistance in *Colocasia* cv. 'Sree Reshmi' and very high resistance in C-9 (a popular variety among farmers). In cv. C-9, no nematode could be detected either from root or tubers. Cv. 'Dodare' was found to be completely resistant to both M. *incognita* and *M. javanica*, while cv. 'Samra' is described as moderately susceptible to these two species.

Resistance has been found in some cultivars/varieties. C-9 (Tamnarakannan) grown extensively in Central Travancore is immune to this nematode. 'Sree Reshmi', a released high-yielding variety is resistant, whereas 'Sree Pallavi', a released high-yielding variety, is highly susceptible to the nematode.

The cultivars 'Mana Ulaulu' and 'Piko Ulaulu' had less nematode reproduction in their roots and corms and produced higher yields. They could become part of a breeding program for root-knot nematode resistance.

6. *Positioning of Control Methods in Terms of the Development Cycle of the Plant* (Fig. 4.42)
   (a) *Before Preparing the Ground*
      - Rotate crops, avoiding cultivating taro for 2 years in a row. Avoid first (previous) crops encouraging the development of nematodes (mainly Solanaceae for *Meloidogyne*).
      - Plant cover crops with a nematode-suppressant/nematicide effect as a first crop, enabling the reduction of nematode populations in the soil.
      - A number of plant species with a nematicide effect may be recommended but must be validated locally since their antagonist action is often limited to certain species of nematode (*Meloidogyne* or *Pratylenchus*), and their effectiveness also depends on the variety of plant species used (Table 4.14).
      - Cover crops may be used as fallow crops cultivated in a mixture (cocktail) or as a pure crop. The cocktail has the benefit of a broader spectrum anti-nematode effect. The disadvantage is that it is more difficult to manage, to avoid a natural self-sowing of the seeds of the different species with different cycles.
      - A pure crop avoids these disadvantages if cutting is done before seed production, but the spectrum of antagonist action on nematode species is narrower. The crop must be cut before seeds are produced and buried in the soil.

   (b) *When Preparing the Ground*
      - Plowing with solarization (sterilization of the soil with sun's heat) under transparent plastic sheet and disinfection of the soil with steam on restricted surfaces constitute an adapted solution to limit nematode populations in the soil.
      - An increase in the organic matter content via manure or compost helps to limit nematodes.

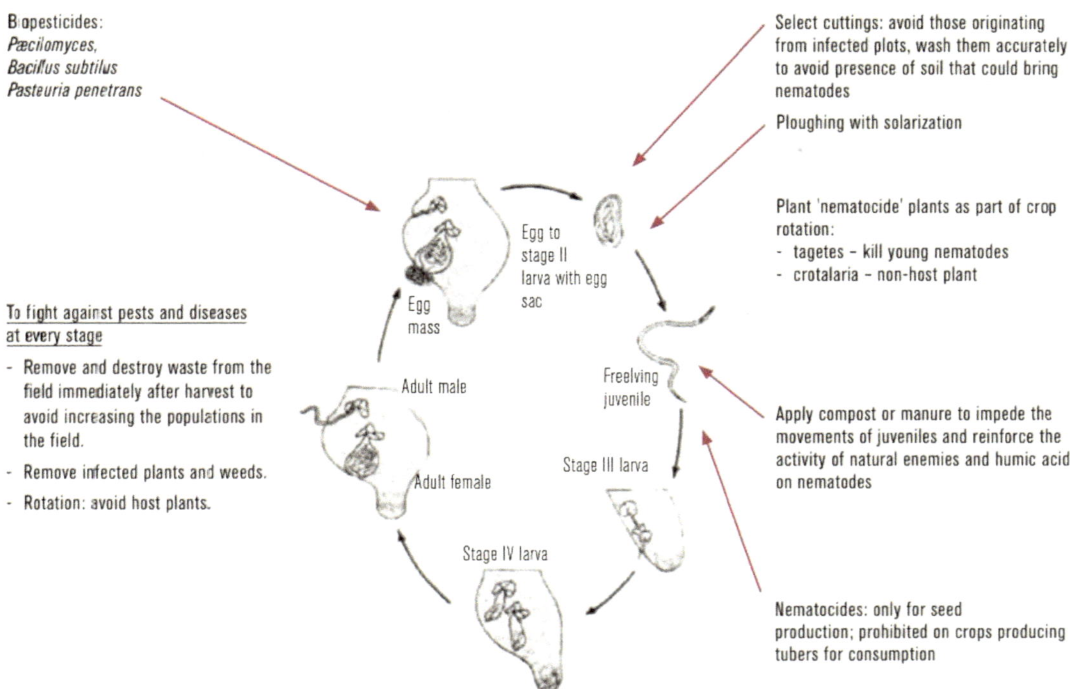

Fig. 4.42 Positioning of control methods in terms of the development cycle of the plant

Table 4.14 Species available as a first crop with nematode-suppressant action

Scientific name	French/English name	Remarks
Tagetes erecta	African marigold	Cover crop as first or associated crop
T. patula	French marigold	Notably the Nemanon variety
T. minuta	Mexican marigold	
Arachis hypogaea	Groundnut	Crop
Cajanus cajan	Pigeon pea	Crop
Calopogonium sp.	Calopogon sp.	Leguminous cover crop
Crotalaria juncea	Crotalaria cover crop	Strong action against *Pratylenchus coffeae* as first crop or associated crops
Macroptilium atropurpureum	Siratro	Leguminous cover crop, antagonist of *Meloidogyne* sp.
Mucuna atterrima	Mucuna noire/black mucuna	Cover crop
Panicum maximum var. *trichoglume*	Guinea or buffalo grass	Cover crop
Sesamia indica	Sesame	Crop
Vigna unguiculata	Cowpea	Crop

Fig. 4.43 Symptoms of lesion nematode infected plants (*left*) and corms (*right*) on taro

(c) *At Planting*
- Rigorous choice of seeds. Eliminate all contaminated material (nematode galls, lesions, or splits).
- Apply nematicides by dipping tubers and as soil treatment when planting for seed production only.

(d) *Throughout the Plant Cycle*
- Eliminate weeds—potential hosts for nematodes.

(e) *After the Harvest*
- Sort and separate the infected tubers from those that appear healthy. Keep them separate.

4.4.2 Lesion Nematode, *Pratylenchus coffeae*

The root lesion nematode *Pratylenchus coffeae* is commonly found associated with the roots of taro. However, only in Japan has this nematode been found to be the cause of a disease.

4.4.2.1 Distribution

Pratylenchus coffeae has been recorded associated with taro in American Samoa, Fiji, and Papua New Guinea and associated with other host plants in Cook Islands, Federated States of Micronesia, Kiribati, Niue, Samoa, Tonga, and Vanuatu.

4.4.2.2 Symptoms

Associated symptoms include root decay, reduced number of cormels, stunting, and death in taro (Fig. 4.43). The nematode (Fig. 4.43) cause localized necrosis (darkened areas of dead tissue) on roots and corms. It is unlikely that symptoms will be seen on plant, except for localized necrosis of feeder roots.

Plants tend to be slow growing and pale green due to a lack of water and nutrients. Yields decrease as the amount of root parasitism increases throughout the growing season. Initial root lesions are small and water soaked but quickly become black and necrotic (Fig. 4.43). These lesions predispose plants to bacterial and fungal infection.

4.4.2.3 Biology

Adults and larvae enter, move within, and leave taro roots during their life cycle. The female lays eggs within the root, with or without fertilization. Eggs either hatch within the root or in the soil if the root decays before hatching. The first larval stage and molt occurs in the egg. The second stage larvae move around in the soil or root, depending on where they were hatched and go through three more molts to the adult stage. The life cycle of the lesion nematode is relatively slow, taking 45–65 days to complete: it is probably near the minimum in the warm, wet climate of American Samoa.

4.4.2.4 Management

1. *Regulatory Methods*: Strict quarantine measures should be followed when germplasm is moved internationally, in order to prevent further spread of *P. coffeae*.
2. *Cultural Methods*: Necrotic tissues should be removed from the tops (approximately 40 cm of petiole and 1–2 cm of corm) and suckers, which should then be washed with running water and allowed to dry before planting. Care should be taken to ensure that no infected materials or trimmings are taken into the field.

   Crop rotations will reduce populations, but, because of its wide host range, damage from *P. coffeae* may not always be avoided. In Japan, crop rotation is regarded as a better strategy for control of *P. coffeae*.
3. *Chemical Methods*: Nematicides (e.g., dazomet, oxamyl, and fenamiphos) are commonly applied as pre-planting treatments to reduce nematode numbers. Their use on taro in the Pacific Island countries is thought to be limited.

4.4.3 "Miti Miti" Disease, *Hirschmanniella miticausa*

A rot of taro corms, known locally in the Solomon Islands as "miti miti" disease, is caused by *H. miticausa*.

4.4.3.1 Distribution

H. miticausa has been found causing the disease in the Solomon Islands (Bridge 1978; Bridge et al. 1983; Bridge and Page 1984; Gowen 1985; Mortimer et al. 1981; Patel et al. 1984) and Papua New Guinea (southern highlands) (Bridge et al. 1983; Bridge and Page 1984).

4.4.3.2 Symptoms

Symptoms include root and corm infection, sometimes causing plants to wilt. Most nematodes (Fig. 4.44) will be found in corms. The nematodes feed endoparasitically in the corm tissues, producing irregular red or brown necrotic zones which normally precede complete secondary brown soft rot of the basal portions of the corms (Fig.). Corms show irregular zones of dry brown rot, originating from base of corm, initially confined to vascular tissue. Healthy tissue alongside rot is red, giving corms appearance of uncooked fatty meat ("miti miti"). Symptoms are usually only apparent when corms are harvested and cut open to observe unique characteristics (dry rot and miti miti symptoms).

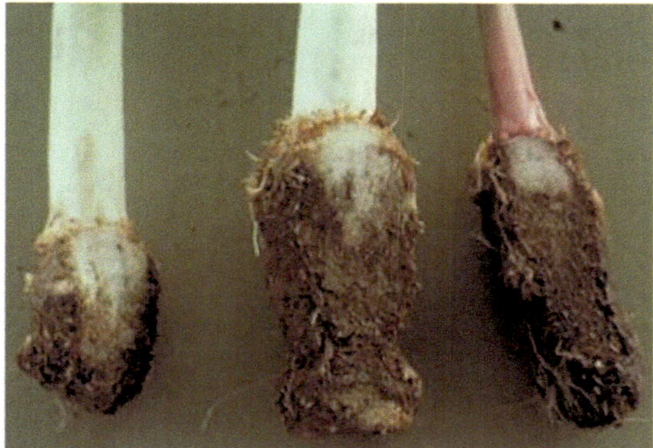

Fig. 4.44 Symptoms of miti miti disease

Often, the basal parts of the corms are completely decayed by secondary brown soft rots. Rots are often not apparent until the taro is harvested, although some chlorosis and wilting of leaves may occur. This is more pronounced in wetland taro where infected plants become stunted and die.

4.4.3.3 Management

1. *Regulatory Methods*: The restricted occurrence of this disease (it is confined to Papua New Guinea and Solomon Islands) means that strict quarantine measures should be followed when germplasm is moved internationally. Guidelines suggest that transfers should only be with sterile, pathogen-indexed plantlets growing in a tissue culture medium.
2. *Cultural Methods*: At a local level, the main source of infection in new land is planting material infested with *H. miticausa*. Therefore, the most effective and practical control measure is to "clean" planting material. Nematodes will be removed if outer leaves and corm are trimmed back to white, healthy tissue, leaving a few centimeters without roots. It is best to avoid planting infected corms or cormels as this increases the risk of spreading nematodes.

Where taro is grown intensively with continuous production, it is not always possible or practical to clean up the planting material or use other hygiene measures. In this case, the use of crop rotation with nonhost plants is the better option. However, little is known about the length of survival of *H. miticausa* in the soil between consecutive crops of taro.

4.5 Future Thrusts

- Multidisciplinary approach shall be adopted to tackle some of the problems limiting the productivity of tuber crops like taro leaf blight.
- Studies conducted on taro leaf blight (TLB) resistance using elicitor from the pathogen *Phytophthora colocasiae* has to be studied in depth and put to use in screening taro lines for TLB resistance. Possibilities of using the glycoprotein for inducing resistance (cross protection) in susceptible varieties also may be studied using a wide array of taro germplasm.
- Popularization of IDM package developed for taro leaf blight developed by the Regional Centre of CTCRI, Bhubaneswar, and recommended by AICRP (Tuber Crops) in the taro growing areas of Eastern and NEH regions.
- Pest problem like taro leaf blight still continues to be serious limiting factor affecting the yield of taro. Priority may be given to develop varieties with resistance/tolerance to the above disease.
- Emerging pests and diseases due to climate change and introduction of invasive pests and pathogens are also main challenges. Appropriate technologies to contain such biotic stresses have to focus on refined eco-friendly methods. The important fungal disease like taro leaf blight is disseminated through planting materials and soil. Efforts to provide healthy planting material by employing diagnostic techniques to identify pathogens are in progress, and this needs to be strengthened.

References

Adegbola MOK (1993) Determination of the most suitable frequency and time of fungicide spraying schedule for the adoption in the control of *Phytophthora* black pod (pod rot) disease of Cocoa. In: Proceedings of the 11th international cocoa research conference, Yamoussoukro, Cote d'Ivoire, pp 18–24

Adejumo TO (1997) Identification, incidence, severity and methods of control of the causal organism of false smut disease of cowpea (*Vignia unguiculata* L Walp. Ph.D. thesis, University of Ibadan, Nigeria, 201 pp

Agyekum E (2004) The effect of plant density and the relative time of planting on Colocasia/Rice intercropping system. MSc thesis, Faculty of Agriculture and Renewable Natural Resources, Kwame Nkrumah University of Science and Technology, Kumasi, Ghana, p 6

Amosa F, Wati P (1997) Effects of taro/maize intercropping systems on the incidence of and severity of taro leaf blight. In: Taro genetic resources: conservation and utilisation, 1995 Annual Research Report. The Institute for Research, Extension and Training in Agriculture (IRETA) and the School of Agriculture (SOA), University of the South Pacific, Alafua Campus, Apia, Samoa, pp 1–2

Anon (1990) Annual report of the Central Tuber Crops Research Institute, Thiruvananthapuram

Anselm RL (2004) The dasheen (*Colocasia esculenta*) beetle (*Ligyrus ebenus*). Division of Agriculture, Plant Protection and Quarantine Unit Bulletin, Botanic Gardens, Roseau

Antonin P, Bailod M, Linder C, Mittaz C (1997) Chemical and biological control of the two spotted spider mite *Tetranychus urticae* Koch in strawberry production. Rev Suisse Vitic Arboric Hortic 29:179–187

Ashok A, Mehrotra RS (1987) Control of *Phytophthora* leaf blight of taro (*Colocasia esculenta*) by fungicides and roguing. Phytoparasitica 15(5):299–305

Ashok B, Saikia UN (1996) Fungicidal management of leaf blight of *Colocasia*. Int J Trop Agric 14(1–4):231–233

Bejarano-Mendoza CA, Zapata M, Bosques A, Rivera-Amador E, Liu LJ (1998) *Sclerotium rolfsii* como componente del complejo patológico causante del mal seco de la yautía (*Xanthosoma sagittifolium*) en Puerto Rico. J Agric Univ P R 82(1–2):85–95

Bergquist RR (1974) Effect of fungicide rate, spray interval, timing of spray application, and precipitation in relation to control of *Phytophthora* leaf blight of taro. Ann Bot 38(154):213–221

Bhagavan BVK, Naga Lakshni R, Prasadji JK, Babu Ratan P, Madhava Rao D (2013) Studies on the reduction of storage rots in *Colocasia* seed cormels. In: Misra RS, Neduchezhiyan M (eds) Aroids: opportunities and challenges. Allied Publishers Pvt Ltd, New Delhi, pp 282–287

Bridge J (1978) Nematodes. In: Pest control in tropical root crops. PANS Manual No. 4, Centre for Overseas Pest Research, Ministry of Overseas Development, London, pp 163–168, 192–193

Bridge J, Page SLJ (1984) Plant nematode pests of crops in Papua New Guinea. J Plant Prot Tropics 1:99–109

Bridge J, Mortimer JJ, Jackson GVH (1983) *Hirschmanniella miticausa* n sp (Nematoda: Pratylenchidae) and its pathogenicity on taro (*Colcasica esculenta*). Rev Nematol 6:285–290

Coffey MD (1991) Strategies for integrated control of soil borne *Phytophthora* species. In: Lucas JA, Shattock RC, Shaw DS, Cooke LR (eds) Phytophthora. Cambridge University Press, Cambridge, UK, 447 pp

Crop Protection Reference (1995) 11ed, C & P Press, New York

Fatuesi S, Tauili'ili P, Taotua F, Vargo A (1991) Cultural methods of pest control on taro (*Colocasia esculenta* Schott) in American Samoa. Micronesica Suppl 3:123–127

Frais T, Garcia ER (1981) Effectiveness of some micro-organisms antagonistic to *Phytophthora palmivora* (Butl) in controlling black pod rot of cocoa. Rev Mexicana Fitopatol 1(3):16–20

Giacometti DC, León J (1994) Tannia. Yautia (*Xanthosoma sagittifolium*). In: Hernaldo JE, León J (eds) Neglected crops: 1492 from a different perspective, Plant production and protection series no. 26. FAO, Rome, pp 253–258

Gollifer DE, Jackson GVH, Newhook JF (1980) Survival of inoculum of the leaf blight fungus *Phytophthora colocasiae* infecting taro, *Colocasia esculenta* in the Solomon Islands. Ann Appl Biol 94:379–390

Gowen SR (1985) Report on nematode survey in the Solomon Islands and Vanuatu. Report to UK Overseas Development Administration, London, January 1985

Grandison GS (1996) Plant-parasitic nematodes of American Samoa. South Pacific Commission, Noumea

Hunter D, Brunt J, Delp C (2001) AusAID/SPC Taro genetic resources: conservation and utilization. A bibliography of taro leaf blight. Secretariat of the Pacific Community, Noumea, pp 2–15

Jackson GVH (1978) Alomae and Bobone diseases of taro, Advisory leaflet no. 8. South Pacific Commission, Noumea

Jackson GVH (1996) Strategies for taro leaf blight research in the region. In: Proceedings of the taro leaf blight seminar. Alafua, Western Samoa, 22–26 November, 1993. South Pacific Commission, New Caledonia, Noumea, pp 95–100

Jackson G (1999). Taro leaf blight. Pest Advisory Leaflet No. 3, Plant Protection Service, Secretariat of the Pacific Community, Suva, Fiji Islands, 2 pp

Jackson GVH, Gollifer DE, Newhook FJ (1980) Studies on the taro leaf blight fungus *Phytophthora colocasiae* in Solomon Islands: control by fungicides and spacing. Ann Appl Biol 96:1–10

Jatala P, Bridge J (1993) Nematode parasites of root and tuber crops. In: Luc M, Sikora RA, Bridge J (eds) Plant parasitic nematodes in subtropical and tropical agriculture. CAB International, Wallingford, pp 137–180

Joshua SA (2010) Identification of the major foliar fungal disease of *Colocasia esculenta* (L) Schott and its management in the Kumasi Metropolis. MSc (Plant Pathology) Thesis, Department of Crop and Soil Sciences, Kwame Nkrumah University of Science and Technology, Kumasi, Ghana, 68 pp

Kirby MF (1987) Root-knot nematodes, Advisory leaflet no. 9. South Pacific Commission, Noumea

Martin FN, Hancock JG (1985) Chemical factors in soils suppressive to *Pythium ultimum*. In: Ecology and management of soil-borne plant pathogens, CA, pp 113–116

McGregor AJ, Moxon JE (1985) Potential for biological control of tent building ants associated with *Phytophthora palmivora* (Butl) pod rot of cocoa in Papua New Guinea. Ann Appl Biol 107:271–277

Mohandas C, Palaniswami MS (1990) Effect of *Meloidogyne incognita* on three popular cultivars of *Colocasia esculenta* (Abstr). National symposium on recent advances in the production and utilization of tropical tuber crops, ISRC, Central Tuber Crops Res Inst, Thiruvananthapuram

Mortimer JJ, Bridge J, Jackson GVH (1981) *Hirschmanniella* sp, an endoparasitic nematode associated with miti miti disease of taro corms in the Solomon Islands. Plant Prot Bull FAO 29:9–11

National Agriculture Marketing and Development Corporation (2003) The dasheen (*Colocasia esculenta*) beetle (*Ligyrus ebenus*). A guide to identification and control. Ministry of Agriculture, Land and Marine Resources, Trinidad

Nirula KK (1959) Root-knot nematode on *Colocasia*. Curr Sci 28:125–126

Nzietchueng S (1984) Root rot of *Xanthosoma sagittifolium* caused by *Pythium myriotylum* in Cameroon. In: ER Terry, EV Doku, OB Arene and NM Mahungu (eds) Tropical root crops: production and uses in Africa. Proceedings of the second triennial symposium of the international society for tropical root crops. Douala, Cameroon, pp 185–188

Odigie EE, Ikotun T (1982) *In-vitro* and *in-vivo* inhibition of growth of *Phytophthora palmivora* (Butl.) by antagonistic microorganisms. Fitopatol Bras 7(2):157–169

Onwueme IC, Charles WB (1994) Tropical root and tuber crops: Production, perspectives and future prospects, FAO plant production and protection papers (no. 126). FAO, Rome, pp 153–154

Ooka J, Brennan BM (2000). Crop profile for taro in Hawaii. College of Tropical Agriculture and Human Resources, University of Hawaii-Manoa, 2 pp

Pacumbaba RP, Wutoh JG, Sama AE, Tambong JT, Nyochembeng LM (1992) Isolation and pathogenicity of rhizosphere fungi of cocoyam in relation to cocoyam root rot disease. J Phytopathol 135:265–273

Patel NZ, Saelea J, Jackson GVH (1984) Breeding strategies for controlling diseases of taro in Solomon Islands. In: Proceedings of the sixth symposium of the international society for tropical root crops, Lima, Peru, 21–26 February 1983

Rajasekhara Rao K (2013) Corm-borer *Aposonyx chalybaeus* (Hope) – a serious pest of taro in eastern Himalayan region of India. In: Misra RS, Neduchezhiyan M (eds) Aroids: opportunities and challenges. Allied Publishers Pvt Ltd, New Delhi, pp 229–232

Saborío F, Umaña G, Solano W, Ureña G, Muñoz G, Hidalgo N, Brenes A (2004a) Mejoramiento genético del tiquisque (*Xanthosoma sagittifolium*) contra el Mal Seco. Memoria REDBIO 2004. Talleres. www.redbio.org. 21 Sept 2005

Saborío F, Umaña G, Solano W, Amador P, Muñoz G, Valerin A, Torres A, Valverde R (2004b) Induction of genetic variation in *Xanthosoma* spp. In: International Atomic Energy Agency (eds) Genetic improvement of under-utilized and neglected crops in low income food deficit countries through irradiation and related techniques. Vienna, Austria, pp 143–154

Sarkar PK, Tarafdar J (2013) Sustainable management of red spider mite (*Tetranychus urticae*) in swamp taro (*Colocasia esculenta* var. *stoloniferum*) in Gangetic Plains of West Bengal. In: Misra RS, Neduchezhiyan M (eds) Aroids: Opportunities and challenges. Allied Publishers Pvt Ltd, New Delhi, pp 219–228

Sriram S, Misra RS (2007) Biological control of taro (*Colocasia esculenta* (L) Schott) leaf blight caused by *Phytophthora colocasiae* (Racib) and storage losses with rhizobacteria. J Biol Control 22(2):181–188

Sriram S, Misra RS, Sahu AK, Maheswari SK (2003) Rhizobacteria: Potential biological control agent against taro leaf blight pathogen, *Phytophthora colocasiae* (Racib). J Root Crops 29(1):50–53

Srivastava AS, Verma RS, Pandey RC (1971) Comparative efficacy of different nematicides tested against root-knot nematode infesting *Colocasia antiquorum*. Labdev J Sci Technol 9B:142–144

Tambong JT, Sapra VT, Garton S (1998) *In vitro* induction of tetraploids in colchicine-treated cocoyam plantlets. Euphytica 104:191–197

Tambong JT, Poppe J, Höfte M (1999) Pathogenicity, electrophoretic characterization and in planta detection of the cocoyam root rot disease pathogen, *Pythium myriotylum*. Eur J Plant Pathol 105:597–607

Tilialo R, Greenough D, Trujillo EE (1996) The relationship between balanced nutrition and disease susceptibility in Polynesian taro. In: Proceedings of a workshop on mineral nutrient disorders of root crops in the Pacific, Nuku'alofa, Kingdom of Tonga, 17–20 April 1995. ACIAR Proceedings No. 65, pp 105–109

Torres-Portuguez SE (1996) Estudio preliminar de la utilización del complejo orgánico sobre el "mal seco" en cultivo de tiquisque blanco (*Xanthosoma sagittifolium*). In: Congreso Nacional Agronómico y de Recursos Naturales. San José, Costa Rica, Memoria 2:111

Ugwuanyi JO, Obeta JAN (1996) Fungi associated with storage rots of cocoyams (*Colocasia* spp.) in Nsukka, Nigeria. Mycopathologia 134:21–25

Vargo A (1990) A rapid rural appraisal of taro production systems in Micronesia, Hawai'i and American Samoa. Pacific Agriculture Development Office, University of Hawai'i at Manoa, Honolulu

Waterhouse DF, Norris KR (1987) Biological control: pacific prospects. Inkata Press, Melbourne

Watt G (1989) Dictionary of the economic plants of India. Supdt Goot Print Calcutta 2:509–513

Yams, *Dioscorea* spp. 5

Abstract

Distribution, economic importance, symptoms/damage, host range, biology, survival and spread, and management of insect and mite pests, fungal, bacterial, viral, and nematode diseases of yams (*Dioscorea* spp.) using regulatory, physical, cultural, chemical, biological, host resistance, and integrated pest management methods are discussed.

Keywords

Yams • *Dioscorea* spp. • Pests • Diseases • Nematodes

5.1 Introduction

Yam is the common name for some plant species in the genus *Dioscorea* (family Dioscoreaceae) that form edible tubers. These are perennial herbaceous vines cultivated for the consumption of their starchy tubers in Africa, Asia, Latin America, the Caribbean, and Oceania. There are many cultivars of yam.

Yams are monocots, related to lilies and grasses. Native to Africa and Asia, yam tubers vary in size from that of a small potato to over 60 kg. There are over 600 varieties of yams and 95 % of these crops are grown in Africa.

Yam tubers can grow up to 1.5 m in length and weigh up to 70 kg and 7.5–15.0 cm high (Fig. 5.1). The vegetable has a rough skin which is difficult to peel, but which softens after heating. The skins vary in color from dark brown to light pink. The majority of the vegetable is composed of a much softer substance known as the "meat." This substance ranges in color from white or yellow to purple or pink in mature yams.

The true yam is a versatile vegetable. It can be barbecued, roasted, fried, grilled, boiled, baked, and smoked and when grated it is processed into a dessert recipe.

5.1.1 Major Cultivated Species

There are many cultivars of yam throughout the humid tropics. The most economically important are discussed below (Calverly 1998).

5.1.1.1 *Dioscorea rotundata* and *D. cayenensis*

Dioscorea rotundata, the "white yam," and *D. cayenensis*, the "yellow yam," are native to Africa. They are the most important cultivated yams. In the

P.P. Reddy, *Plant Protection in Tropical Root and Tuber Crops*,
DOI 10.1007/978-81-322-2389-4_5

Fig. 5.1 Yam vine and tubers

past they were considered two separate species but most taxonomists now regard them as the same species. There are over 200 cultivated varieties between them.

White yam's tuber is roughly cylindrical in shape, and the skin is smooth and brown and the flesh usually white and firm (Fig. 5.2). Yellow yam is named after its yellow flesh, a color caused by the presence of carotenoids. It looks similar to the white yam in outer appearance; its tuber skin is usually a bit firmer and less extensively grooved. The yellow yam has a longer period of vegetation and a shorter dormancy than white yam.

The Kokoro variety is important in making dried yam chips (Dumont and Vernier 2000).

They are large plants; the vines can be as long as 10–12 m. The tubers most often weigh about 2.5–5.0 kg each but can weigh as much as 25 kg. After 7–12 months of growth, the tubers are harvested. In Africa most are pounded into a paste to make the traditional dish of "pounded yam" (Kay 1987).

Fig. 5.2 White yam tuber

5.1.1.2 *Dioscorea alata*

Dioscorea alata, called "water yam," "winged yam," and "purple yam," was first cultivated in Southeast Asia. Although not grown in the same quantities as the African yams, it has the largest distribution worldwide of any cultivated yam, being grown in Asia, the Pacific Islands, Africa, and the West Indies (Mignouna et al. 2003). Even in Africa, the popularity of water yam is second only to white yam. The tuber shape is generally

Fig. 5.3 Purple yam freshly harvested and sliced for cross-sectional view

Fig. 5.4 Segment of a Japanese mountain yam

cylindrical but can vary (Fig. 5.3). Tuber flesh is white and watery in texture.

In the Philippines it is used as an ingredient in many sweet desserts. In Vietnam, it is used mainly as an ingredient for soup.

Dioscorea alata was brought to Hawaii by the early Polynesian settlers and became a major crop in the nineteenth century when the tubers were sold to visiting ships as an easily stored food supply for their voyages (White 2003).

5.1.1.3 *Dioscorea opposita*

Dioscorea opposita, "Chinese yam," is native to China. The Chinese yam plant is somewhat smaller than the African, with the vines about 3 m long. It is tolerant to frost and can be grown in much cooler conditions than other yams. It is now grown in China, Korea, and Japan.

It was introduced to Europe in the nineteenth century when the potato crop there was falling victim to disease and is still grown in France for the Asian food market.

The tubers are harvested after about 6 months of growth (Fig. 5.4). Some are eaten right after harvesting and some are used as ingredients for other dishes, including noodles, and for traditional medicines (Kay 1987).

5.1.1.4 *Dioscorea bulbifera*

Dioscorea bulbifera, the "air potato," is found in both Africa and Asia, with slight differences between those found in each place. It is a large vine, 6 m or more in length. It produces tubers; however, the bulbils which grow at the base of its leaves are the more important food product. They are about the size of potatoes (hence the name "air potato"), weighing from 0.5 to 2.0 kg (Fig. 5.5).

Fig. 5.5 Air potato

Some varieties can be eaten raw, while some require soaking or boiling for detoxification before eating. It is not grown much commercially since the flavor of other yams is preferred by most people. However, it is popular in home vegetable gardens because it produces a crop after only 4 months of growth and continues producing for the life of the vine, as long as 2 years. Also the bulbils are easy to harvest and cook (Kay 1987).

5.1.1.5 *Dioscorea esculenta*

Dioscorea esculenta, the lesser yam, was one of the first yam species cultivated. It is native to Southeast Asia and is the third most commonly

Fig. 5.6 Lesser yam tuber

cultivated species there, although it is cultivated very little in other parts of the world. Its vines seldom reach more than 3 m in length and the tubers are fairly small in most varieties (Fig. 5.6).

The tubers are eaten baked, boiled, or fried much like potatoes. Because of the small size of the tubers, mechanical cultivation is possible, which, along with its easy preparation and good flavor, could help the lesser yam to become more popular in the future (Kay 1987).

5.1.1.6 *Dioscorea trifida*

Dioscorea trifida, the "cush-cush yam," is native to the Guyana region of South America and is the most important cultivated New World yam. Since they originated in tropical rain forest conditions, their growth cycle is less related to seasonal changes than other yams. Because of their relative ease of cultivation and their good flavor, they are considered to have a great potential for increased production (Kay 1987).

5.1.2 Production

Yam crop begins when whole seed tubers or tuber portions are planted into mounds or ridges, at the beginning of the rainy season. The crop yield depends on how and where the sets are planted, sizes of mounds, interplant spacing, provision of stakes for the resultant plants, yam species, and tuber sizes desired at harvest. Small-scale farmers in West and Central Africa often intercrop yams with cereals and vegetables. The seed yams are perishable and bulky to transport. Farmers who do not buy new seed yams usually set aside up to 30 % of their harvest for planting the next year. Yam crops face pressure from a range of insect pests, fungal and viral diseases, as well as nematodes (IITA 2010).

Yam typically grows for 6–10 months and is dormant for 2–4 months, depending on the species. The growth and dormant phases correspond, respectively, to the wet season and the dry season. For maximum yield the yam requires a humid tropical environment, with an annual rainfall of over 1,500 mm distributed uniformly throughout the growing season. White, yellow, and water yams typically produce a single large tuber per year, generally weighing 5–10 kg (Calverly 1998).

Yam production requires high labor. The crop has low yield per hectare compared to crops such as cassava (manioc) or sweet potato. It is not an efficient food staple given the relatively large amount of planting material that is required and its long growing season. Yam's labor requirement exceeds that of other comparable crops. Yam is also difficult to preserve and store over extended periods of time. The cost per 1,000 cal of yam is four times greater than those of other root and tuber crops (Oke 1990).

For these reasons and problems of storing harvested yam, the costs of yam production are high and yam crops are slowly losing ground to cassava and other food staples.

Despite these high costs and low nutrient density, when compared to other tubers and roots, low-technology yam farming produces the highest amount of food calorie and protein annually per hectare per season, on average. Given this nutritional value of yam and its high cultural acceptance in certain parts of Africa, there is an interest in developing knowledge that can improve yam agriculture.

In 2010, yams are farmed on about five million hectares in about 47 countries in tropical and subtropical regions of the world with a production of 48.7 million tons, 95 % of which was produced in Africa. The biggest yam harvest, globally, was in 2008, when the world produced 54 million metric tons of yam (IITA 2010). The top 10 yam-producing countries are presented in Table 5.1.

The world average annual yield of yams was 10.2 t per hectare in 2010. The most productive

Table 5.1 Top 10 yam producers (2012)

Rank	Country	Production (tons)
1	Nigeria	38,000,000
2	Ghana	6,638,867
3	Ivory Coast	5,674,696
4	Benin	2,739,088
5	Togo	864,408
6	Cameroon	520,000
7	Central African Republic	460,000
8	Chad	420,000
9	Papua New Guinea	345,000
10	Colombia	361,034

Source: FAOSTAT (2012)

yam farms in the world were in Colombia, where nationwide average annual yield was 28.3 t per hectare (FAO 2011). These are average national yields; certain farms report yields significantly above 30 t per hectare, for example, for yellow yam; other farms report yields less than 1 t per hectare (Linus Opara 2003).

Despite the high labor requirements and production costs, consumer demand for yam is very high in certain subregions of Africa, making yam cultivation quite profitable to certain farmers (IITA 2010).

There are approximately 600 species of *Dioscorea* in the world, mostly wild, mainly in a tropical environment. Only a handful of them have been domesticated and are regularly cultivated. Of those species cultivated in ACP countries, the most common are listed in the Tables 5.2 and 5.3 below.

In India, greater yam and lesser yam are under cultivation. They are grown in all the states but the major yam-producing states are Kerala, West Bengal, Bihar, Orissa, Tamil Nadu, Assam, Rajasthan, Gujarat, and Maharashtra. It is mostly cultivated in Southern India in homestead gardens. In some pockets of Tamil Nadu, Andhra Pradesh, Gujarat, and Orissa, it is cultivated in large areas (Table 5.4). There is an urgent need to introduce high-yielding varieties in yams and improve the production.

Prain and Burkill (1936) reported the occurrence of about 50 different *Dioscorea* species in India, largely in the west, east, and northeast regions. Many of the *Dioscorea* species serve as a

Table 5.2 Cultivated species of yam

Yam species	Main areas of cultivation	Most common commercial names
D. rotundata	West Africa, Latin America, the Caribbean	White yam, puna (Ghana)
D. alata	Oceania, Southeast Asia, Africa, the Caribbean, Latin America	Water or greater yam, cush-cush (GB)
D. cayenensis	West Africa, Latin America, the Caribbean	Yellow yam
D. trifida	Guyana, Brazil, Central America, the Caribbean	Cush-cush, yampi, Indian yam
D. esculenta	Southeast Asia	Lesser yam, sweet yam
D. opposita	China, Temperate Asia, France (Blois)	Chinese yam, French yam

Table 5.3 Major food yam species in different regions of world

	Africa	Asia	America
Major economic spp.	*D. rotundata*	*D. alata*	*D. trifida*
	D. cayenensis	*D. esculenta*	

"life-saving" plant group for the marginal farming and forest-dwelling communities, during periods of food scarcity (Arora and Pandey 1996). Henry et al. (1989) also reported that the *Dioscorea* spp. are widely distributed in Tamil Nadu.

Most of the species are ethnically important as a secondary staple and a delicacy for tribal people in the interior areas in Western Ghats. Considerable variability is noticed in edible wild yams (*D. pentaphylla*, *D. oppositifolia*). The *Dioscorea* species, viz., *D. wightii*, *D. spicata*, and *D. belophylla*, are endemic to these areas. Balakrishnan et al. (2003) studied the distribution of wild species and their conservation by Kattunaikka, a predominant tribe in Western Ghats. The study revealed that about 21 different types of *Dioscorea* are present in Wayanad district of Kerala which includes a part of the mountain chain of the Nilgiris, Silent Valley, and Kodagu region in the Western Ghats, a biodiversity hot spot. Among the varieties known to Kattunaikka tribe, *D. hamiltonii*, *D. belophylla*,

Table 5.4 Yam-growing states, regions, and production systems in India

State	Production system	Area/region
Kerala	Upland rain fed, sandy soil	Kollam, Alappuzha districts
	Upland rain fed, laterite soil	Kollam, Pathanamthitta districts
	Lowland irrigated	Karavoor region in Kollam district
Tamil Nadu	Plains irrigated	Virudhunagar district
Andhra Pradesh	Plains irrigated	East Godavari district
Odisha	Plains rain fed/ irrigated	Ganjam, Khurda districts

and *D. oppositifolia* are predominantly seen in the interior evergreen and moist deciduous forests. *D. intermedia* is seen in dry deciduous forests while *D. wightii* in rocky grasslands; *D. pentaphylla*, *D. wallichii*, and *D. bulbifera* on wayside bushes; and *D. pubera* in marshy areas. Among the various species, *D. hamiltonii*, *D. belophylla*, *D. oppositifolia*, and *D. pentaphylla* var. *pentaphylla* are the most acceptable ones for culinary purpose, but other than *D. pentaphylla* all are very rare. The tribals collect *Dioscorea* from almost all landscapes but most frequently from the forests and other such unmanaged habitats.

5.2 Insect, Mite, and Vertebrate Pests

The major insects known to attack yam are beetles, termites, and scale insects. They feed on the yam tuber, reducing tuber quality and quantity and leaving holes that serve as entry point for pathogens. Wounded or damaged tubers are susceptible to decay (Adeniji 1970b). The greater yam beetle, *Heteroligus meles*, is widespread in tropical Africa, while the lesser yam beetle, *H. appius*, occurs in southern Nigeria (Coursey 1967). The yam scale insect, *Aspidiella hartii*, infests tubers, and sometimes foliage, causing poor growth (Akinlosotu and Kogbe 1986). Stored tubers are particularly susceptible to attack and large numbers of scale insects cause shriveling. Scale insects are widespread in Africa, Asia, Central America, the Pacific Islands and the West Indies (CABI 1966).

Several other insect pests affect yam either in storage or in field depending on locality (Osagie 1992; Bridge et al. 2005). These insects include larvae of three Lepidoptera spp., viz., *Euzopherodes vapidella* (Sauphanor and Ratnadass 1985), *Decadarchis minusculata* (Plumbley and Rees 1983), and *Dasyses rugosella* (Dina 1977); a Coleoptera *Araecerus fasciculatus* (Plumbley and Rees 1983); a Coccidae *Aspidiella* spp.; and a Pseudococcidae *Phenacoccus* spp. (Nwankiti et al. 1988). Losses due to insect attacks can be estimated at 50 % after several months of storage in Ivory Coast (Osagie 1992).

5.2.1 Scale, *Aspidiella hartii*

5.2.1.1 Distribution

It is widespread in distribution and present in the Old World tropics and Pacific.

5.2.1.2 Damage

The insects are white in color and form a hard scale covering over the tubers during storage. Infestation by scales leads to poor germination and affects cooking quality. Adult and nymphal scale insects (whitish yellow in color) are present on the tubers (Fig. 5.7) and sometimes on the aerial portions of the plant which do not thrive. Stored tubers may be thickly covered by scales and shrivel. The problem is perpetuated by the use of infested tubers as planting material.

This species sometimes grows until it totally covers tubers during storage in the form of small and more or less whitish scales. This does not cause a loss of volume but may cause germination to be delayed or even stopped.

5.2.1.3 Management

1. *Cultural Methods*: Scale insects on tubers may be controlled by using clean planting material. Yams should not be planted after ginger, as this crop also supports the pest.

Fig. 5.7 Yam tubers infected with scale insects

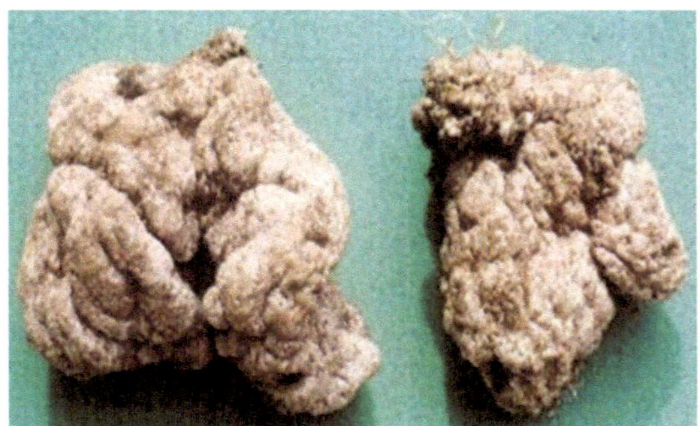

Fig. 5.8 Dipping of yam tubers in 0.05 % monocrotophos for 10 min for control of yam scale (*left*, treated; *right*, control)

Fig. 5.9 Aphid on yam

Scale insects are controlled by dipping the tubers in 2 % petroleum ether extracts of cassava or yam bean seed oil with 0.01 % surfactant for 10 min and shade drying.

Storing the tubers in sand resulted in 92.3 % control on the incidence of scale insects.

2. *Chemical Methods*: Dipping the tubers in 0.3 % aldrin or in a mixture of 5.6 ml diazinon, 20 % EC, and 425 ml white oil in 13.5 L of water or in Monocrotophos (0.05 %) for 10 min before planting has been recommended (Fig. 5.8). Fumigation of tubers with carbon disulfide has also been suggested.
3. *Biological Methods*
   Parasitoids: The nymphs of white scale are often parasitized by an aphelinid, *Physcus comperi*, and an encyrtid, *Adelencyrtus moderatus*. Yam scale insects are frequently checked by two parasitoids *Coccobius (Physua) comperi* and *Adelencyrtus moderatus*. Both are solitary endoparasitoids; the former was more abundant in the field and storage.

5.2.2 Aphid, *Aphis gossypii*

5.2.2.1 Damage

Aphids cause damage to young seedlings. Nymph and adults suck the sap from leaves (Fig. 5.9) and stems of young vines. As a result, new growth is adversely affected. Yellowish speckling and curling of leaves can be seen on infested plant. In severe cases sooty mold develops, reducing vitality of the leaves.

5.2.2.2 Management

1. *Cultural Methods*: Spray neem oil 0.5 % or NSKE 5 %.
2. *Chemical Methods*: Spray Monocrotophos at 1.5 ml + wettable sulfur at 3 g + Teepol at 0.5 ml/l of water.
3. *Biological Methods*: Conserve natural enemies like parasitoids and predators. *Aphelinus mali*, *Syrphophagus aphidivorus*, *Cheilomenes sexmaculata*, *Ischiodon scutellaris*, and *Verticillium* sp. are effective biocontrol agents.

Fig. 5.10 Coffee bean weevil

5.2.3 Coffee Bean Weevil, *Araecerus fasciculatus*

5.2.3.1 Damage

Araecerus fasciculatus damages tubers of yams and aroids. Infested tubers are unfit for consumption as well as for seed purpose. The beetle lays eggs on tubers and grubs bore inside forming zigzag galleries, feed the contents, and convert them into powder. Pupation is inside the tuber and adult emerges by making holes on the tuber surface. The infestation causes weight loss and quality loss.

The damage is caused by the hairy yellowish larvae measuring up to 6 mm in length at maturity. The brownish-red adult (Fig. 5.10) is a good flyer. Its attacks are generally confined to areas around injuries caused by harvest or caterpillars.

5.2.3.2 Management

Storing uninfested tubers in white sand or saw dust prevents infestation. Seed tubers can be stored in a mixture of sand and carbaryl dust (10 %) in the ratio 100:1.

5.2.4 Yam Beetle, *Heteroligus meles*

5.2.4.1 Distribution and Economic Importance

Heteroligus spp. are the commonest and most widely distributed. In the old Asaba province of Delta state, Nigeria, yam beetle damage was estimated at 20 %. More recent studies showed that yam beetle damage ranged from 31 to 51 % resulting in 23–60 % loss in tuber yield.

5.2.4.2 Damage

A major pest in West Africa, adults migrate by flying to other farms/plots; they feed on the tuber, thereby leaving holes on them. Tubers lose appearance and quality and become prone to rotting during storage. The yam beetle is a pest that starts as a curled grub that changes into a beetle. Yam beetles may spoil the tubers by eating and making holes into them. Then the tuber rots. Adult beetles eat the planting sets and plants may wilt and die. The holes in tubers reduce market value. High infestation causes extensive tuber damage which often results in total economic loss as it predisposes the tubers to rot organisms (Onwueme and Sinhad 1991).

Heteroligus is a blackish-brown insect reaching over 30 mm in size. Damage in the form of big holes in tubers a few centimeters deep is caused by adults in the fields, from germination until harvest. Larvae (*H. meles*) develop on the roots of other plants (grasses) in wet areas near yam fields or directly on the yam roots (*H. appius*).

5.2.4.3 Management

Dusting of holes or soil surface with aldrin, endosulfan, or grammalin gives effective control of beetles.

A trial was conducted to evaluate the effectiveness of aldrin, carbofuran, and wood ash for the control of *H. meles.* Tuber yield was highest with carbofuran-treated sets (19.67 t ha^{-1}) and the beetle-inflicted holes were least (2) (Table 5.5) (Law-Ogbomo and Egharevba 2006).

If the yam beetle is infesting the tubers, wash the planting pieces in lead arsenate and Bordeaux mixture. This will stop the grubs from eating into the tubers and spoiling them. Follow crop rotation with nonhost crop.

Farmers are advised to plan their yam planting in such a way to avoid the tuberization period coinciding with the peak population density of the yam beetle in July and August since

Table 5.5 Effects of carbofuran, aldrin dust, and wood ash on yield and beetle damage

Treatment	Yield (tons/ha)*	No. of holes on tuber*
Carbofuran	19.70^{a}	2.5^{c}
Aldrin dust	14.47^{bc}	11.5^{ab}
Wood ash	16.20^{b}	5.0^{b}
Control	14.31^{c}	14.5^{a}

*Means followed by common letter(s) are not significantly different at 5 % probability level

more attack is likely to occur during this period. Very late planting can reduce infestation but affects yields.

5.2.5 Mealybugs, *Ferrisia virgata*, *Planococcus citri*, *P. dioscoreae*, *Pseudococcus citriculus*, *Rhizoecus* sp., *Geococcus coffeae*, and *Phenacoccus gossypii*

5.2.5.1 Damage

Tuber surface is covered with nymphs and adults with its powdery meal. They also infest the tubers of *Amorphophallus*, *Colocasia*, and *Xanthosoma*. Infestation by mealybugs leads to poor germination and affects cooking quality.

These Homopterae develop mainly during storage, forming a sort of white powder near the top of the tubers. They can cause complete necrosis of sprouts preventing the use of tubers as seedlings. Mealybugs, *Geococcus coffeae*, *Phenacoccus gossypii*, *Planococcus citri*, and (in Papua New Guinea) *P. dioscoreae*, feed on the tubers and roots of yams and can multiply to considerable numbers, causing shriveling of the stored tubers.

Yam mealybugs desaps the leaf and the high incidence of this pest causes drying of leaves and withering.

5.2.5.2 Management

Only clean and healthy material should be planted (again the dusting treatment recommended for yam beetles should be used), and, if aerial parts of the plant are affected, spraying with malathion or malathion plus an oil emulsion (e.g., Triona or Albolineum) is recommended.

Storing uninfected tubers in white sand or saw dust prevents infestation by scales and mealybugs. Seed tubers can be stored in a mixture of sand and carbaryl dust (10 %) in the ratio 100:1. If pest infestation is severe, dimethoate (0.05 %) can be used in checking further spread. Malathion, chlorpyrifos, and methyl parathion at 0.5 % concentration completely controlled adult mealybug (*Rhizoecus amorphophalli*) 2 days after treatment.

Treatment with cassava-based biopesticides, cassava seed extract, and also 2 % cassava or yam bean seed oil provides proper control of the pest.

The predators *Scymnus* spp., *Allograpta javana*, *Leucospis* sp., *Geitonia* sp., *Triommata coccidivora*, and *Mallada* sp. have been recorded.

The parasitoids *Aenasius advena*, *Anagyrus qadrii*, and *Blepyrus insularis* parasitize 0.4, 1.4, and 25.1 %, respectively, of the *F. virgata* collected. *Acerophagus* sp. is also found to be parasitic on *F. virgata*.

To traditionally combat mealybugs, the following methods are used. Infested tubers are immediately isolated at harvest to avoid contamination or are watered every morning until the day of planting or sprinkled with the ashes of the roots of *Nauclea latifolia*.

5.2.6 Defoliating Caterpillars, *Loxura atymnus*, *Theretra nessus*, and *Tagiades gana*

5.2.6.1 Damage

Caterpillars first eat the leaf blade and then attack the stems, sometimes causing severe damage only on the species *D. alata* (Fig. 5.11).

5.2.6.2 Management

Rotate yam species and varieties on a single plot of land. Manually eliminate the first damaged leaves. Spray a contact insecticide for caterpillars.

5.2.7 Cricket, *Gybnogryllus lucens* (Orthoptera: Gryllidae)

5.2.7.1 Distribution

The pest is distributed in West Africa.

Fig. 5.11 Defoliating caterpillar

Fig. 5.12 Adult yam weevil

5.2.7.2 Damage

Tubers and vines are eaten by this insect. Young plants may die as a result of the attack.

5.2.7.3 Management

Soil insecticides as used for *Heteroligus meles* will control *Gybnogryllus lucens*.

5.2.8 Yam Weevil, *Palaeopus costicollis* (Coleoptera: Curculionidae)

5.2.8.1 Distribution

The pest is distributed in Cuba and Jamaica.

5.2.8.2 Damage

Damage in the form of big holes in tubers a few centimeters deep is caused by adults in the fields, from germination until harvest. Larvae develop on the roots of other plants (grasses) in wet areas near yam fields or directly on the yam roots.

Tuber sets used for planting may fail to develop if heavily attacked. The tubers, especially the head ends, are tunneled by small (4 mm long) dull black red-legged weevils (Fig. 5.12) and their legless larvae. The market value of the crop which is subsequently affected is further reduced by localized rotting around the damaged parts of the tubers. This insect has also been recorded on sweet potato.

5.2.8.3 Biology

The studies on biology of the weevil revealed that the life cycle was completed in 6–10 weeks on three yam species.

5.2.8.4 Management

1. *Cultural Methods*: Avoid locating yam fields too close to swampy areas where larvae reproduce. Crop rotation should be practiced and clean planting material should be used for planting.
2. *Chemical Methods*: Treat seedlings with insecticides. The treatments recommended for the control of *Heteroligus meles* should control this weevil.

   Both malathion and pirimifos methyl significantly ($P < 0.05$) reduced oviposition and feeding of *P. costicollis* adults on yam tubers of *D. cayennensis*.
3. *Integrated Methods*: A number of cultural and chemical strategies such as application of the mini sett technology, hot water treatment, and chemical treatment of yam sets at planting, fallowing, intercropping, crop rotation, and harvesting on time were identified for the preharvest aspect of an IPM program for *P. costicollis*.

5.2.9 Yam Moth, *Euzopherodes vapidella*

5.2.9.1 Damage

Pyralid moth worms attack tubers in lofts by mining, especially during the first 4 months of storage.

D. alata tubers are attacked more, with their higher water content favoring the development of insects. Moth larvae (Fig. 5.13) appear later and tend to use old moth mines. They progress rapidly and there may be total destruction of stocks after only 1 month.

5.2.9.2 Biology

The mean fecundity per female was 79.0 ± 1.7 eggs with an average oviposition period of 2.2 days. The eggs measuring 0.48 ± 0.02 mm in length and 0.28 ± 0.03 mm in width were laid singly, in chains and in batches. The percentage hatchability was 75. The duration of development for the different stages of *E. vapidella* was 3.5 ± 0.5 days for the egg, 14.5 ± 0.8 days for the total larval stage, 1.5 ± 0.8 days for the pre pupa, and 7.5 ± 0.5 days for the pupa stage. Total duration of development from egg to adult was 27.0 ± 1.5 days. Unmated females lived longer (6.5 days) than unmated males (5.0 days), mated females (5.0 days), and mated males (3.0 days). The adult females were always bigger than the males. The wingspan of the females ranged from 13.8 to 16.8 mm and in males from 11.0 to 13.8 mm. The body length for females ranged from 7.0 to 9.0 mm and was 6.2–7.5 mm for males (Fig. 5.13). Relative humidity of 37 % and below did not favor *E. vapidella* development with no adult emergence, while 58 % RH and above favor development of the moth with significant adult emergence.

5.2.9.3 Management

The following control measures are recommended:

- Sorting of injured or damaged tubers and storing them separately.
- Cleaning and disinfection of storehouses.
- Treatment of stored tubers with insecticide spray before storage.
- Repeat the treatment one month later on damaged tubers (after sorting).
- Storage of yam tubers at low temperatures (but higher than 12 °C to avoid damage to tubers) retarded the development of yam moths.

Fig. 5.13 Adult yam moth

5.2.10 Chrysomelids, *Crioceris* (= *Lilioceris*) *livida* and *Lema armata*

5.2.10.1 Distribution

The pest is distributed throughout West Africa and east to the Kilimanjaro area.

5.2.10.2 Damage

Larvae are shiny, black, and covered in mucus and may reach 1 cm. Larvae and adults feed on the young leaves and shoots. They are defoliating (Fig. 5.14). These larvae cover themselves with their own excreta. They can cause major damage locally, especially during the first months of the crop, but without any real economic impact.

5.2.10.3 Biology

Pink elongate eggs 2 mm long are laid on the underside of the leaves in groups of about six. As they develop they darken to reddish brown. The larvae are thick and fleshy. Adults are basically reddish brown to black, with light yellow to brown elytra on which are dark brown to black spots.

Fig. 5.14 *Left*, *Crioceris* (= *Lilioceris)* larvae; *right*, *Lema* adult

5.2.10.4 Management

1. *Cultural Methods*: Handpicking and destruction of larvae have been recommended.
2. *Chemical Methods*: Carbaryl, dichlorvos, or endosulfan may be used.
3. *Biological Methods*: *Lema lacordairei* is often attacked by a tachinid parasitoid *Pseudalsomyia* sp. The field parasitism is 7.9 %. Adult female oviposited on young grubs (usually second instars). The parasitoid grows along with the host and completes the egg and larval period within 5–6 days. More than one parasitoid develops from a single host.

5.2.11 Maruca, *Diaprepes abbreviatus* and *D. famelicus*

5.2.11.1 Damage

The larvae cause significant local damage to the tubers (Fig. 5.15). The adults (Fig. 5.15), 20 mm long, eat the leaf blades and sometimes the tubers as well. These polyphagous insects also attack macabo and cassava as well as citrus fruits.

Fig. 5.15 Adult of Maruca

5.2.11.2 Management

The following control measures are recommended:

- Sorting of injured or damaged tubers and storing them separately.
- Cleaning and disinfection of store houses.
- Treatment of stored tubers with insecticide spray before storage.
- Repeat the treatment one month later on damaged tubers (after sorting).

5.2.12 Mealworm, *Tenebrio guineensis* (Tenebrionidae)

5.2.12.1 Damage

The larvae and the adults eat into the surface of the tubers and then bore large areas 5–10 mm deep. Despite the significant size of the insects, the damage progresses quite slowly.

5.2.12.2 Management

The following control measures are recommended:

- Sorting of injured or damaged tubers and storing them separately.
- Cleaning and disinfection of store houses.
- Treatment of stored tubers with insecticide spray before storage.
- Repeat the treatment 1 month later on damaged tubers (after sorting).

5.2.13 Sawfly, *Senoclidia purpurata* (Hymenoptera: Tenthredinidae)

5.2.13.1 Distribution

The pest is distributed in Papua New Guinea.

5.2.13.2 Damage

Symptoms include defoliation by numbers of larvae feeding together, often in rows, on large leaves, beginning at the apex of the leaf and moving backward toward the leaf stalk (Fig. 5.16). In serious infestations adult sawflies are commonly seen flying around the garden.

5.2.13.3 Biology

The Tenthredinidae usually lays eggs in young shoots and leaves. Larvae are similar in appearance to Lepidoptera larvae but with more than five pairs of abdominal limbs. Pupation takes place in the soil.

5.2.13.4 Management

Carbaryl applied at recommended rate will kill sawfly larvae.

Entomopathogenic nematodes (*Steinernema carpocapsae* and *Heterorhabditis bacteriophora* at 200 infective juveniles/larvae) are effective against sawfly. *S. carpocapsae* is responsible for a mortality of 92.7 % larvae in 48 h, while *H. bacteriophora* takes 72 h for this.

5.2.14 Termites, *Coptotermes* sp., *Amitermes evuncifer*, and *Protermes minutus*

5.2.14.1 Damage

Termites can attack the tubers during storage (Fig. 5.17), with the infestation possibly beginning in the field. The damage, which may be significant within a few weeks, is often difficult

Fig. 5.16 Sawfly on yam leaf and its damage on leaves

Fig. 5.17 External and internal termite damage of yam tubers

Fig. 5.18 Red spider mite adult

to detect when the colonies use only one gallery for penetration.

5.2.14.2 Management

Planting cassava on ridges reduced the incidence of termites and significantly ($P<0.05$) increased tuber yield. NPKMg (12:12:17:2) suppressed termites in yam and improved yield significantly. Delay in harvesting yam tubers increased termite infestation.

5.2.15 Red Spider Mites, *Tetranychus cinnabarinus* and *T. neocaledonicus*

5.2.15.1 Damage

Red spider mites (Fig. 5.18) are found to infest the luminary base adjoining petiole. Necrotic areas develop on the upper surface of the leaf lamina. Feeding injury predisposes the plant for fungal infection.

5.2.15.2 Management

1. *Chemical Methods*: Spraying of wettable sulfur at 3 g + Teepol at 0.5 ml/L of water can control this pest.
2. *Biological Methods*
   (a) *Predators*: *Stethorus gilvifrons*, *Scolothrips indicus*, predatory phytoseiid mites and spiders are found preying on spider mites.

5.2.16 Rodents

5.2.16.1 Damage

Wherever cassava, yams, taro, or sweet potatoes are grown, they shall be attacked and eaten by rodents. The damage may be seen immediately when the roots are eaten at ground level or may not be apparent until the crop is harvested if rats have burrowed and eaten away the roots from beneath. Losses of the order of 5–15 % in West and Central Africa and up to 50–60 % in the Cayman Islands in the West Indies were reported. Rodent attack may continue after harvest in farm and storage. The species responsible for damage in storage include *Bandicota bengalensis* in the Indian subcontinent and *Mastomys natalensis* and *Rattus exulans* in the Far East and South Pacific region.

5.2.16.2 Management

There are three main methods to control rodents in the field. Trapping has been employed since the earliest times, and much skill has been developed in its use. Poisoning and environmental alteration have the biggest contribution to make to increasing crop yields.

5.3 Diseases

5.3.1 Anthracnose, *Glomerella cingulata* (*Colletotrichum gloeosporioides*)

Anthracnose is widespread throughout tropical countries and is the most prevalent fungal disease infecting yam worldwide (Adebanjo and Onesirosan 1986).

5.3.1.1 Distribution

The disease is distributed in the Caribbean, East Africa, the Far East, India, and West Africa.

5.3.1.2 Symptoms

1. *Aboveground Symptoms*: Symptoms vary according to the age of the leaf, the amount of rain, and the variety of yam. Mild infection leads to very small brown spots on young leaves. These spots become larger as the leaves approach full size, and they may develop pale yellow margins (Fig. 5.19). Sometimes the leaf spots run together to form large irregular blotches, the centers of which may fall out giving a "shot hole" effect. Infected leaves usually fall off. In some varieties, the veins on the underside of the leaf become black where they have been infected. If this occurs as the leaves are expanding, they may become cup shaped and twisted. The "wings" on the stems may also become infected.

Long periods of rain favor epidemics of the disease because the fungal spores (conidia) are spread by rain splash. Young foliage is more susceptible to anthracnose. Therefore, if periods of high rainfall coincide with the stage in crop development where a lot of young leaves are present, the disease can quickly spread throughout the crop. In this case, rapidly expanding black lesions occur on the leaves and stems and the shoots die (Fig. 5.19). This is particularly common when vines have reached the tops of their supporting poles, and the new growth droops near to older infected parts of the plant. Mature leaves, by contrast, develop numerous, brown, pin-point spots which do not usually expand nor penetrate to the other side of the leaf. Both large and pin-point spots can occur on the same leaf (Fig. 5.19).

Fig. 5.19 Symptoms of anthracnose on yam leaves and shoots

Large numbers of spores on mature leaves of some varieties lead to a superficial blackening of the leaf surface. Occasionally, when a leaf has been disturbed from its normal orientation to expose the whole or part of its lower surface, the exposed part of that surface becomes blackened or bronzed instead of the upper surface (Fig. 5.19). Stems may develop similar symptoms of superficial but intense blackening.

Stems are also attacked, but less frequently. If the infection reaches the stele, death of the plant can result. Attacks on the stem generally occur as a result of direct penetration near the terminal bud which leads to rapid death of the plant. Direct infection can also occur at ground level.

The reduction in yield caused by anthracnose can be particularly serious in the more humid regions. *D. alata* is more susceptible but *D. esculenta* is more resistant. *D. cayenensis* and *D. rotundata* are only slightly susceptible, and even in humid forest regions, their leaves only develop a few brown spots and stems are not attacked.

When dieback occurs on young plants, a few of the lower leaves may survive, but usually the whole vine dies resulting in no or poor yields. New shoots sometimes grow from the planting piece and plants become multistemmed in contrast to uninfected plants which usually only have one or two stems. Because each stem produces a tuber, affected plants may have several small tubers instead of the normal one or two.

2. *Tuber Symptoms in Storage*: In the Caribbean, the fungus causes orange-brown tuber rots known as "dead skin" (Fig. 5.20). The rots develop in the flesh of the tuber just under the corky layer or skin. Small blisters occur on the surface of the yam and the skin is easily removed from the layer beneath. Later, deeper rots develop, and eventually all that is left is a wrinkled shell surrounding a rotting core. It is not yet known if these rots occur in Pacific countries. It is likely that they do, but they may have been mistaken for rots caused by the nematode, *Pratylenchus coffeae*.

Fig. 5.20 Small tubers produced by yam anthracnose-infected plants (*bottom*) compared to big tuber produced by healthy plants (*top*)

5.3.1.3 Survival, Development, and Spread

Although the pathogen is not able to survive in soil for more than a few weeks, it is able to survive between growing seasons on crop debris. Therefore, survival from one season to the next may occur but is unlikely to be important where growers practice crop rotation and plow-in crop debris. The fungus, *C. gloeosporioides*, infects many crops and weeds. It is possible that spores from these plants also affect the yam crop.

If tuber rots occur in the Pacific as they do in the Caribbean, then the infected planting material is likely to be the most important way that new crops become diseased. Small immature tubers, derived from early shoot death, may be a major source of infection of the shoots as they develop.

The pathogen overseasons on crop debris as acervuli. Infection is favored by wet and humid conditions. Many acervuli of the *Gloeosporium* type which become erumpent later are produced on lesions. Acervuli remain devoid of setae when humidity is over 95 % but at lower humidity sterile setae mixed with conidiophores convert the acervuli to the *Colletotrichum* type. In very dry conditions, only setae are formed. Spores are formed in large numbers on the leaf spots and are splashed in rain and or carried by dripping dew to adjacent and lower leaves and stems. Spores are mainly dispersed by rain splash and the disease is therefore most severe when there are heavy rains. Once infection is established in a crop, subsequent development of yam anthracnose depends on rainfall and host variety. Severe outbreaks develop on susceptible varieties following rainstorms or cyclones.

5.4 Management

The two most important disease management strategies are to grow resistant varieties and to ensure that planting sets are pathogen-free. Crop debris should be removed as it is a major source of inoculum.

1. *Regulatory Methods*: The global distribution of yams is an important element of international crop improvement programs. However, the movement of yam from one country to another can spread new and potentially devastating diseases. Introduction of new yam germplasm should be limited to small quantities for scientific purposes under the supervision of quarantine officials. Preferably, the material should be transferred as sterile, pathogen-tested plantlets growing on tissue culture medium (FAO/IPPGR 1989). It is important that the plantlets have been tested to ensure that they are pathogen-free. This is done by growing the material in the country of origin or intermediate quarantine station for one crop cycle. Only plants that show no signs of virus (or other pathogen) infection should be released.
2. *Cultural Methods*: Management methods include use of crop rotation, fallowing, removal of debris, planting of healthy material, and destruction of infected cultivars. Plowing immediately after harvest also helps to reduce the inoculum.

Cultural control measures such as the removal of weeds that may be alternative hosts, planting barrier crops of maize, avoiding damage to tubers at harvest, and early staking are likely to reduce disease development.

Tubers should be selected from uninfected plants and stored in a cool dry place during dormancy. Sets should be carefully inspected and any showing areas of rot rejected. The sets should be treated with a broad- spectrum fungicide to eliminate surface-borne fungi and cutting knives should be treated frequently with bleach.

Time of planting is important and should be planned so that plants reach maximum height on supporting poles before the main rainy season. Early planting cannot be used to avoid the disease entirely but only to delay the onset of severe damage to the foliage.

3. *Chemical Methods*: Spraying maneb (0.2 %) and carbendazim (0.25 %) reduces disease severity. Spraying maneb at 3.32 g a.i./L at 10 days' interval is effective. Benomyl alone or applied alternately with mancozeb on a weekly cycle has also been shown to give good control and increase yields. Benomyl + propineb are also effective. Zineb, ferbam, and ziram sprays were shown to give 50 %, 41 %, and 30 % yield increases, respectively.
4. *Biological Methods*: *Trichoderma harzianum* and a native isolate were found effective against *C. gloeosporioides* causing yam anthracnose. *T. viride*, *T. harzianum*, and *Pseudomonas fluorescens* were found effective against yam anthracnose.
5. *Host Resistance*: The use of anthracnose-resistant cultivars (e.g., 'TDA 291', 'TDA 297') bred and released by the International Institute of Tropical Agriculture (IITA) has been advocated (IITA 1993).

   Resistant varieties have also been found in the Caribbean, e.g., Kinabayo, Plimbite, Belep, and Oriental. These varieties are being introduced into the Pacific Islands through regional tissue culture laboratories to ensure they are free from virus.
6. *Integrated Methods*: Soil and tuber treatment with 50 g of *Trichoderma* mixed in fresh cow dung slurry per kg tuber before planting along with foliar application of carbendazim at 0.05 % three times at 15 days' interval after initiation of symptoms was most effective for the management of greater yam anthracnose.

Fig. 5.21 Symptoms of *Cercospora* leaf spot on yams

5.4.1 *Cercospora* Leaf Spot, *Cercospora brasiliensis*, *C. carbonacea*, *C. pachyderma*, *Phaeoramularia dioscoreae*, and *Pseudocercospora contraria*

5.4.1.1 Distribution

The disease is worldwide in distribution on all species of yams.

5.4.1.2 Symptoms

Several *Cercospora* spp. and some species which were formerly classified as *Cercospora* are recorded as causal agents of leaf spots on yams. Generally the symptoms are brownish spots which darken with age and coalesce as the disease progresses. *Phaeoramularia dioscoreae* causes small brown spots which sometimes coalesce. *C. carbonacea* is the most commonly recorded causal agent and causes large blackish-brown to brown spots which may spread over the whole leaf (Fig. 5.21).

5.4.1.3 Development and Spread

The fungus overseasons on crop debris. Conidia are produced on both surfaces of infected leaves and are spread by rain. The disease is more common during warm and wet weather.

Fig. 5.22 *Curvularia* leaf spot on yam leaf

5.4.1.4 Management

Two to three sprays of 0.05 % benomyl or carbendazim at 15 days' interval starting after the first appearance of the disease was found effective in decreasing the disease incidence (24.61 % as compared to 62.64 % in control) and increasing the tuber yield (19.75 t/ha as compared to 7.72 t/ha).

Spraying of maneb or captan at 2 g/L of water every 10 days was found effective in the management of the disease.

5.4.2 *Curvularia* Leaf Spot, *Curvularia eragrostidis*

The leaf spot of yam (*Dioscorea cayenensis*) is one of the most frequent and severe diseases in all yam-growing areas of northeast Brazil. The disease causes a reduction of about 35–40 % of the weight of the commercial tuber (Michereff et al. 1994).

5.4.2.1 Symptoms

The symptoms are necrotic leaf spots (Fig. 5.22) which reduce the photosynthetic area and cause losses of 35–40 % of tuber weight (yield reduction).

5.4.2.2 Management

1. *Biological Methods*: Under greenhouse conditions the antagonists IF-82 and IF-88

strains of *Bacillus subtilis* reduced disease severity. IF-82 strain of *B. subtilis* was the best biocontrol agent for the yam leaf spot disease. *Bacillus subtilis* inhibited pathogen conidial germination, with average inhibition levels of 99.2 %. Under greenhouse conditions the antagonists were applied at three different time intervals relative to *C. eragrostidis* inoculation: 3 days before, at the same time, and 3 days after. *B. subtilis* applied at the same time as pathogen inoculation reduced disease severity to the extent of 75 %. *B. subtilis* showed the best persistence of antagonistic action, with an average of 96.3 %. *B. subtilis* was the best biocontrol agent for the yam leaf spot disease in this study (Michereff et al. 1994).

5.4.3 Tuber Rot, *Fusarium oxysporum*, *F. solani*, *Rhizopus nodosus*, and *Botryodiplodia theobromae*

5.4.3.1 Distribution

The *Fusarium* tuber rot is distributed in China, India, Ivory Coast, Japan, Nigeria, and Puerto Rico. The *Rhizopus* tuber rot is distributed in Ivory Coast and India.

5.4.3.2 Symptoms

Nematode infestation facilitates the entry of *Fusarium* spp. into the tubers. Infected tissues develop a brown-colored dry rot. Often the infection is not noticed until after harvest or until the tubers are cut opened. If the tubers are cut transversely, zonations of infection can be seen. Brown dry rot caused by *Fusarium* occurs centrally, and there may be a blackish zone around this where there is considerable damage and secondary bacterial infection (Fig. 5.23).

Rhizopus nodosus causes a soft rot which particularly affects *D. esculenta*; only weak infection occurs in *D. alata*. The tubers shrink by up to three quarters of their volume and the skin becomes wrinkled.

When the tubers are cut, browning of the tissues is observed and thick colored liquid can be squeezed out of infected tissues. The infection spreads throughout the tuber, only the vascular region remaining uninfected. Eventually the tuber becomes mummified as the infected tissues dry out. A dry rot is also attributed to *R. nodosus* which has also been shown to be a wound parasite.

Fig. 5.23 Yam tubers showing symptoms of a tuber rot disease

Botryodiplodia theobromae is one of the most important causal agents of tuber rot and can be associated with wet rot, soft rot, and brown dry rot. Infection often occurs in conjunction with nematode or insect attack. Lesions caused by nematodes on yam tubers facilitate invasion by disease pathogens. The resulting decay process often destroys the entire tuber in the ground and particularly during storage (Bridge 1982). The more extensive internal decay of tubers known as wet rot or soft rot or watery rot is associated with fungal and/or bacterial pathogens (Adeniji 1970a, b). This general decay of tubers, which is a serious problem in stored yam, is increased when tubers are wounded or damaged by man, rodents, and insects or mainly by nematodes (Adeniji 1970a, b). Nematodes and fungi are often found in combination in the transitional stage between dry rot and wet rot although nematodes tend not to be found when yam tubers are completely decayed (Adeniji 1970a, b).

5.4.3.3 Management

1. *Cultural Methods*: Cut tubers used as planting material should be allowed to dry in order to prevent the entry of the pathogen. The incidence of infection can be greatly reduced by treating the cut surfaces with Bordeaux mixture or coating with lime, chalk, or wood ashes. Crop rotation helps to reduce the disease incidence. Also avoid planting yams in nematode-infested soil. Several measures such as the use of crop rotation, fallowing, and planting of healthy materials and the destruction of infected crop cultivars have been recommended. Minimizing physical damage of tubers during postharvest operations or handling. If wounding cannot be entirely prevented, tubers may be placed in an environment favorable for rapid healing of wounds.
2. *Chemical Methods*: Dipping the tubers in systemic fungicides such as captan, benomyl, and thiabendazole has been found effective. A 5 min dip in 3,000 ppm benomyl followed by dusting the cut ends with 0.3 % benomyl in talcum powder effectively controlled tuber rot. A 30 min dip in 1,000 ppm benomyl or thiabendazole gave 80–90 % control of preharvest rotting in the field. The use of synthetic chemicals such as sodium orthophenylphenate, borax, captan, thiabendazole, benomyl, and bleach (sodium hypochlorite) has been found to significantly reduce storage rot in yam.

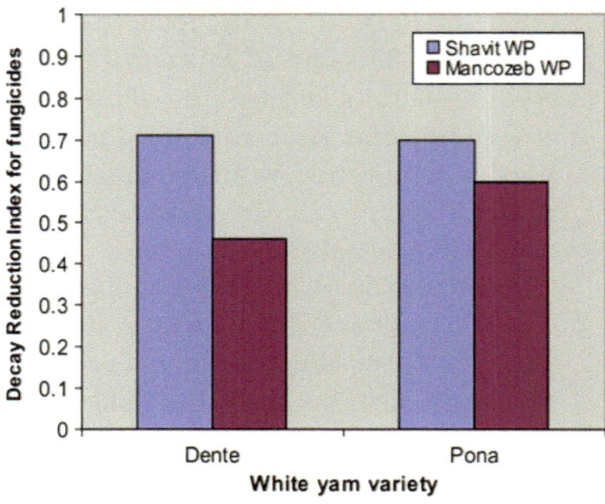

Fig. 5.24 Decay reduction index for the two fungicides in controlling rot of white yam varieties 14 weeks of storage

Surface coating of white yam tubers with chemical fungicides (Shavit WP and mancozeb metalaxyl WP) prior to storage in pits or barns would reduce the incidence of rot diseases and increase shelf life of yam tubers. Shavit WP was more potent than mancozeb metalaxyl WP (Fig. 5.24).

3. *Biological Methods*: Application of *Bacillus subtilis*, isolated from the epiphytic microflora of yam tuber, showed a drastic reduction in spoilage fungi of yams during 5-month storage period (Okigbo 2002, 2005).

The set treatment (30 g/L for 15 min.) and soil application (1 kg/ha) of *Trichoderma viride* recorded the lowest tuber rot incidence (7.33 %) and highest harvest index (66.9 %). Application of *T. viride* effectively controls the normal tuber surface mycoflora throughout 6 months' storage, greatly reducing rotting.

Inoculation of white yam (*Dioscorea rotundata*) with conidia of *T. viride* and subsequent

Fig. 5.25 *Left*, large, brown to black leaf spots with well-defined to diffuse margins observed on yam; *right*, veinal necrosis of yam leaves associated with rust disease

storage of the tubers under the ambient environmental conditions of a traditional yam barn resulted in drastic reduction in the frequency of occurrence of the normal tuber surface mycoflora (*Aspergillus niger*, *Botryodiplodia theobromae*, and *Penicillium oxalicum*) over a 4-month storage period (Okigbo and Ikediugwu 2008).

5.4.4 Collar Rot, *Rhizoctonia* sp.

5.4.4.1 Symptoms

This fungus gets into the roots of tubers of the yams and rots them. There may be signs of brown marks of the disease low down on the stem near the soil. Dark-colored spots on this part of the stem are called "collar rot." This fungus is carried by the tuber. If there are any marks of this disease on the leaves, they are on one side of the leaf only.

5.4.4.2 Management

The most important way of stopping this disease is not to use tubers from diseased plants as planting material. The big tubers are probably good for planting.

5.4.5 Rust, *Goplana dioscoreae*

5.4.5.1 Symptoms

1. *Leaf Symptoms*: Disease symptoms were observed primarily on yam foliage. Petioles and stems also exhibit symptoms. Extensive spotting of leaves was observed at all locations and for most yam varieties.

One of the principal symptom types observed on the variety 'Kehp en Dol' consisted of large, brown to black leaf spots with well-defined to diffuse margins (Fig. 5.25). These spots were roughly circular to oval to irregularly shaped. They had tan to brown, textured centers and irregular margins and ranged in diameter from 1–2 to 25 mm. Individual lesions were surrounded by chlorotic (yellow) halos. These rapidly expanding lesions often coalesced into extensive, blighted areas on leaves and were associated with significant veinal necrosis (Fig. 5.25). Many leaves were either completely destroyed or having from approximately 40–70 % necrotic leaf area. Eventually, leaves begin to curl and change color (yellow and brown) and fall off the vine.

Fig. 5.26 *Left*, "bumpy" leaf spot symptom associated with yam rust disease observed on the yam variety 'Kilimenip en Kokonep'; *right*, significant defoliation due to yam rust disease

The leaf spots tend to expand asymmetrically and leaf veins invaded more rapidly than surrounding leaf tissues. These leaf vein infections are first visible as black veins and later black veins filled with and bearing erumpent (i.e., breaking through the leaf epidermis) rust uredia (i.e., rust spore-bearing structures that contain the infective urediniospores).

A second, "bumpy" leaf spot was associated with the yam variety 'Kilimenip en Kokonep' from Alohkapw (Fig. 5.26). Leaves had small, raised "bumps" (2–4 mm diameter) on the adaxial leaf surface. These green (not necrotic), raised bumps (1–2 mm in height) corresponded with the presence of similar sunken areas on the abaxial leaf surface. Orange-colored fungal spores were abundant within these sunken areas, whereas few or no spores were observed in/on the raised bumps on the upper side of the leaf. Older, mature lesions of this "bumpy" type were blackened and dry.

Depending on the stage of disease and other factors, the observed defoliation attributable to this disease ranged from approximately 5 to nearly 100 % for individual plants (Fig. 5.26).

2. *Petiole Symptoms*: Petiole lesions were numerous, small (1–2 mm length), longitudinal, black spots, running parallel with petiole veins.

Fig. 5.27 Severe stem symptoms associated with yam rust. Stem symptoms are rare in association with *Goplana dioscoreae*

3. *Stem Symptoms*: Stem lesions consisted of numerous, small (2–5 mm diameter) black spots, often coalescing into larger lesions. As the numerous stem lesions coalesce, large blighted areas developed on stems (Fig. 5.27). Stems often became completely necrotic and black. Severely blighted stems died back to the base of the plant.

4. *Yield Suppression*: Yield suppression was significant. Some farmers at some locations reported that there will be no yam harvest for some varieties as a result of this epidemic. For example, a farmer in the municipality of Madolenihmw reported that last year's yam harvest was 500 lb, whereas this year he expects to harvest nothing. Similar scenarios were encountered throughout the course of sampling for this disease at various locations.

5.4.5.2 Epidemiology

The rust was favored by high or frequent rainfall and high relative humidity. Spread of the disease among yam plants (i.e., dispersal of urediniospores) is favored by windy conditions required for significant spore dispersal. Spores of rust fungi are dispersed within and among plants by wind, windblown rain, and mists and by splashing or running water.

5.4.5.3 Management

1. *Chemical Methods*: This is not a very attractive or feasible alternative. Yams are grown almost exclusively using organic methods in Pohnpei, the use of chemical fungicides for disease control is discouraged, and the availability of fungicide products is very limited and prohibitively expensive for the average yam farmer. In addition, the practice of allowing yam vines to scale the heights of tall breadfruit trees makes fungicide sprays of yam foliage very difficult if not practically impossible.
2. *Host Resistance*: Plant resistant varieties (Fig. 5.28). Use resistant varieties interplanted with susceptible varieties. The yam varieties 'Kilimenip Namuu Pwetepwet' (*Dioscorea alata*) and 'Neir' (wild yam variety, *Dioscorea nummularia*) are found resistant to rust.

The possibility exists that susceptible and resistant yam varieties may be crossed to produce rust-resistant progeny with desirable agronomic properties (taste, color, appearance, yield, etc.).

3. *Cultural Methods*
   (a) *Humidity Management*: Dry conditions are not favorable for infection and disease development and pathogen dispersal. A number of options exist for reducing levels of relative humidity within and around yam canopies.
   (b) *Weed Control*: Tall weeds within and around yam plantings create high humidity in the yam plant canopy. Infection and disease is favored by high humidity.
   (c) *Cultivation Method:* Select a cultivation method that increases distance among

Fig. 5.28 Stems and leaves of rust-free yam 'Kehp en Dol' suggesting possible disease resistance or immunity

plants and allows adequate sunlight and air movement to reach the yam canopy.

(d) *Adequate Soil Drainage*: Continually wet soil in and around yam plantings creates high humidity through evaporation into the yam canopy. Improved soil drainage can result in lower levels of rust disease.

(e) *Wind and Canopy Management*: Clear unnecessary foliage of surrounding plants to allow fresh air circulation in the yam canopy. This will reduce humidity levels and allow for more rapid drying of leaf and stem surfaces.

(f) *Scattered Plantings*: Disease spread within and among plants can be reduced by not planting large populations of susceptible yams in one area or location or in one trellis.

(g) *Sanitation*: Sanitation involves the removal and destruction of diseased plant parts. This practice reduces the amount of airborne inoculum (infective spores). Defoliated (symptomatic) yam leaves lying on the ground should be removed on a regular basis and either buried or burned to destroy viable spores in/on the leaves before they can be dispersed to susceptible plants.

(h) *Remote Planting Sites*: Remote planting sites have a lower probability of receiving "spore showers" from other farms.

(i) *Full-Sun Cultivation*: Cultivation of yams in full sun produces leaves and foliage that are less tender. In full sun, yam leaves tend to dry off rapidly after rains. Leaves that often remain wet for several hours are prone to rust infection and disease development.

(j) *Elimination of Alternate or Noneconomic Hosts*: It is possible that there is an alternate host in addition to yam in the life cycle of the rust fungus that causes this disease on yam. For example, while sampling wild yams on Sokehs ridge, an identical rust disease (the same symptoms and similar-appearing fungus spores and structures) was observed on an unidentified, viny climbing plant that resembles yam. The plant was identified as *Dioscorea bulbifera*, the "air potato." Further testing and study are required to determine the identity of the rust fungus responsible for the disease symptoms on *D. bulbifera*.

4. *Biological Methods*: A species of fungal-feeding mite was observed in most of the rust lesions, feeding on the rust urediniospores. It is possible that large populations of this mite can or will afford some levels of disease control at present and in the future.

5.4.6 Neck Rot, *Corticium rolfsii*

5.4.6.1 Distribution

The disease is distributed in Australia, Ivory Coast, Jamaica, and Sierra Leone.

5.4.6.2 Symptoms

This is one of the most important foliar diseases of yam and causes serious damage when stems are affected. The foliar lesions are very different on the various species of yam.

On *Dioscorea alata*, spots are circular and grayish beige in the center bordered with light to purplish brown surrounded by a yellowing halo and can reach 2 cm in diameter. Sometimes the spots run together and often the centers fall out (Fig. 5.29). On the underside of leaves, small reddish-brown sclerotia can be observed in the center of the spots. If the humidity is high, white hyphae begin to grow from the sclerotia and

Fig. 5.29 Leaf spot disease on yam leaves

radiate toward the edge. A bright greasy-looking border shows the area affected by the fungus. Many sclerotia, which are white at first, then reddish brown and finally brown, appear on the underside of necrotic leaves.

On *D. esculenta*, *D. cayenensis*, and *D. bulbifera*, the necrosis is first seen as a grayish-beige spot limited by a dark beige ring surrounded by a yellow halo. Then the whole of leaf takes on a beige or light coffee color with dark bands. The sclerotia appear early and in great number on the underside of the spots and then, sometimes, on the surface. The petiole can become affected after the leaf.

Leaves are infected directly and individually without in contact with the soil which harbors *C. rolfsii*. Rain splashing can result in the transfer of enough sclerotia to ensure infection. When the infection starts at the base of the stem, the first symptom is browning of the lower leaves. Later the lower part of the stem turns black and the aerial parts fade. When the tuber is dug up, total disorganization of the tissues in the neck (the part where the tuber joins the stem) is observed.

Foliar infections are only observed during the rainy season and are not usually very serious. Development of the disease is halted by the start of the dry season.

Neck infection is more serious and causes death in *D. bulbifera* and *D. alata*. However, young stems of *D. alata* possess some resistance as they can produce new tissue which prevents the stem from being girdled.

5.4.6.3 Management

If foliar *C. rolfsii* infection of yams has been observed, it is wise to rotate with resistant cultivars of other crops (other than groundnut, cotton, maze, cassava, and rice which are susceptible). Simple cultural practices can avoid or reduce the damage. Training the vines up supports to keep the leaves clear of the soil can appreciably reduce the chance of infection.

Once neck infection has appeared, it is too late to apply effective control measures. Use of resistant species, *D. esculenta*, instead of the more susceptible species could be encouraged in areas where infection is widespread.

5.4.7 Yam Mosaic Virus

Yam mosaic virus (YMV) is considered to cause the most severe losses in yams. It is known that the most economically important virus diseases of yam so far characterized are caused by members of the potyvirus group, but there is inadequate information on the number and variability of these viruses. It is sap transmissible with difficulty to a small range of hosts and transmitted by aphids in the nonpersistent manner. It causes economically important losses in yams in tropical areas (Africa and America).

5.4.7.1 Distribution

It is reported from several African countries: Ivory Coast, Nigeria (Mohamed and Terry 1979), and Togo (Reckhaus 1980) and probably also present in the Caribbean area.

5.4.7.2 Symptoms

Symptoms associated with yam virus diseases include severe leaf chlorosis, green veinbanding, curling, mottling, green spotting, flecking, mosaic, shoestringing, interveinal chlorosis, stunting, and distortion (Figs. 5.30 and 5.31). These symptoms, which mainly affect the foliage, lead to a reduction in the photosynthetic ability of the infected plant with deleterious effects on the tuber yield, quality, and, in some instances, death of the plants (Odu et al. 2001).

5.4.7.3 Particle Morphology

Yam mosaic virus with flexuous filamentous particles is about 785 nm long, consisting of one species of single-stranded RNA and one species of coat protein (Fig. 5.32).

5.4.7.4 Transmission

The virus is transmitted in a nonpersistent manner by the aphids *Aphis gossypii*, *A. craccivora*, *Rhopalosiphum maidis*, and *Toxoptera citricidus*. Aphids acquire virus after a 5 min access period and infect plants after an inoculation access period of less than 15 min. No transmission has been detected through the seed of *Dioscorea cayenensis* (Thouvenel and Fauquet 1979).

Fig. 5.30 Yam plants infected with yam mosaic virus showing mottling symptom in *Dioscorea esculenta* (*left*) and mosaic symptom in *D. alata* (*right*)

Fig. 5.31 Leaves of yam: *left*, chlorotic mosaic and leaf curling (YMV); *middle*, vein clearing and necrotic mosaic (YMV); *right*, shoestringing and stunting (YMV, badnavirus)

Fig. 5.32 Virus particles from a purified preparation

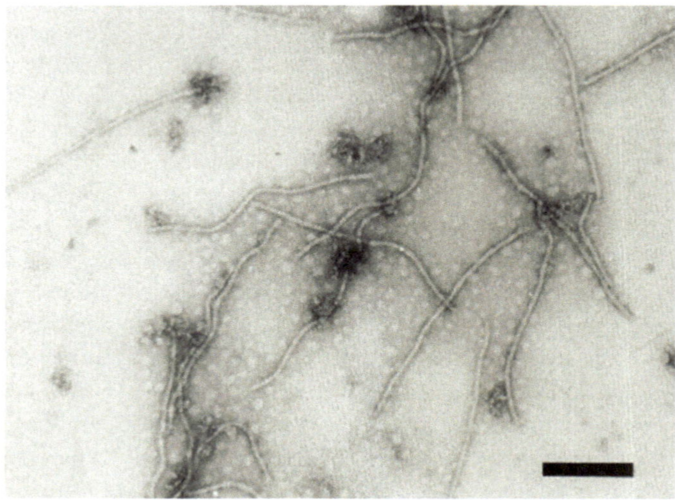

5.4.7.5 Host Range

YMV is known to infect several species of Dioscoreaceae, particularly *Dioscorea alata*, *D. cayenensis*, *D. esculenta*, and *D. rotundata*.

5.4.7.6 Diagnostic Species

- *Dioscorea cayenensis*: Systemic symptoms are apt to differ considerably in different leaves of the same plant: mosaic, veinbanding, green spotting or flecking, curling, and mottling.
- *D. esculenta*: Mosaic and veinbanding.
- *D. preussii*: Mosaic and veinbanding.
- *D. composita* and *D. floribunda*: Not infected.
- *Nicotiana benthamiana*: Mottling on inoculated and systemically infected leaves.
- *N. megalosiphon*: Systemic chlorotic spotting on the whole plant.

5.4.7.7 Management

Since YMV spreads from one generation to and season to the next, one of the main methods to limit the spread of the virus is to ensure that only disease-free planting materials are produced. These can be plantlets derived directly from tissue culture (Fig. 5.33) or from mini-tubers produced in vector-proof conditions (Mantell 1980).

Fig. 5.33 Mosaic virus-tree yam plant produced through meristem-tip culture

5.5 Nematodes

Among the various constraints to production of yam, nematode pests are of significant importance (Bridge et al. 2005). In West Africa, the yam nematode, *Scutellonema bradys*, is a major causal agent of dry rot (Adesiyan and Odihirin 1977; Hahn et al. 1989; Coyne et al. 2005), while in Central America, *Pratylenchus coffeae* causes similar damage to that of *S. bradys* (Acosta and Ayala 1976). *Meloidogyne* spp. can also cause in-field and postharvest losses to yam, although information for *Meloidogyne* spp. is less conclusive (Bridge et al. 2005). A survey of parasitic nematodes on root and tuber crops in Uganda (Coyne et al. 2003) established that both *Meloidogyne* spp. and *Pratylenchus sudanensis* were common on yam. It has also been shown that yams could be severely damaged by *Meloidogyne* spp. in Uganda, substantially affecting newly introduced, high-yielding cultivars of *D. rotundata* (Mudiope et al. 1998).

The yam nematode (*S. bradys*) and the root-knot nematodes (*Meloidogyne* spp.) are the most serious nematode pests of yam and are widespread in West Africa (Caveness 1992). Various *Pratylenchus* spp. are also known to infest yam and have been reported in Puerto Rico and in several countries of the South Pacific (Kermarrec et al. 1991). Nematodes cause lesions beneath the tuber skin which are yellow at first, developing into dark brown dry rots which may cover the tuber surface in heavily infested tubers. Infection often starts before harvest and continues in storage, leading to a loss of food and planting material for the next season's crop (Siddiqi 1972). Major sources of losses in storage can be attributed to an interaction between nematodes (*S. bradys*, *Meloidogyne* spp., and *Pratylenchus* spp.), fungi, and bacteria and moderated by physical factors of the environment such as temperature and humidity (Bridge 1972).

Radopholus similis was associated with yam in Papua New Guinea (Bridge and Page 1984), in Fiji (Butler and Vilsoni 1975), and in the Solomon Islands (Bridge 1988) causing dry rot disease, similar to that caused by *P. coffeae* and *S. bradys*. *Aphelenchoides besseyi* was also found associated with drying and blackening of the foliage and cracking of the tuber of *D. trifida*, with internal decay in Guadeloupe (Kermarrec and Anais 1973). *Paratrichodorus porosus* was reported associated with blackening, cracking, and corkiness of the tuber tips of Chinese yam, *D. opposita*, in Japan (Niashizawa 1973).

5.5.1 Yam Nematode, *Scutellonema bradys*

Scutellonema bradys has been recorded on yams all over the tropics and is recognized as an economically major nematode pest in West Africa. An infestation results in high quantitative losses, as well as loss in the marketable value of the affected tubers.

The dry rot of yam tubers is caused by *S. bradys*. Nadakal and Thomas (1967) reported the occurrence of dry rot of *Dioscorea alata* and the association of *S. bradys* from Kerala. *Dioscorea alata*, *D. cayenensis*, *D. esculenta*, and *D. rotundata* are good hosts of the yam nematode. Sesame and cowpea support high root populations and melon can increase soil populations.

5.5.1.1 Economic Importance

The primary importance of *S. bradys* on yams is in the direct damage it causes to the tubers resulting in dry rot disease (Jatala and Bridge 1990). The nematodes produce a marked reduction in the quality, marketable value, and edible portions of tubers, and these reductions are more severe in yams that have been stored. Weight differences between healthy and diseased tubers harvested from the field have been estimated to be 20–30 % in the Ivory Coast (Bridge 1982) and 0 to 29 % in Nigeria (Wood et al. 1980). Weight reduction due to moisture loss is more likely to occur in late harvested tubers left in dry soil (Bridge 1982). Water loss from tubers continues during storage and is significantly greater in tubers infected with *S. bradys* compared with healthy tubers (Adesiyan et al. 1975). Nearly 47 % of all yam tubers on sale in Nigerian markets were infested with *S. bradys* (Bridge 1973) and both dry rot and wet rot diseases of tubers have been observed in all Nigerian yam barns and markets sampled (Adesiyan and Odihirin 1977). When dry rot is followed by wet rot in stored yams, losses of whole tubers can be as high as 80–100 %.

In West Africa, some 11 million tons of yams are lost annually because of damage in storage initiated by *S. bradys*. With West African yam exports rising, there is a heightened risk of *S. bradys* spreading to potato-producing countries in southern Africa, where potato production already suffers from other pests and diseases.

5.5.1.2 Distribution

Scutellonema bradys is widely distributed in the tropics, especially in yam-growing areas. It has been reported from the West African countries of Nigeria, Ivory Coast, Senegal, Gambia, Ghana, Benin, and Togo and from the central African country of Cameroon. It also occurs in Cuba, Jamaica, Guatemala, Puerto Rico, Guadeloupe, Haiti, and Martinique in the Caribbean and from Venezuela, Brazil, and India (Figs. 5.34 and 5.35) (Bridge et al. 2005).

5.5.1.3 Symptoms

Scutellonema bradys is a migratory endoparasite of roots and tubers, confined usually to the outer 1–2 cm of the tuber, feeding intracellularly in yam tuber tissues. The damage results in ruptured cell walls, loss of cell contents, formation of cavities (Bridge 1973), tuber dry rot (Bridge et al. 2005), tuber decay, and higher rates of desiccation (Nwauzor and Fawole 1981) (Fig. 5.36).

The eggs are most often laid in plant tissues but they are almost always to be found in the soil. Very large populations can build up in the tuber. A tuber disease commonly associated with *S. bradys* is referred to as "dry rot," where the nematodes feed within the tuber causing tissue breakdown and producing cavities. Initially small yellowish lesions develop beneath the periderm and as the infection spreads these turn into areas of dark-brown- or black-colored rot (Fig. 5.19). External symptoms are slight to deep cracks on the tuber skin and malformation of the tuber. The nematodes cause small cracks on the surface of tubers, with the proximal parts (head) affected most. Under these lesions are brownish-black necrotized

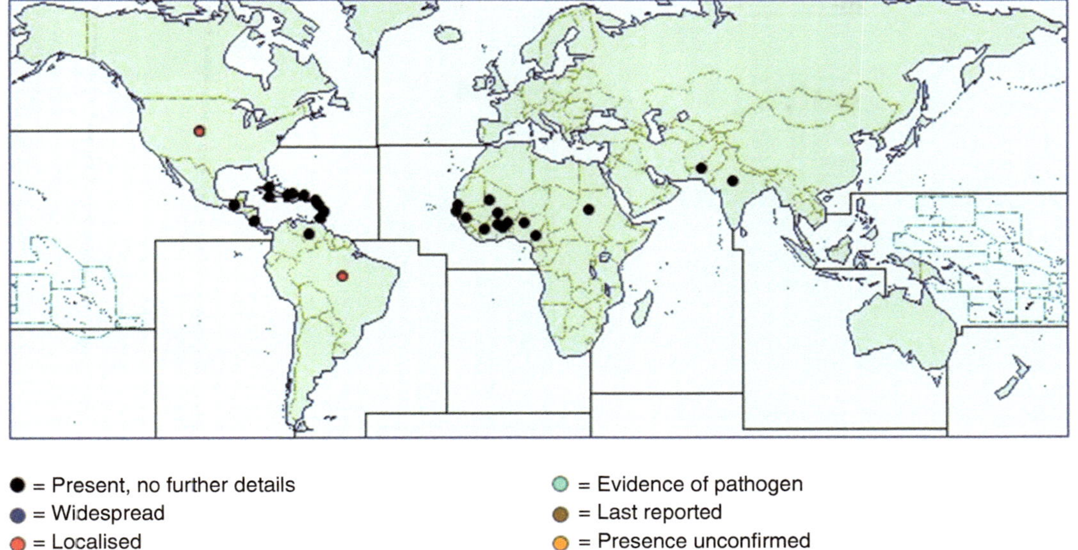

Fig. 5.34 Distribution of *Scutellonema bradys* in the world

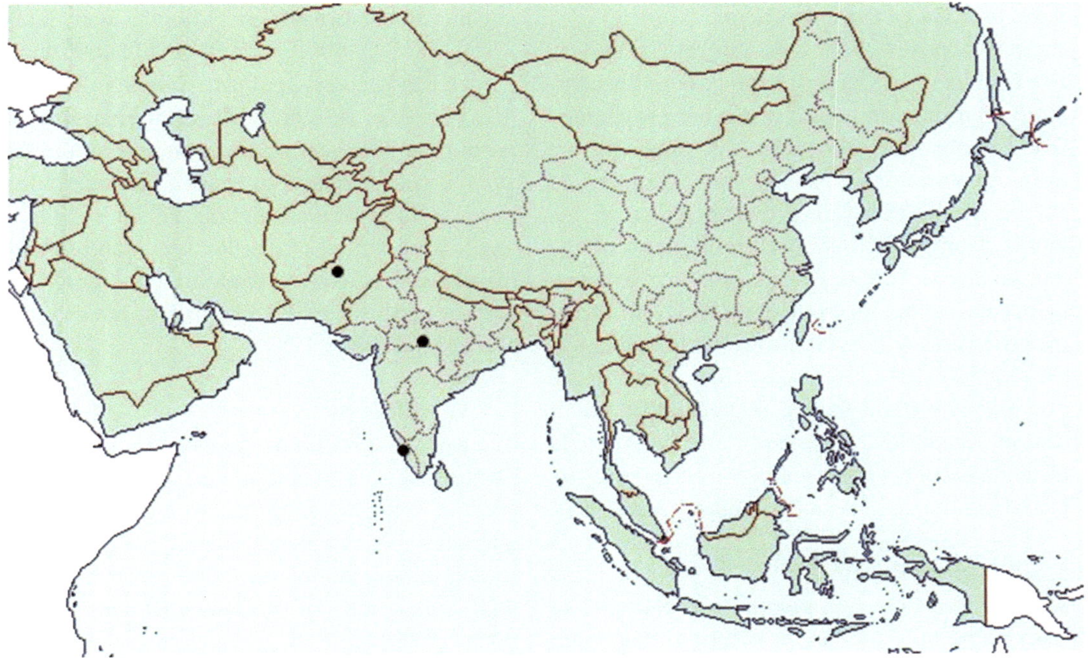

Fig. 5.35 Distribution of *Scutellonema bradys* in India

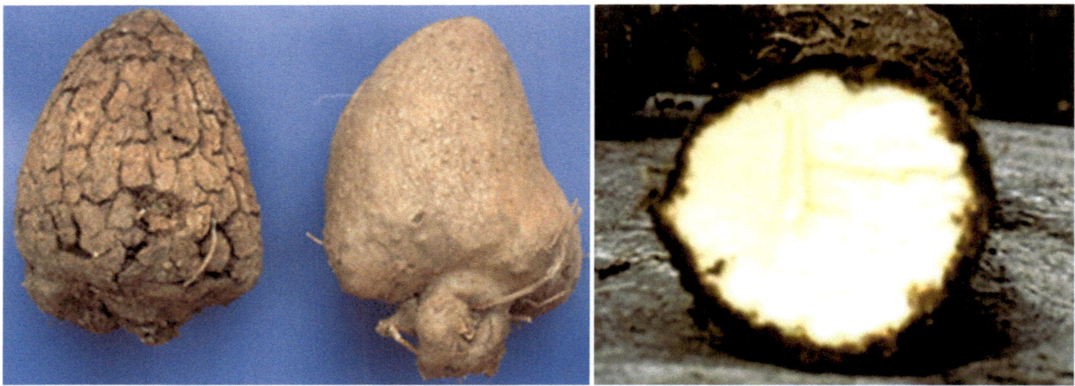

Fig. 5.36 *Left*, yam nematode-infested and healthy tubers; *right*, dry rot nematode on tubers

areas whose size depends on how long ago the damage occurred. The species *D. rotundata* is generally affected most.

Nematode damage to the tubers results in considerable reduction in the edible portion and marketable value of the tubers. Moisture loss is also significantly greater from diseased tubers during storage.

5.5.1.4 Risk of Introduction

S. bradys poses a considerable phytosanitary risk because of its survival and ease of dissemination within yam tubers. Yams are propagated from whole tubers or pieces of tuber and are thus the principal means of dissemination of *S. bradys* in the yam-growing areas of the world. Comparatively low populations of the nematodes in tubers do not produce external symptoms of damage (Bridge 1973) and therefore the risk of dissemination by this means is greater. Infested seed tubers rather than soil are probably the main source of nematode inoculum in yam fields. The spread of the nematode in infested tubers has been highlighted between islands in the Caribbean (Kermarrec et al. 1987).

5.5.1.5 Host Range

All the *Dioscorea* spp. grown as food crops are susceptible to *S. bradys*. In addition, other yam species known to be affected by *S. bradys* are *D. bulbifera*, *D. trifida*, and *D. transversa* (Ayala and Acosta 1971). Two wild *Dioscorea* spp. growing in forests in Nigeria and Cameroon have been shown to be natural hosts (Bridge et al. 2005). A wide range of other crops and some weeds have been reported to support low root populations of *S. bradys* including yam bean (*Pachyrhizus erosus*), pigeon pea (*Cajanus cajan*), okra (*Hibiscus esculentus*), tomato (*Lycopersicon esculentum*), sorghum (*Sorghum bicolor*), Loofah (*Luffa cylindrica*), and roselle (*Hibiscus sabdariffa*) (Adesiyan 1976). It occurs also in cassava (Missah and Peters 2001), taro (*Xanthosoma* sp., *Colocasia esculenta*), and sweet potato (*Ipomea batatas*) (Kermarrec et al. 1987). These alternative hosts permit the yam nematodes to survive in the soil even in the absence of yams. However, only cowpea (*Vigna unguiculata* subsp. *unguiculata*), watermelon (*Citrullus* spp.), and sesame (*Sesamum indicum*) in addition to yam have been found to actually increase populations of *S. bradys* (Bridge 1982).

5.5.1.6 Biology

The life cycle is simple. Eggs are laid in soil or plant tissues (roots and tubers) where they hatch and the juveniles develop into adults by subsequent molting. All stages seem to be infective.

Dense populations can build up in the tubers with a maximum of 62,000 nematodes/10 g of tuber recorded in Nigeria (Bridge 1972) but 100,000 nematodes were also reported to be found in the infested tubers in Nigeria (Bridge 1982). The *S. bradys* populations are affected by storage conditions and increase at twice the rate in tubers stored at 22–32 °C and relative humidity 40–85 %, when compared to tubers stored at 16–18 °C (Adesiyan 1977).

5.5.1.7 Histopathology

The nematodes were found in the intercellular spaces of the tissues lying within the periderm where extensive cell destruction was noticed.

5.5.1.8 Survival and Spread

Sizeable populations of the nematode are maintained in the absence of yams probably on other host plants. Yams are propagated from whole tubers or pieces of tuber which are the principal means of spread of *S. bradys*. Infested seed tubers rather than soil are probably the main source of nematode inoculum in yam fields.

5.5.1.9 Interaction with Other Pathogens

The more extensive internal decay of tubers known as "wet rot," "soft rot," or "watery rot" is associated with fungal (*Botryodiplodia theobromae* and *Fusarium* spp.) and bacterial (*Erwinia* sp.) pathogens. This general decay of tubers, which is a serious problem in stored yams, is increased when tubers are wounded or damaged. The damage caused by nematodes can predispose the tubers to invasion by decay organisms resulting in complete rotting of the tubers.

5.5.1.10 Management

Management of *S. bradys* can be achieved by one or more of the following measures: (1) controlling nematodes in field soil by cultural or chemical means, (2) use of planting material that is naturally free of nematodes or treatment of seed material (tubers and sets of yams) prior to planting to reduce or eliminate nematodes from propagative material, and (3) in the case of yams, treatment of tubers after harvesting to prevent storage losses (Jatala and Bridge 1990).

1. *Physical Methods*: Hot water treatment consists of heating water to temperature of 50–55 °C and submerging tubers for 45 min, which provides good control of *S. bradys* without damaging *D. rotundata* tubers (IITA 2005). The hot water treatment is effective but far from practical for farmers, due to the need for temperature control, labor, and fuel and simply due to the huge volume of yam material needed to be treated. Other factors, such as yam species, cultivars, nematode densities, depth of infestation, and age of tubers, also conflict with efficiency (Bridge et al. 2005). However, it is feasible for small-scale operations and for establishing nematode-free planting material for research experiments.
2. *Cultural Methods*
   (a) *Phytosanitation and Clean Planting Materials*: Phytosanitation is the first method used for nematode control on yam. It is necessary to separate infested tubers before storage and planting in order to prevent establishment of nematode infection. Healthy planting materials can also be obtained by using tissue-cultured planting material (Dropkin 1980; Speijer et al. 2000). In yam, use of nematode-free planting material is a practical and economic means of preventing damage by *S. bradys* and their dissemination. Aerial tubers of the yam *D. bulbifera* and some forms of *D. alata*, which are used for propagation, should, however, be completely free of nematodes. A number of yam species, such as *D. alata*, *D. rotundata*, and *D. dumetorum*, can also be produced from vine cuttings (Coursey 1967). Although these methods of propagation are not practical for producing yam tubers, they can be used to produce nematode-free seed tubers (IITA 2005). The use of "microsets" or "minisets," cut from mature tubers (IITA 2007), can be used to provide clean planting material if the mother seed yam tubers selected is free of nematodes.

Even true seed can be used for propagating *D. rotundata* (Sadik and Okereke 1975). Although these methods of propagation are not a practical means of producing ware tubers, they can be used to produce nematode-free seed tubers (Jatala and Bridge 1990).

Some traditional practices in Africa, such as the use of wood ash on yam tubers or mixing cow dung in yam mounds before planting, are reported to decrease nematode numbers (Adesiyan and Adeniji 1976).

(b) *Agronomic Practices in the Field*: Agronomic practices such as fallow, crop rotation with nonhost or cover crops are efficient for nematode control. In Cuba, keeping fallow land free of all host plants is a suggested means of reducing damage by *S. bradys* to yams (Decker et al. 1967) but this is unlikely to be economic or practical in most situations. A fallow of 8–12 months as well as crop rotation with nonhosts can reduce the nematode population in the soil (Adesiyan 1976). However, high land pressure and the cultivation of perennials prevent the successful application of fallow. When it is practical, rotation of yam with nonhosts or poor hosts such as groundnut (*Arachis hypogaea*), tobacco (*Nicotiana tabacum*), or cotton will limit damage by *S. bradys*. However, crop rotation to effectively reduce *S. bradys* populations seems to be nonrealistic because of its absence or low density in the soil.

Soil populations of *S. bradys* will be reduced if a nonhost or poor host crops, such as groundnut, chili pepper, tobacco, Indian spinach (*Beta vulgaris* var. *bengalensis*), cotton, maize, or sorghum, are grown prior to yams (Adesiyan 1976).

Using *Mucuna* sp. and *Crotalaria* sp. as cover crops has been reported to reduce soil nematode populations, particularly *Meloidogyne* spp. and *S. bradys* (Claudius-Cole et al. 2004). Other cover crops such as *Tagetes* spp., *Stylosanthes* spp., *Centrosema* spp., and *Aspilia* spp. have been recommended to lower nematode populations and restore fertility for yam production in Nigeria (Atu et al. 1983). Using cover crops in crop rotation is not efficient to control *Meloidogyne* spp., since these nematodes have a wide host range; moreover, as these crops are not edible, farmers are deterred from using them as a rotation crop.

(c) *Organic and Mineral Fertilizers*: It was reported that the application of mulch or organic matter might result in increased plant vigor and probably to some level, tolerance to nematodes. The use of cow dung, mixed in yam mound before planting at a rate of 1.5 kg per mound (1,886.3 kg/ha), was reported to increase yields of tubers and significantly decreased nematodes numbers (Adesiyan and Adeniji 1976). The use of *Azadirachta indica* powder at a rate of 2.5 t/ha has also been reported to decrease nematode populations as well as to increase soil fertility and tuber yields (Onalo et al. 2001).

For inorganic fertilizer use, superphosphate fertilizers were reported to reduce *S. bradys* populations in tubers of *D. alata* to a low level but did not increase yam yield (Baimey et al. 2006). The traditional practice of using wood ash on yam tubers before planting is reported to decrease nematode numbers (Adesiyan and Adeniji 1976).

NPK fertilizers can reduce *S. bradys* populations in tubers of *D. alata* to a very low level. N alone can increase both populations of *S. bradys* and the percentage of infested tubers of *D. rotundata*, whereas P alone can reduce percentage of infested tubers. These results support observations by farmers in certain yam-growing areas of Nigeria that yams fertilized with N alone do not store well, but yams fertilized with mixtures that contain P store longer.

3. *Chemical Methods*: Nematicide treatment of planting material however, to generate healthy seed yam free of nematodes, can have a major effect and proved to be highly economical (IITA 2006). Bare root-dip treatment of diseased yams in fensulfothion (Dasanit), thionazin (Nemafos), and DBCP (Nemagon) at 1,250 ppm for 15 min. and 625 ppm for 30–60 min was found effective in eliminating the nematode infection (Ayala and Acosta 1971). Significant increases in yield have been obtained by soaking tuber pieces of *D. alata* infected with *S. bradys* for 30 min. in 1,000 ppm aqueous solutions of carbofuran and oxamyl.

Application of aldicarb, oxamyl, carbofuran, miral, or isazophos each at 2 kg a.i./ha two weeks after planting reduced soil populations of *S. bradys* to very low levels with remarkable yield increases recorded.

4. *Host Resistance*: Yam cv. 'Florido' of *Dioscorea alata* did not seem to be susceptible to nematode attack (Ayala and Acosta 1971). One cultivar of yellow yam, *D. esculenta*, and one of *D. dumetorum* have shown some resistance to *S. bradys* (Bridge 1982; Kwoseh 2000).
5. *Biological Methods*
   (a) *Antagonistic Bacteria*: Varghese and Mohandas (2004) reported *Pasteuria penetrans* was effective against the yam nematode, *S. bradys.* Eight to ten spores were found attached in different parts of the body (4–5 spores in the head region and 3–4 spores in the esophageal region). *P. penetrans* appears to be a potential biocontrol agent against yam nematode.
   (b) *Plant Growth-Promoting Rhizobacteria (PGPR)*: Rhizobacteria such as *Rhizobium* spp. and *Pseudomonas* spp. not only have a positive effect on plants by promoting their growth, but in addition they show a repellent effect toward nematodes (Aalten et al. 1998; Hallmann et al. 2001).
   (c) *Arbuscular Mycorrhizal Fungi (AMF)*: An efficient practice to increase soil fertility and decrease nematode population in the soil, as well as on yam tubers, could be to increase the levels of soil microorganisms and particularly AMF spore density and diversity or inoculation of selected AMF species or isolates (Cardoso and Kuyper 2006; Smith and Read 2008).

The presence of AMF tended to lead to improved growth of yam, especially *D. alata* cvs., as compared with the non-AMF control plants. When challenged with the yam nematode *S. bradys*, plantlets of the two *D. alata* cultivars preinoculated with *G. mosseae* and cv. 'TDr97-00551' preinoculated with *G. dussii* yielded significantly higher tuber weights compared to non-AMF control plantlets, indicating a relatively high degree of functional specificity of the AMF species.

6. *Integrated Methods*

A combination of hot water treatment and phytosanitary measures as well as crop rotations was experimentally used in IITA-Ibadan, West Africa, for successful yam nematode control (IITA 2005).

5.5.2 Lesion Nematode, *Pratylenchus* spp.

Pratylenchus coffeae, commonly called "lesion nematode," causes a marked reduction in yield and dry rot on stored yams. The symptoms on the affected tubers are similar to that of *S. bradys* (Bridge 1982).

Pratylenchus sudanensis is morphologically similar to *P. pseudopratensis* (Coyne et al. 2003) and associated with cracked tubers (Mudiope et al. 2004). Host range studies of *P. sudanensis* in Sudan showed that 20 plant species such as cotton, sorghum, and pigeon pea (*Cajanus cajan*) acted as favorable hosts while groundnut (*Arachis hypogaea*) and wheat (*Triticum turgidum*) were considered as poor hosts (Saadabi 1985).

5.5.2.1 Distribution

Pratylenchus coffeae has been reported as a pest of yam in Puerto Rico, Jamaica, British Solomon Islands, the Pacific, and Central America (Thompson et al. 1973). *Pratylenchus sudanensis* was recently observed on yam in Uganda (Coyne et al. 2003), where it was reported as dominant nematode on yam (Mudiope et al. 2001). *Pratylenchus brachyurus* has been found in yam tubers, roots, and yam soil in Nigeria (Caveness 1967), Ivory Coast (Miège 1957), Guatemala (Jenkins and Bird 1962), Fiji, and Tonga (Bridge 1988) but appears to be of limited importance.

5.5.2.2 Symptoms

Symptoms of damage to yam tubers by *P. coffeae* are similar to those caused by *S. bradys*. The nematodes are restricted to the outer layers of tuber tissue, cause severe necrosis or dry rot and cracking of the skin, and are associated with complete deterioration of tubers.

The lesion nematode lives endoparasitically and is typically a root parasite, but it also attacks underground stems and tubers. *Pratylenchus coffeae* is assumed to have a life cycle of 3–4 weeks on yam and causes dry rot of the peridermal and subdermal regions similar to that caused by *S. bradys* (Thompson et al. 1973). The nematode is concentrated in the apical portion within 6 mm up to 15 mm into yam tuber tissue. Dry rot can extend 1–2 cm into the outer tissues of *D. rotundata* tubers (Acosta 1974) but has been estimated to extend up to 5 cm in *D. alata* tubers (Bridge and Page 1984). *Scutellonema bradys* and *P. coffeae* can induce a synergistic effect on yam growth in pot experiments (Acosta and Ayala 1976) but were together exclusively present in tubers harvested from fields in India (Castognone-Sereno and Kermarrec 1988).

5.5.2.3 Biology

P. coffeae is a migratory endoparasite of yam roots and tubers. The life cycle is completed in 3–4 weeks on yams.

5.5.2.4 Host Range

P. coffeae is a parasite of *D. alata*, *D. cayenensis*, *D. rotundata*, *D. bulbifera*, and *D. trifida*. In addition to yams, *P. coffeae* has an enormous host range covering almost all plant families.

5.5.2.5 Management

1. *Physical Methods*: Immersion of *D. rotundata* tubers in hot water maintained at 46–52 °C for 15–30 min has given good control of *P. coffeae*.
2. *Cultural Methods*: Using plant material which is free of nematodes is an effective means of controlling or reducing damage by *P. coffeae*. Collection of central or distal tuber pieces, which generally contain least *P. coffeae*, is recommended for propagative material.
3. *Chemical Methods*: Immersion of *D. rotundata* tubers in DBCP and Fensulfothion at 1,250 ppm for 15–30 min effectively reduced *P. coffeae* present to a depth of 6 mm in the tuber tissues.

Aldicarb as a single application at planting at 5.4 kg a.i./ha can give 72 % control of *P. coffeae* and significantly increase high-quality tuber yields of *D. rotundata*. Significant increases in yield of *D. rotundata* have also been obtained by a combination of foliar and seed tuber treatments with oxamyl.

4. *Host Resistance*: *D. alata* cv. 'Florido' is not susceptible to attack by *P. coffeae* (or *S. bradys*). *D. esculenta* is possibly less susceptible to *P. coffeae* because of its different growth habit.

5.5.3 Root-Knot Nematodes, *Meloidogyne* spp.

The root-knot nematode is another important nematode occurring on yams and is present worldwide. Yams have been found infested by *M. incognita*, *M. javanica*, *M. arenaria*, and *M. hapla*.

5.5.3.1 Distribution

Meloidogyne spp. has been found on yam in Africa (Ghana, Ivory Coast, Nigeria, Uganda, and Ethiopia), the Caribbean (Jamaica, Martinique, Puerto Rico, Trinidad), the Pacific (Fiji, Kiribati, Papua New Guinea, Western Samoa), Brazil, Guatemala, and Japan (Bridge et al. 2005). They are well distributed all over the world. The most widely spread species are *M. incognita* and *M. javanica* (Bridge et al. 2005). They can become abundant on yam above all in areas where *S. bradys* is not present. A general survey conducted by Coyne et al. (2005) in West Africa showed that the *Meloidogyne* spp. is associated with yam across the whole region. In spite of their widespread occurrence and abundance, root-knot nematodes are considered of limited importance on yam in West Africa (IITA 2005).

5.5.3.2 Economic Importance

M. javanica populations of 30,000 nematodes/plants can reduce yields of *D. opposita* by over 50 %. It is estimated that there is a reduction of 39–52 % in the price of galled tubers compared to healthy ones (Nwauzor and Fawole 1981). The proportion of galled tubers collected from yam barns and markets in Nigeria can be as high as

90 % for *D. alata* and 70 % for *D. rotundata* (Adesiyan and Odihirin 1978), although in general the proportion of affected yams is much lower (Coyne et al. 2005).

Nwauzor and Fawole (1981) recorded losses of 25–75 % due to *Meloidogyne* spp. infection on yam within a storage period of 16 weeks in Nigeria.

5.5.3.3 Symptoms

Yam seedlings infested with root-knot nematodes can be severely stunted and chlorotic; very young seedlings may be killed by severe infestation. Foliar symptoms, such as early yellowing, leaf fall, and inhibited vine growth, have been observed on yam infested with *Meloidogyne* spp. (Nwauzor and Fawole 1981). The roots swell in size at the site of infection producing the characteristic and typical knots on the roots (Fig. 5.37). There is sometimes a proliferation of tuberous roots on these galls (hirsutism). *D. alata* is more sensitive to this type of nematodes than *D. rotundata*.

Tuber formation was severely affected, often infested tubers are smaller, and most of the tuber forming roots does not develop. Rotting of tubers has been reported.

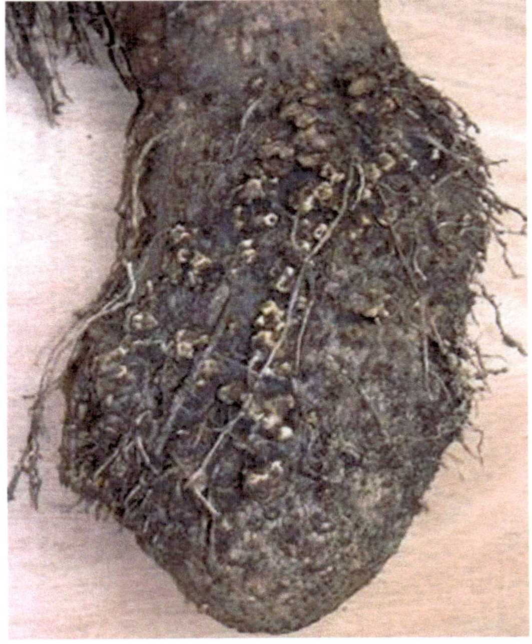

Fig. 5.37 Root-knot nematode-infected yam tuber

The root-knot nematode infests the roots and tubers causing reduction in size of tubers, irregular wartlike projections on the surface, and branching of tubers with irregular shape. This nematode also continues to inflict postharvest damage after harvest in store.

5.5.3.4 Biology

The life cycle of *M. incognita* in *D. rotundata* or *D. alata* tubers is 35 days under controlled conditions at 28 °C (Nwauzor and Fawole 1981). In *D. alata*, most nematodes are concentrated to a depth of 2 mm with none beyond the 8 mm depth. In *D. rotundata* they are concentrated at depths between 4 and 6 mm with few at 14 mm.

5.5.3.5 Host Range

Susceptible yam hosts of *M. incognita* are *D. alata*, *D. cayenensis*, *D. esculenta*, *D. bulbifera*, *D. composita*, *D. floribunda*, *D. trifida*, *D. prehensilis*, *D. spiculiflora*, and *D. rotundata*. Hosts of *M. javanica* are *D. alata* and *D. rotundata*.

5.5.3.6 Survival and Spread

Where *Meloidogyne* juveniles and/or eggs survive in stored tubers, they will be spread in propagative material. However, *Meloidogyne* spp. have extremely wide host ranges and damaging populations will come from field soil having survived on other weed hosts or be introduced into yam fields on infested seedlings of other crops.

5.5.3.7 Management

1. *Physical Methods*: Hot water treatment of corms at 45–51 °C for 30 min. eliminates root-knot nematode infection.
2. *Cultural Methods*: The root-knot nematode can be controlled by growing sweet potato cv. 'Shree Bhadra' as a trap crop.

   Intercropping yam miniset (40,000/ha) with groundnut SAMNUT 10 at (160,000/ha or 120,000/ha), nematode damage of seed yams will be reduced by about 60 % (Nwauzor 1996).

   Use of nematode-free tubers for propagation will effectively reduce the incidence and damage.

3. *Chemical Methods*: Application of carbofuran at 3 kg a.i./ha gave effective control of root-knot nematodes. Granular oxamyl at 3 or 6 kg a.i./ha applied at planting and at 3- and 4-week intervals can control *M. javanica* on *D. rotundata*. In the presence of both *M. javanica* and *Pratylenchus brachyurus*, tuber yields can be increased by over 40 % when granular oxamyl at 3 kg a.i./ha applied at planting is combined with subsequent applications of calcium nitrate or ammonium sulfate incorporated at 3–4-week intervals, each at 60 kg N. These treatments also reduce the incidence of rot in stored yams associated with nematodes.
4. *Host Resistance*: The only yam species consistently found to be resistant to attack by *M. incognita* is the cluster yam, *D. dumetorum*. *D. alata* cv. 'Obunenyi' is reported to be resistant to *M. incognita*, and *D. cayenensis* can be resistant to *M. incognita* and *M. javanica*. Yam cvs. 'Sree Latha' and 'Sree Kirthi' were found resistant to root-knot nematodes.
5. *Biological Methods*: At harvest tuber galling was significantly lower on plantlets inoculated with both arbuscular mycorrhizal fungi (AMF) species and *Meloidogyne* spp. compared to *Meloidogyne* spp. inoculation alone. Plants inoculated with *Meloidogyne* spp. had no roots at harvest and highly galled tubers, compared to tubers harvested from plantlets inoculated with AMF species and *Meloidogyne* spp. (Fig. 5.38). AMF inoculation suppressed the densities of *Meloidogyne* spp. both in yam roots and in soil. For individual yam cultivars, only *Glomus dussii* significantly suppressed *Meloidogyne* spp. soil density on cv. 'TDa98-01183.' However, application of either AMF species (*Glomus mosseae* or *G. dussii*) improved tuber quality of plants inoculated additionally with *Meloidogyne* spp. compared to nematode inoculation alone (Fig. 5.38).

When challenged with *Meloidogyne* spp., cv. 'TDa98-01183' plantlets preinoculated with either AMF species (*Glomus mosseae* or *G. dussii*) yielded heavier tubers than the non-mycorrhizal control plantlets. *Scutellonema bradys* densities on yam plantlets preinoculated with AMF were generally suppressed, although no differences were observed in visible damage scores, which remained low or absent across treatments. However, *Meloidogyne* spp. densities and galling damage tended to be similar with or without AMF. Thus, with regard to nematode suppression, AMF appears to counter *S. bradys* attack more than an attack by *Meloidogyne* spp.

Fig. 5.38 Quality of yam tubers and roots at harvest 6 months after planting following arbuscular mycorrhizal fungi inoculation at planting and *Meloidogyne* spp. inoculation 2 months after planting. L to R, control (non-inoculated), *G. mosseae* + *Meloidogyne* spp., *G. mosseae*, and *Meloidogyne* spp.

Table 5.6 Pests and diseases of yams and recommended control strategies

Pest/disease	Incidence and characteristics	Control strategy
Yam beetle, *Heteroligus meles*	A major pest in West Africa; adults migrate by flying to other farms/plots; they feed on the tuber, thereby bore holes on them. Tubers lose appearance and quality and become prone to rotting during storage. Plants may wilt and die. The holes in tubers reduce market value	Dust the sets with insecticide. Dusting of holes or soil surface with aldrin, endosulfan, or grammalin. Very late planting can reduce infestation but affect yields
Chrysomelid beetle, *Lilioceris livida*	Cause major damage locally, especially during the first months of the crop, but without any real economic impact	Removal of larvae by hand or by spraying carbaryl
Scales, *Aspidiella hartii*	Adult and nymphal scale insects (whitish yellow in color) are present on the tubers	Removal of pests with a brush and treatment with diazinon + white oil/ malathion prior to planting
Mealybugs, *Planococcus citri*	These homopterae develop mainly during storage, forming a sort of white powder near the top of the tubers. They can cause complete necrosis of sprouts preventing the use of tubers as seedlings	Using a clean planting stock
Anthracnose, *Colletotrichum gloeosporioides*	Occurs in all yam-producing areas; infection from inoculum borne within the planting sett or in the soil; results in blackening and dieback of the leaves and shoot. It is seen as small, black spots between the leaf veins. Some infections can coalesce to form massive blighted lesions	Soak the planting material with recommended copper fungicides prior to planting and drench the planting holes with the fungicide after planting. Spray the vines with maneb or benomyl upon shoot emergence. Encourage healthy plants by ensuring adequate levels of nutrients. Sanitize the field by raking and removing fallen leaves prior to cultivation to reduce the source of the inoculum (spores of the fungus). Practice crop rotation. Intercrop with corn helps to move the spores up and away from the yam leaves reducing the infection rate. Use windbreaks on the windward side of the crop. Inspect the field continuously and rogue out infected plants, especially at the windward side of the field. Use resistant cultivar (e.g., 'TDA 291', 'TDA 297')
Nematodes, *Scutellonema bradys*, *Pratylenchus coffeae*, *Meloidogyne* spp.	They reside in the tuber and remain active during storage. Wounded areas provide entry for decay-causing bacteria and fungi	Crop rotation and fallowing and planting with healthy materials. Soil fumigation with nematicides may be uneconomical
Yam virus complex	Occurs throughout the West African yam zone. Reduces yields considerably	Use virus-free planting materials or resistant cultivars, thermotherapy, and meristem-tip culture
Tuber rots, soft rots caused by *Penicillium* spp., *Fusarium oxysporum*, and *Botryodiplodia theobromae*. Dry rots caused by *Rosselinia* and *Sphaerostilbe*. Other fungi: *Rhizopus nodosus*, *F. solani*	Infection in the field can persist and lead to rotting during storage	Plant with disease-free material. Crop rotation. Minimize physical damage of tuber during postharvest operations. Treat the sett or tuber with systemic fungicide or alkaline material such as Bordeaux mixture. Provide adequate aeration and inspect stored tuber regularly

5.6 Pests and Disease Management

Yams are susceptible to a variety of pests and diseases during growth as well as postharvest. Attack by the yam beetle and microorganisms such as nematodes and yam virus is the most devastating. The major postharvest disease is tuber rots caused mostly by fungi. Table 5.6 summarizes the common pests and diseases, their characteristics, and potential remedies. Fumigation is generally carried out using methyl bromide, and the application must be checked with appropriate regulatory authority and with the importer if produce is destined overseas. Fencing, poisoning, and trap setting are common methods for controlling rodents.

5.7 Future Thrusts

- Yam (*Dioscorea alata*) accessions Da10, Da 11, Da 60, Da 62, Da 64, and Da 157 which are showing field tolerance to anthracnose can be utilized for breeding resistant cultivars.
- Pest problems like anthracnose in greater yam and dry rot nematode in white yam, still continue to be serious limiting factors affecting the yield of yams. Priority may be given to develop varieties with resistance/tolerance to the above pests.
- Detailed studies should be undertaken in developing resistant varieties to yam anthracnose disease using biotechnological methods (pathogen-derived and elicitor-induced methods).
- Refinement of IDM technologies developed for the management of anthracnose in greater yam needs to be undertaken.
- Safe control measures and storage techniques are to be developed to ensure longer storage life and better viability of planting materials of *Dioscorea*. There is a need for development of environmental friendly control measures through use of biopesticides, attractants, and repellants.
- Emerging pests and diseases due to climate change and introduction of invasive pests and pathogens are also main challenges. Appropriate technologies to contain such biotic stresses have to focus on refined eco-friendly methods. The important fungal disease like yam anthracnose is disseminated through planting materials and soil. Efforts to provide healthy planting material by employing diagnostic techniques to identify pathogens need to be strengthened.
- Effective technology has to be developed in order to control viral diseases in planting materials of yams. Thermotherapy and chemotherapy studies should be intensified.
- There is a need to intensify the breeding efforts toward developing nematode-resistant/tolerant varieties of white yams.

References

Aalten PM, Vitour D, Blanvillain D, Gowen SR, Sutra L (1998) Effect of rhizosphere fluorescent *Pseudomonas* strains on plant-parasitic nematodes *Radopholus similis* and *Meloidogyne* spp. Lett Appl Microbiol 27:357–361

Acosta N (1974) Depth penetration of phytoparasitic nematodes in yam tubers. Nematropica 4:7–11

Acosta N, Ayala A (1976) Hot water and chemical dips for nematode control in tubers of *Dioscorea rotundata*. J Agric Univ Puerto Rico 60(3):395–402

Adebanjo A, Onesirosan PJ (1986) Surface-borne infection of *Dioscorea alata* tubers by *Colletotrichum gloeosporioides*. J Plant Protect Tropics 3:132–137

Adeniji MO (1970a) Influence of moisture and temperature on yam decay organisms. J Phytopathol 60:1698–1699

Adeniji MO (1970b) Fungi associated with storage decay of yam in Nigeria. Phytopathology 60:590–592

Adesiyan SO (1976) Host range studies of the yam nematode *Scutellonema bradys*. Nematologica 22:289–297

Adesiyan SO (1977) Penetration and multiplication of *Scutellonema bradys* in yams (*Dioscorea* spp.). Nematol Mediterr 5:313–317

Adesiyan SO, Adeniji MO (1976) Studies on some aspects of yam nematode (*Scutellonema bradys*). Ghana J Agric Sci 9:131–136

Adesiyan SO, Odihirin RA (1977) Plant parasite nematodes associated with yam tuber in mid-western state, Nigeria. Nig J Plant Protect 3:178–179

Adesiyan SO, Odihirin RA (1978) Root-knot nematodes and pest of yams (*Dioscorea* spp.) in southern Nigeria. Nematologica 24:132–134

Adesiyan SO, Odihirin RA, Adeniji MO (1975) Economic losses caused by the yam nematode, *Scutellonema bradys*, in Nigeria. Plant Dis Rep 59:477–480

Akinlosotu TA, Kogbe J (1986) Studies of the incidence of yam scale *Aspidiella hartii* on *Dioscorea* spp. and its chemical control. J Root Crops 14(2):21–23

Arora RK, Pandey A (1996) Wild edible plants of India: diversity, conservation and use. NBPGR, New Delhi

Atu UG, Odurukwe SO, Ogbuji RO (1983) Root-knot nematode damage to *Dioscorea rotundata*. Plant Dis 67:814–815

Ayala A, Acosta N (1971) Observation on yam (*Dioscorea alata*) nematodes. Nematropica 1:39–40

Baimey H, Coyne D, Labuschagne N (2006) Effect of fertilizer application on yam nematode (*Scutellonema bradys*) multiplication and consequent damage to yam (*Dioscorea* spp.) under field and storage conditions in Benin. Int J Pest Manag 52:63–70

Balakrishnan V, Ratheesh Narayanan MK, Anil Kumar N (2003) Ethnotaxonomy of *Dioscorea* among the Kattunaikka people of Wayanad District, Kerala. India IPGRI News Lett 135:24–32

Bridge J (1972) Nematode problems with yams (*Dioscorea* spp.) in Nigeria. PANS 1:89–91

Bridge J (1973) Nematodes as pests of yam in Nigeria. Mededelingen Faculteit Landbouwwetenschappen Gent 38:841–852

Bridge J (1982) Nematodes of yam. In: Miege J, Lyonga SN (eds) Yams. Clarendon Press, Oxford, pp 253–264, Ignamea

Bridge J (1988) Plant parasitic nematode problems in the Pacific islands. J Nematol 20:173–183

Bridge J, Coyne D, Kwoseh CK (2005) Nematode parasites of tropical root and tuber crops (Excluding potatoes). In: Luc M, Sikora RA, Bridge J (eds) Plant parasitic nematodes in subtropical and tropical agriculture, 2nd edn. CAB International Publishing, Cambridge, MA, pp 221–258

Bridge J, Page SLJ (1984) Plant nematode pests of crops in Papua New Guinea. J Plant Protect Tropics 1:99–109

Butler L, Vilsoni F (1975) Potential hosts of the burrowing nematode in Fiji. Fiji Agric J 37:38–39

CAB International (1966) Distribution maps of plant pests. CAB International, Oxfordshire

Calverly (1998) Storage and processing of roots and tubers in the tropics. Food and Agriculture, Association of the United Nations, Rome

Cardoso I, Kuyper TW (2006) Mycorrhizas and tropical soil fertility. Agric Ecosyst Environ 116:72–84

Castognone-Sereno P, Kermarrec A (1988) Association between *Pratylenchus coffeae* and *Scutellonema bradys* in yam tubers under agronomic conditions in the French West Indies. Nematropica 18:155–157

Caveness FE (1967) Shade house host ranges of some Nigeria nematodes. Plant Dis Rep 51:115–119

Caveness FE (1992) Nematological research at IITA 1969–1988. A summary of investigations. In: Lowe J (ed) Plant health management research monograph 2. IITA, Ibadan, p 52

Claudius-Cole AO, Asiedu R, Fawole B (2004) Cultural control of *Scutellonema bradys* on *Dioscorea* spp. In: Akporoda MO, Ekanayake IJ (eds) Proceedings of the international symposium for root and tuber crops, Ibadan, 12–16 November 2001, pp 522–526

Coursey DG (1967) Yams: an account of the nature, origins, cultivation, and utilization of the useful members of Dioscoreaceae. Longmans, Green and Co Ltd, London, 230 pp

Coyne DL, Talwana HAL, Maslen NR (2003) Plant-parasitic nematodes associated with root and tuber crops in Uganda. Afr Plant Protect 9:87–98

Coyne DL, Tchabi A, Baimey H, Labuschagne Rotifa I (2005) Distribution and prevalence of nematodes (*Scutellonema bradys* and *Meloidogyne* spp.) on marketed yam (*Dioscorea* spp.) in West Africa. Field Crop Res 96:142–150

Decker H, Casamayor GR, Bosch D (1967) Observaciones sobre la presencia del nematode *Scutellonema bradys* en el tuberculo die name en la provincea de Oriente (Cuba). Centro, Boletin de Ciencias y Tecnologia, Universidad Central de Las Villas 2:67–75

Dina SO (1977) Occurrence of *Dasyses rugosella* Stainton (Lepidoptera: Tineidae) on stored yam tubers in Nigeria. Trop Sci 19:113–117

Dropkin VH (1980) Introduction to plant nematology. Wiley, New York

Dumont R, Vernier P (2000) Domestication of yams (*Dioscorea cayenensis-rotundata*) within the Bariba ethnic group in Benin. Outlook Agric 29:137

FAO (2011) Production crop data, Yams (2010). Food and Agriculture Organization of the United Nations, Rome

FAO/IPPGR (1989) Safe movement of yam germplasm. In: Brunt AA, Jackson GVH, Prison EA (eds) Food & Agriculture Organisation/International Board for Plant Genetic Resources Technical Guidelines, 19 pp

FAOSTAT (2012) FAO statistical yearbook 2012. Food & Agriculture Organization of the United Nations, Rome

Hahn SK, John C, Isoba G, Ikoun T (1989) Resistance breeding in root and tuber crops at the International Institute for Tropical Agriculture (IITA), Ibadan Nigeria. Crop Prot 8:147–168

Hallmann J, Quadt-Hallmann A, Miller WG, Sikora RA, Lindow SE (2001) Endophytic colonization of plants by the biocontrol agent *Rhizobium etli* G12 in relation to *Meloidogyne incognita* infection. Phytopathology 91:415–422

Henry AN, Kumari GR, Chithra V (1989) Flora of Tamil Nadu, India, vol 2. Sugarcane Breeding Institute, Coimbatore

IITA (International Institute of Tropical Agriculture) (1993) Crop improvement division/Tuber root Improvement Program Archival Reports (1989–1993), Part III yam, *Dioscorea* spp. Ibadan, Nigeria, pp 20–85

IITA (2005) Plant health management annual report, Ibadan, Nigeria

IITA (2006) Plant health management annual report, Ibadan, Nigeria

IITA (2007) Plant health management annual report, Ibadan, Nigeria

IITA (2010) Yam. International institute of tropical agriculture, Nigeria

Jatala P, Bridge J (1990) Nematode parasites of roots and tuber crops. In: Luc M, Sikora RA, Bridge J (eds) Plant parasitic nematodes in tropical and subtropical agriculture. CAB International, Wallingford, pp 137–180

Jenkins WR, Bird GW (1962) Nematode associated with wild yam, *Dioscorea* sp, with special reference to pathogenicity of *Meloidogyne incognita*. Plant Dis Rep 46:858–860

Kay DE (1987) Root crops. Tropical Development and Research Institute, London

Kermarrec A, Anais A (1973) Presencia en Guadalupe de *Aphelenchoides besseyi* Christie 1942 (Nematoda, Aphelenchoididae) en foliage del mapuey: *Dioscorea trifida* L. Turrialba 23:389–393

Kermarrec A, Anais A, Gamiette F, Suard C, Dos Santos F (1991) Evaluation of susceptibilities of yam cultivars (*D. alata, D. cayenensis, D. trifida*) to soil nematodes: *Meloidogyne incognita, Pratylenchus coffeae* and *Scutellonema bradys*. Caribbean Food Crops Society, Antilles Guyane: INRA Proceedings of the twenty fifth annual meeting 1989, pp 225–235

Kermarrec A, Castagnone-Sereno P, Degras L, Anais A, Denon D (1987) Nouvelle distribution of *Scutellonema bradys* (Tylenchida: Hoplolaiminae) dans le Caraibe. Le cas des Antilles Francaises Medelingen van de Faculteit Landbouwwetenschappen Rijksuniversiteit Gent 52:617–624

Kwoseh CK (2000) Identification of resistance to major nematode pests of yams (*Dioscorea* spp.) in West Africa. PhD thesis. The University of Reading, Reading, 196 pp

Law-Ogbomo KE, Egharevba RKA (2006) Establishment, yield and control of *Heteroligus meles* (Bilb.) (Coleoptera: Dynastidae) in *Dioscorea rotundata* Poir on *Amitemes evancifer* (Silv.) infested site using wood ash, carbofuran and aldrin. Asian J Plant Sci 5:320–323

Linus Opara U (2003) Yams: post-harvest operation. INPho post-harvest compendium, FAO, Rome, 22 pp

Mantell SH (1980) Apical meristem-tip culture for eradication of flexuous rod viruses in yam (*Dioscorea alata*). Trop Pest Manag 26:170–179

Michereff SJ, Silveira NSS, Reis A, Mariano RLR (1994) Epiphytic bacteria antagonistic to *Curvularia* leaf spot of yam. Microb Ecol 28:101–110

Miège J (1957) Influence de quelques caractères des tubercules semences sur la levee et le rendement des ignames cultivées. Journal d'Agriculture Tropicale et de Botanique Appliquée 4:315–342

Mignouna HD, Abang MM, Asiedu R (2003) Harnessing modern biotechnology for tropical tuber crop improvement: Yam (*Dioscorea* spp.). Molecular Breeding (Available online)

Missah A, Peters JC (2001) Plant parasitic nematodes associated with soils, intercrops and tubers of yam in Ghana. In: Proceedings of the 7th triennial symposium of the international society for tropical root crops-Africa branch. Cotonou, Benin, 11–17 October 1998, pp 639–644

Mohamed T (1979) Trop Agric Trin 56:175

Mudiope J, Speijer PR, Maslen NR, Adipala E (1998) Evaluation of yam host-plant response to root knot nematodes in Uganda. Afr Plant Protect J 4:119–122

Mudiope J, Coyne DL, Adipala E, Sikora RA (2004) Pathogenicity of the lesion nematode, *Pratylenchus sudanensis* (Loof and Yassin) in pots. In: Proceeding of the VIth Africa crop science conference, Nairobi, Kenya, 12–17 October 2003

Mudiope J, Speijer PR, Maslen NR, Adipala E (2001) *Pratylenchus*, the dominant genus affecting yam (*Dioscorea* spp.) in Uganda. In: Proceedings of the 7th triennial symposium of the international society for tropical root crops- Africa branch, Cotonou, Benin, 11–17 October 1988, pp 637–638

Nadakal AM, Thomas AN (1967) Observations on nematodes associated with dry rot of *Dioscorea alata* L. Sci Cult 33:142–143

Niashizawa T (1973) Pathogenicity of *Trichoderma porosus* to Chinese yam. Jpn J Nematol 3:33–37

Nwankiti AO, Atu UG, Alozie SO, Igbokwe M, Okpokiri A, Nnodu E (1988) Effect of modifying two traditional yam barn structures on storability of four cultivars of yam (*Dioscorea* spp). Trop Agric 65(2):125–128

Nwauzor EC (1996) Use of groundnut (*Arachis hypogaea* L) as an intercrop for nematode control in seed yam production by minisett techniques. Final Report. National Agricultural Research Project (NARP) Priority Research and REFILLS Projects 22–26

Nwauzor EC, Fawole B (1981) Root knot nematodes on yams (*Dioscorea* spp.) in eastern Nigeria. In: Proceeding of the 3rd research planning conference on root-knot nematodes, *Meloidogyne* spp., Region IV and V, Ibadan, Nigeria, pp 161–167

Odu BO, Shoyinka, SA, Hughes JDA, Asiedu R, Oladirin OA (2001) Yam viruses of Nigeria. In: Proceedings of the 7th triennial symposium on international society of tropical root crop –African branch, Cotonou, Benin, 12–16 October 1998

Oke OL (1990) Roots, tubers, plantains and bananas in human nutrition. FAO, Rome

Okigbo RN (2002) Mycoflora of tuber surface of white yam (*Dioscorea rotundata* Poir) and post-harvest control of pathogens with *Bacillus subtilis*. Mycopathologia 156:81–85

Okigbo RN (2005) Biological control of post-harvest fungal rot of yam (*Dioscorea* spp.) with *Bacillus subtilis*. Mycopathologia 159:307–314

Okigbo RN, Ikediugwu FEO (2008) Studies on biological control of post-harvest rot in yams (*Dioscorea* spp.) using *Trichoderma viride*. J Phytopathol 148:351–355

Onalo JI, Asiedu R, Adisiyan SO (2001) Control of the yam nematode (*Scutellonema bradys*) with neem fruit powder. In: Proceedings of the 7th triennial symposium of the international society for tropical root crops-Africa branch, Cotonou, Benin, 11–17 October 1998, pp 634–637

Onwueme IC, Sinhad TD (1991) Field crop production in tropical Africa. CTA, Ede-Wageningen, pp 337–342

Osagie AU (1992) The yam tuber in storage. Post Harvest Research Unit, University of Benin, Benin, pp 107–173

Plumbley RA, Rees DP (1983) An infestation by *Araecerus fasciculatus* (Degeer) (Coleoptera: Anthribidae) and *Decadarchis minuscula* (Walsingham) (Lepidoptera: Tineidae) on stored fresh yam tubers in south-east Nigeria. J Stored Prod Res 19(2):93–95

Prain D, Burkill IH (1936) An account of the genus *Dioscorea* in the East, Part 1. The species which twine to the left. Ann R Bot Gard Calcutta 14(1):1–210

Reckhaus (1980) Ph.D. dissertation. University of Bonn, Federal Republic of Germany, 103 pp

Saadabi AM (1985) Studies on the biology and control of some root-knot nematodes with emphasis on *Pratylenchus sudanensis* n. sp Loof and Yasin, 1970 on cotton in Gezira area. Int Nematol Netw Newslett 2:11–13

Sadik S, Okereke OU (1975) Flowering, pollen grain germination, fruiting, seed germination and seedling development of white yam, *Dioscorea rotundata* Poir. Ann Bot 39:597–604

Sauphanor B, Ratnadass A (1985) Problemes entomologiques lies a la conservation de l'igname en Cote d'lvoire. L'Agronomie Tropicale 40(3):261–270

Siddiqi MR (1972) *Pratylenchus coffeae*. Descriptions of plant parasitic nematodes. Set 1, No. 6. Commonwealth Agricultural Bureaux, Slough

Smith SE, Read DJ (2008) Mycorrhizal symbiosis, 3rd edn. Academic, London

Speijer PP, Ssango F, Vuylsteke D (2000) Evaluation of host plant response to nematodes in *Musa* germplasm in Uganda. Acta Horticult 540:225–232

Thompson AK, Been BO, Perkins C (1973) Nematodes in stored yams. Exp Agric 9:281–286

Fauquet T (1979) Yam mosaic: a new potyvirus infecting *Dioscorea cayenensis* in Ivory Coast. Ann Appl Biol 93:279

Varghese R, Mohandas C (2004) *Scutellonema bradys*, a new host of *Pasteuria penetrans*. Indian J Nematol 34(2):220–221

White LD (2003) Canoe plants of ancient Hawai'i: Uhi. Waitsfield, Vermont

Wood TG, Smith RW, Johnson RA, Komolafe PO (1980) Termites damage and crop loss studies in Nigeria- preharvest losses to yam due to termites and other soil pests. Trop Pest Manag 26:355–370

6 Chinese Potato, *Plectranthus rotundifolius*

Abstract

Distribution, economic importance, symptoms/damage, host range, biology, survival and spread, and management of insect and mite pests and fungal, bacterial, viral, and nematode diseases of Chinese potato (*Plectranthus rotundifolius*) using regulatory, physical, cultural, chemical, biological, host resistance, and integrated pest management methods are discussed.

Keywords

Chinese potato • *Plectranthus rotundifolius* • Coleus • Pests • Diseases • Nematodes

6.1 Introduction

Plectranthus rotundifolius (synonym: *Solenostemon rotundifolius*) (Common names: Hausa potato, Coleus potato, Country potato) is a member of the Family Lamiaceae (Labiatae) (mint family) which consists of heterogeneous assemblage of over 300 plants. The crop is thought to have originated in Kenya or Ethiopia in East Africa from where it spread throughout tropical Africa through the savannas of Togo, Guinea, and then into Southeast Asia including India, Sri Lanka, Malaysia, and Indonesia (Harlan et al. 1976). Chinese potato is a small herb that has prostrate or ascending succulent stems and branches (Fig. 6.1). It attains a height of 15–30 cm and presents a distinctive, distractive, lighten odor due to the presence of volatile oils in the glands or sacs of its leaves (Phungpanya et al. 2012). The flowers are small and may be white, blue, pinkish, or pale violet in color, being borne on distal inflorescences with slender false spikes measuring up to 15 cm in length. These flowers are hermaphroditic, and the fruits consist of four nutlets which rarely develop. In Africa today, cultivation of this crop is mostly limited to Burkina Faso, Eastern Mali, Northern Ghana, and South Africa. In Nigeria, Chinese potato is still popular in the middle belt and Northeastern regions of the country especially around the states of Bornu, Taraba, Nasarawa, Jos Plateau, and Kaduna. On the world scale, the crop is the most widespread of the cultivated Lamiaceae (Labiatae) and is believed to be of more economic importance in tropical Asia particularly in India where it is grown as a monsoon crop as well as in Sri Lanka, Madagascar, Malaysia, Indonesia, and Southern

P.P. Reddy, *Plant Protection in Tropical Root and Tuber Crops*,
DOI 10.1007/978-81-322-2389-4_6

Fig. 6.1 Chinese potato plants with tubers

Thailand (PROTA 2013). In ethnobotany, preparation of boiled leaves of Chinese potato is used to treat such disorders as dysentery, blood in urine, and eye disorders including glaucoma (Schipper 2000).

It is cultivated for its edible tubers (Fig. 6.1) primarily in West Africa, as well as more recently in parts of Asia, especially India, Sri Lanka, Malaysia, and Indonesia (National Research Council 2006). Tubers of the crop provide essential dietary and energy requirements to the populace during the lean periods. The tubers taste similar to Irish potato and trifoliate yam and can be eaten as the main starchy staple or part of it in combination with legumes, rice, and vegetables. It can also be prepared in Indian and African cuisines as boiled, baked, or fried (Schipper 2000). Some workers reported that the tubers of Chinese potato can be used to make aromatic, alcoholic beverages (Schipper 2000), while the leaves sometimes may also be eaten as potherbs (PROTA 2013). Compared to other tuberous staples in Nigeria, the tubers of this crop are rich in protein, vitamins, and minerals. One hundred grams of the raw tubers contain water 76 %, carbohydrate 21 %, protein 1.4 %, fiber 0.7 %, fat 0.2 %, and ash 1.0 % among other important nutrients (PROTA 2013).

Yields averaging 5–15 t/ha have been reported from the crop in Ghana and Nigeria. The potential yield of the crop could be up to 18–20 t/ha (PROTA 2013). A study in South Africa, however, indicated that potential yield from the crop may amount up to 45 t/ha under favorable conditions of rains, soil fertility, and texture (Nkansah 2004). Such optimum yields could be obtained on well-drained light sandy loams, while heavy soils prone to waterlogging are unsuitable for the crop. Chinese potato may be propagated vegetatively by suckers or like cassava by soft woody stem cuttings as well as by tubers as yams and cocoyams. However, stem cuttings are scarcely used. Plantings are done on mounds, ridges, and flats in well-drained locations at the beginning of the rains as sole crop or intercrop with maize, yam, millet, melon, sorghum, and groundnut (PROTA 2013). The ridges are incorporated with farmyard manure, wood ash, cattle dung and urine, or inorganic fertilizers (Olojede et al. 2005).

Chinese potato is an important crop which has contributed immensely in alleviating poverty, hunger, and starvation in various farm families. Properly directed researches to find solutions to the many agronomic and storage-health challenges besetting the crop are imperative toward making the crop in achieving its full quantitative and qualitative yield potentials and contributing immensely to the drive for food security in Nigeria.

Chinese potato is an important minor tuber crop cultivated in India. It is grown in most of the homestead gardens of Tamil Nadu and Kerala. Chinese potato is mainly confined to central Kerala and Malabar in laterite soils (200 ha) and sandy coastal soils sometimes extending to Mangalore district of Karnataka. Due to its wide acceptance as an aromatic tuber vegetable in Kerala and Tamil Nadu, there is a gradual spread of the crop even to black soils in Tamil Nadu as an irrigated commercial crop, and of late it is intensely cultivated in more than 500 ha in South Tamil Nadu.

Tuber yield from the crop have been reported to correlate strongly with amount and regularity of rain (Nkansah 2004). A well-distributed rainfall of 1,000 mm has been reported sufficient for the crop, while excessive rainfall which could lead to waterlogging is detrimental to its tuber development.

6.2 Insect Pests

6.2.1 Tinged Bug, *Monanthia globulifera*

6.2.1.1 Damage

Female bug inserts eggs on leaves and tender foliage. Black, spiny nymphs and lace-winged adults cause damage by sucking the sap from the leaves. Infested leaves develop yellow patches, become brown, shrivel, and dry up from margins in case of heavy infestations.

6.2.1.2 Management

Removal and destruction of heavily infested leaves reduces the bug population. Foliar sprays of acephate or carbaryl or methyl parathion or phenthoate each at 0.05 % offer effective control of the bug.

6.2.2 Spike Borer, *Helicoverpa armigera*

6.2.2.1 Damage

Cream-colored, round eggs are seen on tender foliage. Larvae initially feed on tender leaves and later bore into developing spikes by making holes and feeding on buds and flowers.

6.2.2.2 Management

Two or three sprays of 0.07 % Endosulfan or methyl parathion or Etofenprox each at 0.05 % at 15 days interval control the pest.

6.2.3 Grasshopper, *Chrotogonus* sp.

6.2.3.1 Damage

Nymphs and adult grass hoppers attack the crop during the day. They cause damage by making irregular holes on leaves and cutting tender shoots.

6.2.3.2 Management

Malathion at 0.1 % or carbaryl at 0.2 % sprays offer protection to foliage from grasshopper damage. Spraying of 2 % neem oil checks attack of the pest.

Fig. 6.2 Mealybug infestation on coleus

6.2.4 Mealybug, *Orthezia insignis*

6.2.4.1 Damage

Feed on Chinese potato by piercing plant and sucking out juices. Certain species of mealybugs will also feed on roots of Chinese potato. Honeydew is excreted by bugs, similar to aphids. A cottony-looking waxy substance is secreted by females, most commonly at stem joints (Fig. 6.2). Chinese potato will develop unusually small leaves and its branches will wither.

6.2.4.2 Management

- Rubbing alcohol diluted with water to 25 or 50 % or insecticidal soap may be useful in eliminating them.
- Chemical insecticides are also available.

6.2.5 Leaf and Shoot Folders, *Pycnarmon cribrata, Phostria piasusalis*

6.2.5.1 Management

Leaf and shoot folder is controlled by two beneficial endoparasitoids *Microgaster psarae* on *Pycnarmon cribrata* and *Leptobatopsis indica* on *Phostria piasusalis*. The parasitoids showed preference to second instars. The total developmental period ranged from 8 to 14 days. The field efficacy ranges from 15 to 50 %.

6.3 Diseases

6.3.1 Wilt, *Fusarium chlamydosporum*

6.3.1.1 Symptoms

Symptoms of Chinese potato wilt caused by *F. chlamydosporum* were described by Shyla (1998). In the field, the infected plants were characterized by gradual yellowing, marginal necrosis, and withering of leaves followed by loss in vigor and premature death. Such plants showed discoloration of roots and complete decaying of tap and lateral root system. The bark of such plants easily peeled off. There was extensive sloughing off and shredding of affected bark. Such affected plants were finally killed due to severe root and collar rot. The infected tubers showed rotting and emitting of bad odor (Fig. 6.3).

6.3.1.2 Management

1. *Cultural Methods*: Water stagnation in fields may lead to severe infections of *Fusarium*; therefore, water stagnation in the planted fields should be avoided.

Botanical pesticides like neem cake reduced the percent disease index (40–60 %) and increased the root yield of Chinese potato (Singh et al. 2008; 2009b).

Fig. 6.3 Symptoms of *Fusarium* wilt of Chinese potato (*left* healthy, *right* diseased)

2. *Chemical Methods*: Dipping the terminal cuttings in Carbendazim solution (1 g/l) before planting protects Chinese potato from fungal pathogens.

The chemical Emisan (0.2 %) has been found to protect the plants against *Fusarium* wilt to some extent, but the protection provided to plants inoculated with biocontrol agents was found to be higher (Boby and Bagyaraj 2003). Chemical fungicides (benomyl) reduced the disease incidence (54.54 %) caused by *Fusarium chlamydosporum* on Chinese potato during field study (Singh et al. 2009a, b). Paramasivan et al. (2007) reported that the use of chemical fungicide (Carbendazim) reduced the disease incidence by 18 %. Kulkarni et al. (2007) reported that the lowest population (cfu/g soil) of *F. chlamydosporum* and *R. bataticola* was observed with the use of Carbendazim.

3. *Biological Methods*: *Pseudomonas fluorescens*, mainly considered as a PGPR, can suppress a wide range of plant pathogens including *Fusarium*. Increase in plant height, number of branches, fresh and dry weight of the plant as well as less disease severity, and less root rot were observed due to PGPR treatments (fluorescent *Pseudomonas* sp.) compared to control (only pathogen). Application of PGPR strains (RB50 and RB31) recorded maximum plant height, number of branches, fresh and dry weight of the plants with less disease severity, and root rot index (Mallesh 2008) (Table 6.1).

Singh et al. (2009a, b) reported that inoculation of bio-inoculants (*T. viride*, *G. fasciculatum*, *G. mosseae*, and *P. fluorescens*) was also effective.

4. *Integrated methods*
   (a) *Two Bioagents*: Inoculation with *T. viride* + *G. mosseae* gave the best result in controlling the disease. The same treatment also resulted in maximum growth, yield of Chinese potato. The next best treatment was *P. fluorescens* + *T. viride*

Table 6.1 Biocontrol potentiality of PGPR strains in Chinese potato against Fusarium wilt

PGPR strains	Plant height (cm)	No. of branches	Fresh weight (g)	Dry weight (g)	Disease severity	Root rot index
RB 01	32.5	5.0	97.5	15.6	+	1.5
RB 10	29.5	4.5	93.5	14.5	++	3.0
RB 13	34.0	6.0	107.5	17.0	+	1.5
RB 22	28.0	4.0	94.7	15.0	++	2.5
RB 31	36.5	6.5	116.5	18.5	+	1.0
RB 43	30.0	5.5	95.5	15.3	++	2.0
RB 50	37.0	7.0	120.0	18.9	+	1.0
Control	25.0	3.5	80.0	13.0	+++	3.5

followed by *G. mosseae* + *P. fluorescens* and *T. viride* alone. Application of Emisan resulted in better growth than control, but it was less when compared to single or dual inoculations of biocontrol agents (Boby and Bagyaraj 2003). Paramasivan et al. (2007) reported that the use of bio-inoculants like *T. viride* and *P. fluorescens* reduced the disease incidence by 20–21 %.

(b) *Biological and Cultural Methods*: Combination of *T. viride* + Neemato (neem-based product applied at 500 g/5 m^2) resulted in lowest wilt incidence by 12.76 % (Kulkarni et al. 2007).

Fig. 6.4 *Downy mildew* symptoms on Chinese potato

6.3.2 Downy Mildew, *Peronospora* sp.

6.3.2.1 Symptoms

Downy mildew appears on the leaves, making the plant look dirty because it is brown in color and can also result in curled and twisted leaves. Symptoms include irregular, brown leaf spots or flecking, leaf twist and leaf drop. Leaf spotting can take form of squarish patches bounded by larger veins. The undersurface of the leaf appears "dirty" to the naked eye due to grayish-brown spores (Fig. 6.4).

6.3.2.2 Management

- Diseased plants should be discarded.
- For protectant action against downy mildews, Mancozeb is especially effective, and copper (0.1 % copper oxychloride) is also helpful.
- Systemic or partly systemic materials—Aliette and Alude (phosphorous acid materials) and Compass and Heritage (Strobilurins)—are effective against downy mildews.

6.3.3 Leaf Blight, *Rhizoctonia solani*

Rhizoctonia solani has been reported to cause the leaf blight of Chinese potato (Shukla et al. 1993). Blight disease is common during monsoon or during period of high humidity.

6.3.3.1 Symptoms

The disease was initially expressed as water-soaked areas and the affected tissues soon turned

Fig. 6.5 *Left* – Blight symptoms on Chinese potato plant; *Right* – Root systems of diseased (*left*) and healthy (*right*) plants

into a soft, black, watery mass at the collar region of the plant. The infection was also found on roots and caused decay, which ultimately resulted in collapse of the plant. The infected plant roots showed discoloration followed by rotting of root hairs. Extensive sloughing and peeling off affected bark was also observed. Under conditions of high humidity, the disease was found to spread rapidly (Ramaprasad Shresti 2005). Severe infection results in defoliation and death of the plant (Fig. 6.5).

6.3.3.2 Management

1. *Integrated Methods*
   (a) *Biological and Cultural Methods*: The blight incidence was significantly minimum in the plots where *T. viride* (10 ml spore suspension/plant) combined with Neemato (500 g/5 m^2) were applied compared to other treatments (Ramaprasad Shresti 2005).

The treatment *T. viride* + Neemato (T8) recorded significantly lowest per cent blight incidence (12.76) over control (35.52) followed by *T. harzianum* (T2) (18.87) and *P. fluorescens* (T3) (19.98) (Ramaprasad Shresti 2005).

Colony-forming units of *R. bataticola* were found to be significantly highest in control (21.60 × 10^3/g of soil) followed by FYM (Ramaprasad Shresti 2005) (Table 6.2).

Table 6.2 Management of leaf blight of Chinese potato using different biocontrol agents, organic amendments, and chemicals (Ramaprasad Shresti 2005)

Treatment	% blight incidence[ab]	Cfu[c] *Rhizoctonia bataticola*
T1– *Trichoderma viride* at 10 ml/plant (8 × 10^3 cfu/ml)	21.09 (27.33)	12.20
T2 – *T. harzianum* at 10 ml/plant (8 × 10^3 cfu/ml)	18.87 (25.74)	12.60
T3 – *Pseudomonas fluorescens* at 10 ml/ plant (24 × 10^5 cfu/ ml)	19.98 (26.51)	14.20
T4 – pronto at 5 % soil drench	23.31 (28.84)	15.60
T5 – Neemato at 500 g/5 m^2	21.09 (27.24)	16.40
T6 – carbofuran 3G at 15 g a.i./5 m^2	24.42 (29.57)	17.60
T7 – farm yard manure at 5 kg/5 m^2	25.53 (30.38)	18.80
T8 – *T. viride* at 10 ml/ plant (8 × 10^3 cfu/ ml) + Neemato 500 g/5 m^2	12.76 (20.93)	9.60
T9 – carbendazim at 0.1 % soil drench	21.19 (27.33)	6.80
T10 – propiconazole at 0.1 % soil drench	23.31 (28.84)	7.40
T11 – control	35.52 (36.59)	21.60
CD at 5 %	**3.48**	**2.72**

[a]Observations recorded at harvest (150 days after planting)
[b]Figures in parentheses are arc sin angular transformed values
[c]Cfu – Colony-forming units ×10^3/g of soil (average of 5 replications)

6.3.4 Root Rot, *Sclerotium rolfsii*

6.3.4.1 Symptoms

The earliest symptom of the disease was darkening of the stem at collar region of the plant. The leaves became flaccid and dropped off. White, fan-shaped mycelial strands crept over the stem portion, developing small light to dark-brown sclerotia on the infected portion. The sclerotial initials were white at first and then later turned brown with age. Finally the plant wilted and dried (Fig. 6.6) (Ramaprasad Shresti 2005).

6.3.4.2 Management

1. *Integrated Methods*
   (a) *Biological and Cultural Methods*: Treatment involving field sanitation + dipping stem cuttings in Carbendazim (0.1 %) + one more drench with 0.1 % Carbendazim 30 days after planting (DAP) recorded maximum reduction of disease incidence over control (76.50 and 72.72 % at 45 and 90 DAP, respectively). This was on par with treatment soil application of zinc sulfate at 20 kg/ha + soil drenching with neem cake and *Trichoderma viride* mixture at 50 g/plant, where reduction in disease incidence was 73.73 and 70.11 % over control at 45 and 90 DAP, respectively. The tuberous root yield was also maximum in the above treatments (5.787 and 5.759 t/ha).

Fig. 6.6 Root rot symptoms on Chinese potato (*left* diseased, *right* healthy)

6.3.5 Tuber Rot

6.3.5.1 Symptoms

Recent investigations in Yola, Northeast Nigeria, showed that *Fusarium oxysporum*, *Aspergillus niger*, *Penicillium expansum*, and *Rhizopus stolonifer* were implicated with tuber rot of Hausa potato (Mohammed et al. 2013a, b). Preliminary studies into the cause of rot of tubers of the crop in Southeast Nigeria suggested the association of *Colletotrichum* spp. with rot of the tubers in storage (Nwaneri 2013). However, organisms responsible for the spoilage and rot of Hausa potato have not been fully documented.

6.3.5.2 Management

Tuber rot diseases are controlled using synthetic chemical agents such as captan, borax, naphthalene acetic acid, and orthophenylphenate (Okigbo and Nmeka 2005).

Studies by Mohammed et al. (2013b) in Adamawa state of Northern Nigeria, with ash from *Anogeissus leiocarpus* wood, revealed that it significantly inhibited the development and spread of tuber rot (*Aspergillus niger*, *Fusarium oxysporum*, *Rhizoctonia stolonifer*, and *Penicillium expansum*) on Hausa potato in storage for 4 months.

6.3.6 Bacterial Wilt, *Ralstonia solanacearum*

The bacterial wilt of Chinese potato caused by *R. solanacearum*, biovar N2 race was reported for the first time from Brazil. Bacterial wilt is the major disease of Chinese potato causing heavy losses (>50 %) in South India (Chandrashekara and Prasannakumar 2010).

6.3.6.1 Symptoms

Water-soaked patches with linear streaks on collar region of the infected plants were observed. Leaves became flaccid and drooped quickly, and the plants wilted and dried. The leaves showed roll up symptoms and the whole plant dried up. Wilted plants came off easily with a gentle pull and vascular discolorations were observed. Such tubers when pressed exhibited oozing of bacterial exudates (Fig. 6.7).

Fig. 6.7 Symptoms of bacterial wilt on Chinese potato

Table 6.3 Biocontrol potentiality of PGPR strains on Chinese potato against bacterial wilt

PGPR strains	Plant height (cm)	No. of branches	Fresh weight (g)	Dry weight (g)	Disease severity	Root rot index
RB 01	31.5	4.5	94.0	15.00	++	2.0
RB 10	27.5	4.0	92.5	14.25	+++	3.5
RB 13	32.0	5.0	101.0	16.50	++	2.0
RB 22	26.0	4.0	93.0	14.25	+++	3.0
RB 31	34.0	5.5	107.0	16.90	++	1.5
RB 43	28.0	4.5	93.5	14.50	++	2.5
RB 50	35.0	6.0	110.5	17.50	+	1.5
Control	22.0	3.5	69.5	11.12	++++	4.0

6.3.6.2 Management

1. *Cultural Methods*: Water stagnation in Chinese potato fields may lead to severe infections of bacterial wilt; therefore, water stagnation in the planted fields should be avoided.
2. *Chemical Methods*: Drenching the soil around the roots with streptocycline 300 ppm solution is effective.
3. *Biological Methods*: Application of PGPR strains (fluorescent *Pseudomonas* spp. strains RB50 and RB31) recorded maximum plant height, number of branches, fresh and dry weight of the plants with less disease severity, and root rot index (Mallesh 2008) (Table 6.3).
4. *Integrated Methods*
   (a) *Biological and cultural methods*: Application of 2 kg of *Pseudomonas fluorescens* mixed with 300 kg of compost is also effective.

6.3.7 Impatiens Necrotic Spot Virus

6.3.7.1 Symptoms

Causes brown or yellow spots on leaves, rings, black or brown stem discoloration, and brown leaf veins. It is a virus that causes plants to die. The disease is spread by Western flower thrips (*Frankliniella occidentalis*). It only takes a few of these insects to infect a whole greenhouse. Characteristic ring spots are produced (Fig. 6.8).

Fig. 6.8 Symptoms of impatiens necrotic spot virus

6.3.7.2 Management

- Discard affected plants.
- Eradicate weeds that may be hosts.
- Control thrips population by spraying dimethoate at 2 ml/l.

6.4 Nematodes

6.4.1 Root-Knot Nematodes, *Meloidogyne* spp.

In Chinese potato, the root-knot nematode causes severe damage and huge reduction in yield. Tubers rot under severe infestations. Reduction in dry weight and starch had been recorded.

The root-knot nematode infestation was reported on Chinese potato from Kerala and Orissa. The dry weight of the tubers was reduced by 20 % due to root-knot nematodes. The percentage of starch on fresh weight basis showed drastic reduction (16 %) in the infested tubers. *M. incognita* has been reported to cause a yield reduction of up to 86 % (Senthamarai et al. 2006b), while severe losses also occur in Chinese potato because of *M. arenaria* infestations (Bhandari et al. 2007).

Root-knot nematode affected roots, impairs water and nutrient uptake by the crop, reduces marketable root yield and quality, and results in total crop failure in severe cases (Asawalam and Adesanya 2001). Infestation of root tubers by root-knot nematodes in the field in addition to abrasions sustained during harvesting strongly predisposes tubers to rapid decay during transit and storage (Okigbo 2004; Okigbo and Nmeka 2005; Okigbo et al. 2009).

6.4.1.1 Symptoms

The newly inoculated Chinese potato plant exhibited stunted growth, yellow patches, and severe galling of roots due to *M. incognita*. Tuber formation was found commensurate with the degree of galling and yield reduction was 65 % under field conditions. The pathogenicity study which was carried out under controlled conditions revealed that the effective damage to the plant was caused by 1,000 larvae per plant.

The galls on Chinese potato roots are very big and pronounced (Fig. 6.9). The root-knot nematode damage often leads to crop failure. The infested tubers swell in size with irregular surface and cracking of the skin. Rotting sets in even before harvest, if the infestation is severe (Fig. 6.1). Infested tubers rot after harvest and rarely reach market.

6.4.1.2 Management

1. *Cultural Methods*: Management of the nematode is very simple. As infestation spreads through the planting material, clean planting material will help to contain the spread of the nematode. It is very easy to distinguish infested and healthy planting materials as infested tubers have irregular wartlike projections, irregular shape, and also dried patches on the surface. Such tubers need to be separated and avoided. Heavily infested tubers rot in store and get eliminated by themselves.

Sree Bhadra, a high-yielding variety of sweet potato released by Central Tuber Crops Research Institute, Trivandrum, is identified as a resistant trap crop for the root-knot nematode.

Fig. 6.9 Root-knot nematode-infested plant and roots of Chinese potato

Table 6.4 Biocontrol potentiality of PGPR strains on Chinese potato against root-knot nematodes

PGPR strains	Plant height (cm)	No. of branches	Fresh weight (g)	Dry weight (g)	Root-knot index
RB 01	31.5	4.5	94.0	15.00	2.0
RB 10	27.5	4.0	92.5	14.25	3.5
RB 13	32.0	5.0	101.0	16.50	2.0
RB 22	26.0	4.0	93.0	14.25	3.0
RB 31	34.0	5.5	107.0	16.90	1.5
RB 43	28.0	4.5	93.5	14.50	2.5
RB 50	35.0	6.0	110.5	17.50	1.5
Control	22.0	3.5	69.5	11.12	4.0

The nematodes were able to penetrate the root but giant cell formation was not induced, and as a result, the nematodes die inside the roots. Planting this variety in root-knot nematode-infested field helped in clearing the field free of nematodes and giving good tuber yield. Subsequently growing of susceptible crop like Chinese potato escaped nematode damage (Mohandas 2001).

Continuous cropping of a particular crop results in buildup of the nematodes in fields. Hence, crop rotation is essential. Farmers rotate Chinese potato after rice in Tirunelveli district of Tamil Nadu.

2. *Chemical Methods*: Chemical agents such as methyl isothiocyanate mixtures, oxamyl, thionazin, and carbofuran have been found effective in the management of nematodes. Application of carbofuran 3 % granules at the rate of 20 kg per hectare under wet conditions near the root zone was also found effective.

Treatment with fensulfothion reduced damage to 5.7 % as against 21 % in control. The above treatment also increased the production of healthy tubers to the extent of 47 % (Pillai 1976).

3. *Biological Methods*: Soil application of *Trichoderma harzianum* and *Pseudomonas fluorescens* each at 2.5 kg/ha recorded increased plant growth and yield and reduced *M. incognita* population (Senthamarai et al. 2006a, 2008).

Increase in plant height, number of branches, fresh and dry weight of the plant, as well as less root-knot index were observed due to PGPR treatments (fluorescent *Pseudomonas* sp.) compared to control. Application of PGPR strains (RB50 and RB31) recorded maximum plant height, number of branches, and fresh and dry weight of the plants with less root-knot index (Table 6.4) (Mallesh 2008).

4. *Integrated methods*
   (a) *Biological, Cultural, and Physical Methods*: Integration of soil solarization in the nursery for 15 days with 150 gauge LDPE film and application of *Paecilomyces lilacinus* + neem cake or *P. lilacinus* + *Bacillus macerans* in the main field are the best treatments in increasing plant height (64.3 and 60.3 cm compared to 40.0 cm in control), number of leaves (593.3 and 583.3 compared to 310.0 in control), weight of tubers/plant (560.0 and 546.6 g compared to 350.0 g in control), and tuber yield (11.5 and 11.3 kg/plot compared to 6.9 kg/plot in control) and in reducing root galls (0.3 and 1.0 compared to 50.6 in control), nematode population in soil (25.0 and 30.0/100 ml soil compared to 196.6/100 ml soil in control), and roots (1.0 and 1.6/5 g roots compared to 79.0/5 g roots in control) (Nisha and Sheela 2006).
   (b) *Biological and Cultural Methods*: Dipping of stem cuttings in *P. fluorescens* + soil application of neem cake at 400 kg/ha + growing marigold as intercrop followed by their biomass incorporation during earthing up increased the yield (22.7–30.0 %) and reduced the root-knot nematode population (71.2–73.8 %) superiorly, followed by the integration of *P. fluorescens* + marigold intercrop, which were almost equally effective (Seenivasan and Deevrajan 2008).

6.4.2 Root-Knot Nematode (*Meloidogyne incognita*) and Wilt Disease (*Fusarium chlamydosporum*) Complex

6.4.2.1 Symptoms

Among the different diseases affecting Chinese potato, root-knot and wilt disease complex caused by *M. incognita* and *F. chlamydosporum* was observed in severe form. In the interaction studies, *M. incognita* was the most aggressive pathogen compared to *F. chlamydosporum*. However, simultaneous inoculation of *M. incognita* and *F. chlamydosporum* caused greater reduction in plant growth as well as nematode multiplication (Fig. 6.10). However, with respect to individual pathogen, *M. incognita* caused greater reduction in plant growth compared to *F. chlamydosporum*. Sequential inoculation of *M. incognita* 7 days prior to *F. chlamydosporum* caused reduction in plant growth parameters.

Fig. 6.10 Symptoms of root-knot and wilt disease complex on Chinese potato

The effect of simultaneous inoculation of *M. incognita* with *F. chlamydosporum* on Chinese potato was additive in nature. However, when *M. incognita* was inoculated with *F. chlamydosporum*, the resultant effect was almost equal to the sum of individual effect. These results suggest that the nematode can predispose the coleus to infection by fungus, namely, *F. chlamydosporum*, and can aggravate the disease. Reduction in number of galls per plant and final nematode population was observed in simultaneous inoculation.

6.4.2.2 Management

1. *Biological Methods*: Significant increase in tuber yield was recorded in all the plants treated with PGPR strains (fluorescent *Pseudomonas* sp.). However, highest tuber yield was observed in the treatment RB 50 (6.74 kg) followed by RB 31 and RB 13 (5.45 and 4.60 kg) (Mallesh 2008) (Table 6.5).

Table 6.5 Efficacy of talc formulations of PGPR strains on fresh tuber yield parameters and disease incidence in Chinese potato under field conditions

PGPR strain	Tubers/plant	Total fresh biomass (g)	Root-knot index	% disease incidence	% decrease over control	Tuber yield (kg/plot)
RB 01	15.25	964.44	1.13	16.67	41.05	4.18
RB 10	12.87	823.13	1.60	25.00	11.75	3.25
RB 13	16.63	1,016.74	0.80	20.00	29.40	4.60
RB 22	11.87	811.84	1.53	23.33	17.64	3.36
RB 31	16.50	1,094.54	0.73	13.33	52.94	5.45
RB 43	14.92	841.83	1.33	18.33	41.05	3.95
RB 50	18.00	1,243.02	0.66	10.00	64.70	6.74
Control	10.70	730.44	2.20	28.33	–	3.00
CD at 5 %	**1.52**	**14.66**	**0.65**	**5.02**	**–**	**0.79**

Fig. 6.11 Efficacy of talc formulation of PGPR strains against root-knot nematode on Chinese potato under field conditions (*left* RB 50, *right* control)

Lowest RKI was noticed in RB 50 (0.66), RB 31 (0.73), and RB 13 (0.80) which showed 70.00, 66.81, and 63.63 % reduction in root-knot index compared to control, respectively (Mallesh 2008).

Lowest disease incidence of 10.00 % (64.70 % reduction over control) was recorded in RB 50 followed by 13.33 % in RB 31 (52.94 % reduction over control). In general due to PGPR treatment, there was a reduction of disease incidence in Chinese potato which varied from 64.70 to 11.75 % reduction over control (Fig. 6.11) (Mallesh 2008).

The highest biomass of 1,243.02 g and 212.27 g were recorded on fresh weight and dry weight basis, respectively, in RB 50, followed by 1,094.54 and 190.89 g in RB 31 (Mallesh 2008).

2. *Host Resistance*: Studies on screening of different genotypes of coleus against *M. incognita* and *F. chlamydosporum* disease complex showed that the genotypes, viz., yellow tubers and Rabakavi local, showed resistant reaction. Orange tubers and Nimbanur local genotype showed moderately resistant reaction.
3. *Integrated Methods*
   (a) *Biological and Cultural Methods*: Combined application of plant products (neem seed kernel powder at 5 g/kg of soil) with biocontrol agents (*Paecilomyces lilacinus, Trichoderma viride* + *Pseudomonas fluorescens* at 10 g/kg of soil) was found effective in reducing the number of galls, nematode population, number of egg masses, root-knot index, and root rot index and improving the plant growth parameters as compared to inoculated control.

6.4.3 Root-Knot Nematode and Collar Rot Disease Complex

Rhizoctonia bataticola, *Sclerotium rolfsii*, *F. chlamydosporum*, and *M. incognita* were found to be the most commonly associated fungi with collar rot disease complex (Fig. 6.12) (Ramaprasad Shresti 2005).

6.4.3.1 Symptoms

Wilt symptoms were first recorded at 45 days after inoculation in treatment *M. incognita* inoculated 7 days prior to inoculation of all the fungal pathogens simultaneously (*F. chlamydosporum* + *R. bataticola* + *S. rolfsii*) (Fig. 6.13) (Ramaprasad Shresti 2005).

Fig. 6.12 Root-knot and collar rot disease complex on Chinese potato. (*left* infected, *right* healthy

Fig. 6.13 Sequential inoculation of *Meloidogyne incognita* 7 days prior to inoculation of fungal pathogens. *1* control, *2* nematode → *Fusarium*, *3* nematode → *Rhizoctonia*, *4* nematode → *Sclerotium*, *5* nematode → *Fusarium* + *Rhizoctonia* + *Sclerotium*

6.4.3.2 Management

1. *Biological Methods*: Greenhouse studies using the plant growth promoting rhizobacterial (PGPR) fluorescent pseudomonads (bioformulations containing RB 50 and RB 31 strains) showed significant increase in seedling biomass besides reduction in Chinese potato wilt complex due to combination of pathogens. RB 50 and RB 31 strains decreased root-knot index and incidence of disease complex while increasing biomass and tuber yield. Biochemical analyses in the above treatments showed elevated expression of defense enzymes (peroxidase, polyphenol oxidase, and phenylalanine ammonia lyase) and higher accumulation of phenolic compounds (activity being highest in respect of RB 50 and RB 31 treated plants) compared to respective inoculated checks. The suppression of disease complex in Chinese potato is largely due to the aforementioned biocontrol mechanism as well as induction of systemic resistance by efficient PGPR strains (Lingaraju and Mallesh 2010).
2. *Integrated Methods*
   (a) *Biological, Cultural, and Chemical Methods*: The wilt incidence, nematode population, number of galls, and colony-forming units of *F. chlamydosporum* and *R. bataticola* were significantly minimum in the plots where *T. viride* (10 ml/plant spore suspension) combined with Neemato (500 g/5 m^2) were applied compared to other treatments (Ramaprasad Shresti 2005).

The treatment *T. viride* + Neemato (T8) recorded significantly lowest per cent wilt incidence (12.76) over control (35.52) followed by *T. harzianum* (T2) (18.87) and *P. fluorescens* (T3) (19.98). The number of galls/5 g of root was significantly lowest in treatment *T. viride* + Neemato (T8) (10.13), followed by carbofuran (T6) (14.93) which was on a par with each other. Colony-forming units of *F. chlamydosporum* were significantly highest in the control (19.60) followed by FYM (T7) (Table 6.6) (Ramaprasad Shresti 2005).

Table 6.6 Management of collar rot complex of Chinese potato using different biocontrol agents, organic amendments, and chemicals

Treatment	% wilt incidence[a]	No. of galls/5 g roots	Cfu[c]	
			Fusarium chlamydosporum	*Rhizoctonia bataticola*
T1	21.09 (27.33)[b]	21.13	7.60	12.20
T2	18.87 (25.74)	19.53	8.00	12.60
T3	19.98 (26.51)	18.27	8.00	14.20
T4	23.31 (28.84)	17.33	10.60	15.60
T5	21.09 (27.24)	16.07	12.60	16.40
T6	24.42 (29.57)	14.93	16.20	17.60
T7	25.53 (30.38)	25.67	15.20	18.80
T8	12.76 (20.93)	10.13	6.20	9.60
T9	21.19 (27.33)	23.33	3.60	6.80
T10	23.31 (28.84)	23.00	3.80	7.40
T11	35.52 (36.59)	28.40	19.60	21.60
CD at 5 %	**3.48**	**5.38**	**2.49**	**2.72**

T1, *Trichoderma viride* at 10 ml/plant (8×10^3 cfu/ml); T2, *T. harzianum* at 10 ml/plant (8×10^3 cfu/ml); T3, *Pseudomonas fluorescens* at 10 ml/plant (24×10^5 cfu/ml); T4, pronto at 5 % soil drench; T5, Neemato at 500 g/5 m^2; T6, carbofuran 3G at 15 g a.i./5 m^2; T7, farm yard manure at 5 kg/5 m^2; T8, *T. viride* at 10 ml/plant (8×10^3 cfu/ml) + Neemato at 500 g/5 m^2; T9, carbendazim at 0.1 % soil drench; T10, propiconazole at 0.1 % soil drench; T11, control

[a]Observations recorded at harvest (150 days after planting)

[b]Figures in parentheses are arc sin angular transformed values

[c]Cfu – Colony-forming units $X10^{-3}$/g of soil (average of 5 replications)

6.4.4 Root-Knot Nematode, *Meloidogyne incognita*, and Root Rot *Macrophomina phaseolina* Disease Complex

The productivity of Chinese potato has been hampered by its susceptibility to nematode and root rot disease. Pathogens associated with root rot are *Macrophomina phaseolina* and the root-knot nematode *Meloidogyne incognita*. Severely affected areas were Salem, Attur, and Rasipuram in Tamil Nadu. Due to this disease complex, the yield loss ranged from 50 to 60 %. Therefore, it is necessary to take effective control measures to save the crop from complete devastation.

6.4.4.1 Symptoms

Simultaneous inoculation of *M. incognita* and *M. phaseolina* as well as nematode inoculation followed by fungus 15 days later caused significant reduction in tuber yield and 100 % root rot disease in coleus. The nematode multiplication was adversely affected when fungus was inoculated prior to nematode (Table 6.7) (Senthamarai et al. 2008).

6.4.4.2 Management

The following control measures are recommended:

- Select the Chinese potato cuttings from disease-free plants.
- Chinese potato cuttings should be treated with Carbendazim solution (0.1 %).
- Soil drenching with Carbendazim (0.1 %) or Propiconazole (0.1 %).
- Soil application of FYM at 12.5 t/ha + *Trichoderma viride* at 2.5 kg/ha before planting.
- Application of carbofuran 3G at 1 kg a.i./ha before planting.
- Stem cuttings to be dipped in 0.1 % *Pseudomonas fluorescens* at the time of planting and growing.
- Use drip irrigation to minimize the spread of pathogens from infected plants to healthy plants.

1. *Physical Methods*: Hot water treatment of Chinese potato tubers at 53 °C for 10 min eliminated *Meloidogyne* spp. infection.
2. *Biological Methods*: Dipping of plants in 3 % solution of *Bacillus macerans* + soil drenching with 2 % solution of *B. macerans* 7 days after planting recorded maximum reduction in nematode population and root galling (14 galls/g of roots) and gave maximum increase in yield (Sheela et al. 2004).
3. *Integrated Methods*
   (a) *Biological and Cultural Methods*: INM strategy includes dipping of stem cuttings in *Pseudomonas fluorescens* (strain Pf1) + growing marigold (*Tagetes erecta*) as intercrop and their biomass incorporation during earthing up. Biointensive disease management strategy (BDMS) include soil drenching with *P. fluorescens* (strain PfC6) at 2.5 kg/ha at planting and 30, 60, 90, and 120 days after planting. INMS along with BDMS treatment

Table 6.7 Effect of *Meloidogyne incognita* and *Macrophomina phaseolina* on root rot, root galling, and yield of Chinese potato

Treatment	Tuber yield/plant (g)	No. of galls/plant	% Disease incidence
M. incognita (1 J2/g soil)	44.00	768	0
M. phaseolina (5 g/kg soil)	61.50	0	50
M. incognita (prior) + *M. phaseolina* (15 days later)	34.50	373	100
M. phaseolina (prior) + *M. incognita* (15 days later)	52.50	110	50
M. incognita + *M. phaseolina* (simultaneously)	13.00	316	100
Uninoculated control	84.38	0	0
CD (P = 0.05)	**10.07**	**2.08**	**43.99**

increased the tuber yield by 40.6 % and reduced nematode infestation in terms of mean number of juveniles per 100 ml soil (73.2 %), mean number of adult females per g of root (82.4 %), mean number of egg mass/g root (85.9 %), and mean number of eggs per g of root (87.9 %) with least gall index (1.6). This treatment also decreased the incidence of *Macrophomina phaseolina* root rot disease up to 50.4 %. The effect of this treatment on *M. incognita*, *M. phaseolina*, and root tuber yield of Chinese potato was almost equal to that of the treatment of INMS. But the latter is highly economical with lesser BC ratio than other treatments including standard chemical check, i.e., carbofuran 3G at 1 kg a.i./ha + soil drenching with Carbendazim 0.1 %. Hence, it could be concluded that treatment of INMS, i.e., stem cuttings dipping in 0.1 % *P. fluorescens* + marigold intercropping, can be commercially exploited for the management of *M. incognita* and *M. phaseolina* disease complex in Chinese potato (Seenivasan 2010).

References

Asawalam EF, Adesanya SO (2001) Comparison of nematicidal potential of *Azadirachta indica* on the growth and yield of nematode infested okra. J Aust Agric Environ 3(2):85–90

Bhandari S, Harsh NSK, Singh P (2007) First report on *Meloidogyne arenaria* on *Coleus forskohlii* in India. Indian Forester 133:1709–1710

Boby BU, Bagyaraj DJ (2003) Biological control of root rot of *Coleus forskohlii* Briq. using microbial inoculants. World J Microbiol Biotechnol 19:175–180

Chandrashekara KN, Prasannakumar MK (2010) New host plants for *Ralstonia solanacearum* from India. New Dis Rep 22:6

Harlan JR, Dewet JMJ, Stemler ABL (1976) Origins of African plants domestication. Monton, The Hague

Kulkarni MS, Ramprasad S, Hedge Y, Laxminarayan H, Hedge NK (2007) Management of collar rot complex disease of *Coleus forskohlii* (Wild) Briq. using bioagents, organic amendments and chemicals. Biomed 2:37–40

Lingaraju S, Mallesh SB (2010) Emerging nematode diseases in two horticultural crops: their management through novel approaches. In: National conference on innovations in nematological research for agricultural sustainability – challenges and a roadmap ahead, Tamil Nadu Agricultural University, Coimbatore, p 51

Mallesh SB (2008) Plant growth promoting rhizobacteria, their characterization and mechanisms in the suppression of soil borne pathogens of Coleus and Ashwagandha. Ph D thesis, University of Agricultural Sciences, Dharwad, 206 pp

Mohammed A, Chimbekujwo IB, Bristone B (2013a) Identification and control of fungi associated with the post-harvest rot of *Solenostemon rotundifolius* (Poir) J K Morton in Adamawa State. J Agric Healthc 3(5):136–140

Mohammed A, Chimbekujwo IB, Bristone B (2013b) Effect of different storage methods on development of post harvest rot of *Solenostemon rotundifolius* (Poir) J K Morton in Yola, Adamawa State, Nigeria. J Biol Agric Healthc 3(5):1–5

Mohandas C (2001) Nematode diseases of tuber crops and their management. National congress on centenary of nematol in India – Appraisal & Future Plans. Indian Agricultural Research Institute, New Delhi, pp 35–36

National Research Council (2006) Lost crops of Africa: Volume II: Vegetables. National Academies Press, Washington, DC, pp 269–285

Nkansah GO (2004) *Solenostemon rotundifolius* (Poir), PROTA 2: Vegetables/legumes (CD-ROM). PROTA, Wageningen

Nisha MS, Sheela MS (2006) Bio-management of *Meloidogyne incognita* on Coleus, *Solenostemon rotundifolius* by integrating solarization, *Paecilomyces lilacinus, Bacillus macerans* and neem cake. Indian J Nematol 36:136–138

Nwaneri JA (2013) Preliminary investigation into rots of Hausa Potato in Southeast Nigeria. A seminar paper presented to the Department of PHM, MOUA, Umudike

Okigbo RN (2004) A review of biological control methods for postharvest yams (*Dioscorea* spp.) in storage in South Eastern Nigeria. KMITL Sci Technol J 4(1):207–215

Okigbo RN, Nmeka A (2005) Control of yam tuber with leaf extracts of *Xylopia aethiopica* and *Zingiber officinale*. Afr J Biotechnol 4(8):804–807

Okigbo RN, Anuagasi CL, Amadi JE (2009) Advances in selected medicinal and aromatic plants indigenous to Africa. J Med Plants Res 3(2):86–95

Olojede AO, Nwokocha CC, Eke-Okoro ON, Emehute JKU (2005) Determination of optimum plant population and NPK fertilizer requirement for Hausa potato (*Solenostemon rotundifolius* Poir) production under low altitude agro-ecology. NRCRI annual report (2004), pp 92–93

Paramasivan M, Mohan S, Muthukrishnan N (2007) Management of *Coleus* dry root rot pathogen *Macrophomina phaseolina* by fungal and bacterial antagonist. Indian J Plant Prot 35:133–135

Phungpanya C, Thongpoon C, Pripdeevech P, Machan T (2012) Analysis of aroma volatiles from fresh leaves of *Plectranthus amboinicus* using solid phase micro-extraction (SPME) technique. In: Pure and applied chemistry international conference 2012 (PACCON 2012). 2012, 16701672

Pillai KS (1976) Nematicidal control of root-knot nematode on *Coleus parviflorus*. J Root Crop 2:60–63

PROTA (Plant Resources of tropical Africa) (2013) *Solenostemon rotundifolius* Poir (Synonyms: *Germania rotondifolius* Poir, *Plectranthus rotundifolius* Sims) database. PROTA.org/db/w-wpd/exec/db Retrieved November

Ramaprasad Shresti AY (2005) Studies on collar rot complex of *Coleus forskohlii* (Wild.) Briq. M Sc (Agri.) thesis, University of Agricultural Sciences, Dharwad, Karnataka, India

Schipper RR (2000) African indigenous vegetables: an overview of the cultural species. Natural Resources Institute, Chattam, 214 pp

Seenivasan N (2010) Biointensive management of *Meloidogyne incognita* and *Macrophomina phaseolina* disease complex in medicinal coleus. Indian J Plant Protect 38:186–192

Seenivasan N, Deevrajan K (2008) Integrated approach for the management of root-knot nematode, *Meloidogyne incognita* in medicinal coleus. Indian J Nematol 38(2):154–158

Senthamarai M, Poornima K, Subramanian S (2006a) Bio-management of root-knot nematode, *Meloidogyne incognita* on *Coleus forskohlii* Briq. Indian J Nematol 36:206–208

Senthamarai M, Poornima K, Subramanian S (2006b) Assessment of avoidable yield loss on *Coleus forskohlii* due to *Meloidogyne incognita*. Indian J Nematol 36:296–297

Senthamarai K, Poornima K, Subramanian S, Sudheer J (2008) Nematode-fungal disease complex involving *Meloidogyne incognita* and *Macrophomina phaseolina* on medicinal coleus, *Coleus forskohlii* Briq. Indian J Nematol 38:30–33

Sheela MS, Nisha MS, Mohandas C (2004) Eco-friendly management of nematodes associated with Chinese potato (Coleus), *Solenostemon rotundifolius* (Poir) Morton. In: National symposium on paradigms in nematological research for biodynamic farming, University of Agricultural Science, Bangalore, p 90

Shukla RS, Kumar S, Singh HN, Singh KP (1993) First report of aerial blight of *Coleus forskohlii* caused by *Rhizoctonia solani* in India. Plant Dis 77:429

Shyla M (1998) Etiology and management of a root-rot of Coleus forskohlii. M Sc thesis, University of Agricultural Sciences, Bangalore

Singh R, Paramaeswarn TN, Prakasa Rao EVS, Puttanna K, Kalra A, Srinivas KVNS, Bagyaraj DJ, Divya S (2009a) Effect of arbuscular mycorrhizal fungi and *Pseudomonas fluorescens* on root-rot/wilt, growth and yield of *Coleus forskohlii*. Biocontrol Sci Tech 19:835–841

Singh R, Parameswaran TN, Divya S, Puttanna K, Satyasrinivas KVN, Bagyaraj DJ, Kalra A (2009b) Management of root-rot/wilt of *Coleus forskohlii* Briq. In: CIMAP golden jubilee national symposium on medicinal & aromatic plants "Fifty Years of Research on Medicinal & Aromatic Plants", CIMAP Regional Centre, Bangalore, p 18

Singh R, Parameswaran TN, Prakasa Rao EVS, Puttanna K, Bagyaraj DJ, Srinivas KVNS, Kalra Alok, Kumar Ravi, Ravish BS (2008) Influence of bioinoculants and organic inputs on growth and yield of *Coleus forskohlii* under semiarid tropical organic regime. In: National seminar on developments in soil science, 73rd annual convention, University of Agricultural Sciences, Bangalore, p 292

7 Elephant Foot Yam, *Amorphophallus paeoniifolius*

Abstract

Distribution, economic importance, symptoms/damage, host range, biology, survival and spread, and management of insect and mite pests and fungal, bacterial, viral, and nematode diseases of elephant foot yam (*Amorphophallus paeoniifolius*) using regulatory, physical, cultural, chemical, biological, host resistance, and integrated pest management methods are discussed.

Keywords

Elephant foot yam • *Amorphophallus* • Pests • Diseases • Nematodes

7.1 Introduction

Elephant foot yam (*Amorphophallus paeoniifolius*), a striking aroid from tropical Asia, is extensively cultivated for its edible tubers. It is a strong herbaceous plant with erect, solitary stem usually 1.0–2.5 m in height and having leaves at the top which is highly dissected. The life cycle of the plant is usually 4–6 years. At the end of the plant's life cycle, a large terminal inflorescence is produced. The underground stem is large globose with a depression on top. It is called corm (Fig. 7.1) and at the end of each season, it produces small cormels. The corm has carminative properties. It is also used as an expectorant. Corm is good for piles and is applied externally for treating rheumatism. Fermented juice of the petiole is used to treat diarrhea.

Elephant foot yam (EFY) is a popular tuber crop of humid tropics. Among root and tuber crops, *Amorphophallus* is one of the most important tuber crops that occupy special significance as high value cash crop due to its high production potential and popularity as a vegetable. The genus *Amorphophallus* is indigenous to tropical Asia and Africa, and *A. paeoniifolius* is found widely distributed in thickets and secondary growth forests at low and medium altitude in the Philippines, Malaysia, Indonesia, Sri Lanka, and the Southeast Asia subcontinent. In India, elephant foot yam is grown in an area of 32,012 ha globally with an annual output of 667,323 tons of tubers. The elephant foot yam is traditionally cultivated in Andhra Pradesh, Tamil Nadu, Gujarat, Maharashtra, and Kerala States. It is also being cultivated in nontraditional areas such as West Bengal, Bihar, Jharkhand, and Chhattisgarh

P.P. Reddy, *Plant Protection in Tropical Root and Tuber Crops*,
DOI 10.1007/978-81-322-2389-4_7

Fig. 7.1 Elephant foot yam crop, corms, and flower

Table 7.1 Elephant foot yam growing states, regions, and production systems in India

State	Production system	Area/region
Andhra Pradesh	Plains irrigated and in Lankas (delta lands)	Prakasam, Guntur, Krishna, West Godavari and East Godavari, besides Chittoor district
West Bengal	Plains irrigated	Nadia, 24 Parganas districts
Gujarat	Plains irrigated	Ahmedabad, Surat regions

(Tables 7.1 and 7.2). In elephant foot yam, rich diversity has been reported from Western Ghats. Seven wild species such as *A. hohenackeri*, *A. commutatus*, *A. paeoniifolius* var. *dubius*, *A. bulbifera*, *A. smithsonianus*, *A. bonoccordensis*, and *A. sylvaticus* are located in the Western Ghats in Kerala and Karnataka. Chemical composition of elephant foot yam corms: moisture 60–80 %, starch 25–28 %, sugars 0.5–0.6 %, crude protein 1.0–2.8 %, fat 0.1–0.3 %, ash 0.5–2.6 %, vitamin A 0.8 mg, vitamin B1 0.08–0.09 mg, vitamin B2 0.02–0.03 mg, and vitamin C 5–12 mg. Besides being used as a vegetable, the tubers can also be used for making pickles and flour.

Scientists at the Central Tuber Crops Research Institute (CTCRI), Thiruvananthapuram, under the All India Coordinated Research Project on tuber crops, have developed an acrid-free variety of yam named Gajendra. The variety is endowed with yield potential of 40–50 tons of tubers per hectare. The crop is found to grow well in hot and humid climates. Well-drained, fertile, sandy loam soil is ideal for growing the crop.

Under irrigated conditions, the crop is planted during mid-March (especially in Kerala) and harvested during October, and under rain-fed conditions, the crop is planted during the last week of June in north India. Cut tubers are used for planting. The tubers should be planted 90×90 cm apart in pits filled with decomposed cow dung manure and sandy loam soil. Before planting, it is advisable to treat the tubers with cow dung slurry mixed with *Trichoderma* formulation followed by drying under shade. Weeding and fertilizer applications have to be done twice, 1 week after the crop starts sprouting leaves and again after a month of planting the tubers. A fertilizer dose of 80 kg of urea, 60 kg of phosphorus, and 100 kg of potash should also be applied while planting the tubers.

Mulching with organic waste or polyethylene sheets helps in reducing the weed growth and conserving soil moisture. Depending on the moisture availability, irrigation should be given at regular intervals till the monsoon. Irrigation should be stopped 4–5 months after planting to allow the crop to mature. Care should be taken to prevent water stagnation. The crop matures in 6–7 months. Yellowing of the leaves indicates crop maturity. Banana, coconut, and vegetable crops may be intercropped to generate additional revenue to the farmers. The crop is susceptible to mosaic virus, collar rot, and leaf blight infestations. Use of disease-free planting material, mulching with paddy straw, and spraying 2 %

Table 7.2 State-wise area, production, and productivity of elephant foot yam in India

State	Area (ha)	Production (tons)	Productivity (tons/ha)
West Bengal	11,730	163,560	13.94
Kerala	11,182	279,550	25.00
Andhra Pradesh	3,439	100,829	29.32
Chhattisgarh	898	8,675	9.66
Tamil Nadu	807	20,175	25.00
Bihar	408	17,883	43.83
Gujarat	350	14,400	41.14
Jharkhand	198	8,250	41.67
Other states	3,000	54,000	18.00
Total	*32,012*	*667,323*	*20.85*

Source: Unpublished statistics from Department of Agriculture/Horticulture of various states, 2008–2009

Mancozeb solution mixed with any systemic insecticide at 60 and 90 days after planting should be undertaken to protect the crop. In the case of soilborne infestations, soil application of neem cake before planting the tubers may be effective in control. During harvest, care should be taken to avoid injury to the tubers. Tubers devoid of any physical damage are ideal for marketing.

Fully mature, graded, and cured tubers should only be used for storage as planting material. The storage place should be well ventilated and cool. The tubers should be stored in a single layer. But if the storage place is insufficient, then they can be stored in two layers. Avoid storing the tubers in a heap.

Fig. 7.2 Mealybug infestation on elephant foot yam tuber

7.2 Insect Pests

7.2.1 Mealybug, *Rhizoecus amorphophalli*

During the last two decades, the mealybug, *R. amorphophalli*, has emerged as a noxious pest infesting stored EFY tubers (Rajamma et al. 2002, 2007).

7.2.1.1 Damage

Rhizoecus amorphophalli sucks cell sap from the tubers, and the severely infested deformed tubers find no place in market nor are accepted for cooking. Mealybug multiplication is rapid during the season with high temperature and less humidity, and they spread all over the tubers with white powdery mealy substance disfiguring them (Fig. 7.2) (Palaniswami 1994).

In the absence of mealybug control measures, mealybug numbers increased by 4–5 times during the storage period. The pest affected the quality of the corms and reduced subsequent field establishment and crop growth.

Infested seed materials showed a delay in sprouting, and it was reflected in their growth in first 2 months after planting. At harvest, it affected the yield and infested seed material produced lesser yield (Table 7.3).

Table 7.3 Yield of elephant foot yam raised by using infested and uninfected planting material

Infestation level of seed material	Average weight of seed materials*		Average weight of harvested tubers*	
	kg	t/ha	kg	t/ha
Fully infested	0.51[a]	6.29[a]	2.01±0.27[a]	24.81[a]
Partially infested	0.54[a]	6.66[a]	1.95±0.20[a]	24.07[a]
Uninfected	0.50[b]	6.17[b]	3.15±0.37[b]	38.94[b]

*Values with the same superscript (column wise) do not differ significantly ($P \leq 0.05$)

7.2.1.2 Biology

The mealybug reproduces sexually and the adult female secretes an egg sac of white waxy substance in which eggs are laid. Female nymph molts normally, but male instar produces a cottony puparium around its body and forms a pupal stage from which adult males emerge. On tubers of elephant foot yam, average fecundity and incubation period were 68.30±6.22 and 7.80±0.88 days, respectively. The total life cycle of bug which includes three instars and a pupal stage took 27.10 days for females and 22.40 days for males on tubers of elephant foot yam.

7.2.1.3 Management

1. *Cultural Methods*: Salt (NaCl) solution (1,000 ppm), cow urine, cow dung slurry (2 kg cow dung in 1 L of water), and clay slurry (1 kg of clay in 1 L of water) treatments were effective in reducing mealybug numbers and the associated corm damage. Common salt is cheap, widely available, and easy to use in treating the corms prior to storage. Relative to untreated corms, those treated with salt solution recorded greater emergence when field planted as well as producing vigorous plant growth.

   Application of early instars of mealybugs with lower concentration of cassava seed extract (CSE) (0.5 %) with a few drops of soap solution is effective. Surfactant at 0.1 % was also highly toxic to early instars and the mortality was 50 %. Adult females were not as sensitive as first instar, and treatment with 2 % concentration of CSE only killed all the females.
2. *Chemical Methods*: Spraying dimethoate 0.05 % is effective if infestation is severe. Cent per cent mortality of mealybugs was observed 1 day after treatment with methyl parathion (0.05 %) which was on par with malathion, chlorpyrifos, and quinalfos. Two days after treatment, 100 % mortality was observed with malathion, which was on par with chlorpyrifos and dimethoate (0.5 %).
3. *Biological Methods*: The mealybug in storage was parasitized by an encyrtid parasitoid *Anomalicornia tenuicornis* (Hymenoptera) up to 10–20 %. Approximately 120 numbers of *A. tenuicornis* were emerged from 1 kg of elephant foot yam tuber fully infested with mealybugs. Coccinellid *Cryptolaemus montrouzieri* (2 beetles for 1 kg tuber) also predated successfully on *Rhizoecus* spp. Both these bioagents controlled the EFY mealybugs within 3–4 days.

7.2.2 Scale Insect, *Aspidiella hartii*

7.2.2.1 Damage

Infestation of white scale insect, *Aspidiella hartii*, on tubers of major aroids perpetuates through seed tubers and affects the quality and viability of harvested tubers.

7.2.2.2 Management

Use of pest-free seed tubers and dipping seed tubers before planting in 0.1 % dimethoate is advocated. Scale insect is also controlled by dipping planting material in Monocrotophos (0.5 %).

7.3 Diseases

7.3.1 Collar Rot, *Sclerotium rolfsii*

7.3.1.1 Symptoms

This disease is generally observed in the later part of crop growth, but it can infect the crop at any stage of crop growth. Injury to the collar region during intercultural operations and infestation of roots by *Meloidogyne incognita* (nematode) predisposes the plants to collar rot disease. The disease causes damage by rotting of the col-

Fig. 7.3 *Amorphophallus* collar rot

lar region of the stem (Fig. 6.3) which ultimately topples down and is responsible for heavy reduction in yield and qualitative degradation of the crop. The symptoms include water-soaked patch on the collar region of the pseudo-stem and finally the entire plant falls due to rotting of the collar region.

The white mycelial mat and lot of sclerotia could be seen on the collar region of the affected plant (Fig. 7.3)

Waterlogging, poor drainage, and mechanical injury at the collar region favor disease incidence.

7.3.1.2 Management

1. *Cultural Methods*: Disease can be managed by the use of disease-free planting material, removal of infected plants, improving drainage, and application of neem cake in soil.

Thick mulching with paddy straw or other organic waste reduces the frequency of weeding and other intercultural operations, thereby avoiding injury to the plants. Mulching with paddy straw/polythene sheet reduced the collar rot disease.

Seed treatment and drenching of soil with vermiwash showed lower collar rot incidence (10–30 %). Soil application combined with spraying of vermiwash 90 and 120 DAP resulted in tuber yield increase of 44.0 and 121.4 % over package of practices (POP) recommendations and control, respectively.

Table 7.4 Effect of different dates of sowing on collar rot and corm yield (Singh et al. 2006)

Date of sowing	% collar rot incidence[a]	Yield (t/ha)
February 15	52.08 (46.19)	12.326
March 15	49.99 (44.99)	12.586
March 28	47.22 (43.39)	13.338
April 30	56.25 (48.60)	11.486
May 16	73.33 (59.21)	9.982
CD at 5 %	*8.835*	–

[a]The figures given in parentheses angular transformed values

The minimum collar rot incidence, i.e., 47.22 %, was recorded in the crop sown on March 28, being at par with March 15 and February 15 sown crop (Table 7.4). Maximum yield of the crop which was recorded in March 23 sown crop being at par with February 15, March 15, and April 30 sown crop indicated that the crop sown on early dates get more time for its vegetative growth which may have resulted in good yield. The lowest incidence and severity of collar rot in March 28 sown crop may be another reason for enhancing the crop yield (Singh et al. 2006).

2. *Biological Methods*: An isolate of *Trichoderma* sp. (Td-1) isolated from the parasitized sclerotia of *S. rolfsii* was highly effective in suppressing *S. rolfsii* on elephant foot yam under storage conditions (Mukherjee and Raghu 1997).

Potent biocontrol agents of vermicompost origin against elephant foot yam collar rot were identified as *Bacillus subtilis*, *B. cereus*, *Providencia rettgeri*, and *Trichoderma asperellum*. The bacterial isolates expressed N_2 fixing, P and K solubilizing, and growth promotion properties. Disease suppression varied with the origin of vermicompost. Minimum concentration of 10 % was required for pathogen suppression. Application of vermicompost reduced collar rot incidence by 10–30 %.

3. *Chemical Methods*: Disease can be managed by drenching soil with captan 0.2 %.
4. *Integrated Methods*
   (a) *Biological and Cultural Methods*: Application of neem cake and biological control agents like *Trichoderma viride*, *T. harzianum*, and *Pseudomonas fluorescens* was found effective against amorphophallus collar rot.
   (b) *Biological and Chemical Methods*: Seed corm treatment followed by soil application/drenching with *Trichoderma viride* and Carboxin at 600 g/ha is recommended for the management of collar rot of *Amorphophallus*.

   Combination of corm and soil treatment with *T. harzianum* lowered disease incidence to 12.9 % followed by corm + soil treatment with captan (0.2 %) recording 14.8 % disease incidence against 83.3 % in control. Both *T. harzianum* and *B. subtilis* multiplied more in soil when applied as soil treatment than used as corm treatment. Antagonists and chemical treatment significantly lowered the rhizosphere population of *S. rolfsii* (Table 7.5) (Gogoi et al. 2002).
   (c) *Cultural, Chemical, and Biological Methods*: The maximum reduction in collar rot incidence (75.99 %) was exhibited by the crop treated with *Trichoderma harzianum* (T.h.) + *Pseudomonas fluorescens*

Table 7.5 Effect of corm and soil treatment with antagonists and chemical on the incidence of collar rot of elephant foot yam (Gogoi et al. 2002)

Treatment	% Disease incidence (PDI)[a]	Population density (x 10^3 cfu/g of soil)[a]		Population density of S. rolfsii[a] (viable sclerotia/50 g of soil)
		T. harzianum	*B. subtilis*	
Corm treatment				
T. harzianum	38.8 (38.3)[b]	26.3	–	5.2
B. subtilis	49.9 (44.8)	–	31.6	6.0
Captan	33.3 (35.0)	–	–	5.5
Soil treatment				
T. harzianum	31.4 (32.3)	31.1	–	4.9
B. subtilis	40.7 (37.8)	–	34.1	6.1
Soil drenching				
Captan	27.7 (31.2)	–	–	4.4
Corm + soil treatment				
T. harzianum + *T. harzianum*	12.9 (17.6)	46.2	–	3.5
B. subtilis + *B. subtilis*	33.3 (35.1)	–	43.2	5.0
Captan + captan	14.8 (21.6)	–	–	3.8
T. harzianum + captan	33.3 (35.1)	23.6	–	5.0
B. subtilis + captan	24.0 (28.8)	–	31.7	4.8
Captan + *T. harzianum*	25.9 (27.5)	33.9	–	4.0
Captan + *B. subtilis*	31.4 (33.8)	–	36.5	4.8
Control	83.3 (80.9)	16.9	19.8	8.2
CD. at 5 %	*7.6*	*1.7*	*2.0*	*0.7*

[a]Average of three replications pooled over two seasons
[b]Figures within parenthesis are angular transformed values

Table 7.6 Effect of the application of antagonists, their combinations, and chemicals on collar rot and corm yield (Singh et al. 2006)

Treatment/dose/mode of application	Collar rot		Yield t/ha	% yield increase
	% Disease incidence	% disease control		
Trichoderma harzianum (Th)—100 g/10 L of water—seed treatment	18.75 (25.45)	63.99 (53.12)	21.354	70.83 (57.31)
Pseudomonas fluorescens (Pf)—100 g/10 L of water—seed treatment	27.08 (31.33)	48.00 (43.85)	17.708	41.66 (40.20)
Pf (a)+Pf (b)—(1:1) 50+50 g/10 L of water—seed treatment	22.91 (28.55)	56.00 (48.45)	20.607	64.85 (53.64)
Th+Pf (c)—(1:1) 50+50 g/10 L of water—seed treatment	64.85 (53.64)	68.01 (55.56)	22.743	81.94 (64.85)
Carbendazim—10 g/10 L of water—seed treatment	37.50 (37.72)	27.99 (31.94)	14.756	18.04 (25.13)
Ridomil MZ—25 g/10 L of water—seed treatment	29.16 (32.66)	44.00 (41.55)	16.718	33.74 (35.51)
Mancozeb—30 g/10 L of water—seed treatment	33.32 (33.99)	36.02 (36.88)	15.451	13.60 (29.06)
Pf (a)+Pf (b)+FYM—75+75 g+300 g/pit	25.00 (27.11)	51.99 (46.14)	20.052	60.41 (51.01)
Th+FYM—150 g+300 g/pit	24.58 (20.70)	52.80 (46.61)	21.927	75.41 (60.27)
Pf (c)+FYM—150 g+300 g/pit	25.00 (28.55)	5 1.99 (46.14)	18.836	50.68 (45.39)
Th. + Pf (c)+FYM—75+75 g+300 g/pit	12.50 (20.70)	75.99 (60.66)	23.559	88.47 (70.15)
Check—treated with plain water—seed treatment	52.08 (48.59)	0.05 (1.28)	12.500	0.05 (1.25)
CD at 5 %	*4.191*	*5.004*		*5.11*

The figures given in parentheses are angular transformed values

strain (P.s.c.)+FYM with 88.47 % yield increase, which was significantly higher than the check (Table 7.6). The use of antagonists, their various combinations, and chemicals showed significant reduction in the collar rot incidence by lowering the number of infected plants as compared to check (Singh et al. 2006).

After one application, the maximum reduction in collar rot incidence was recorded with the application of Ridomil MZ+neem cake and *Trichoderma harzianum* which were similar (53.85 %) (Table 7.7). In the case of two applications of fungicides, bioagents, and neem cake, it was found that maximum reduction in collar rot incidence (64.01 %) was achieved due to the application of Ridomil MZ+neem cake. The use of bioagents, fungicides, and neem cake showed significant reduction in disease incidence by lowering the number of infected plant as compared to control.

The highest yield was recorded in the case of neem cake+Ridomil MZ. The yield increase may be due to the lowering down of the inoculum density and capacity which resulted in the improvement of the plant health, with the application of antagonists.

7.3.2 Tuber Rot

Six major pathogens that caused postharvest rotting of elephant foot yam tubers are *Phytophthora colocasiae, Sclerotium rolfsii, Botryodiplodia theobromae, Fusarium,* and *Rhizopus* species. One bacterial pathogen, *Erwinia carotovora*, was found to cause serious damage to elephant foot yam tubers stored at higher temperature with poor aeration.

Table 7.7 Effect of the fungicides, antagonists, and oil cakes as soil application in standing crop against collar rot and corm yield

Treatment/dose	% Disease control with 1 application	% Disease control with 2 applications	Yield (t/ha)	% Yield increase
Neem cake—50 g/plant	38.44 (38.32)	52.00 (46.15)	15.173	23.61 (29.07)
Neem cake + *T. harzianum* (Th)—50 g + 10 g/plant	46.14 (42.79)	56.00 (48.45)	18.211	48.37 (44.07)
Neem cake + [Th + *P. fluorescens* (Pf)]—50 g + 10 g/plant	38.44 (38.32)	60.00 (50.77)	20.277	65.20 (53.85)
Th—10 g/plant—10 g/plant	53.85 (47.21)	56.00 (48.45)	18.593	51.48 (45.85)
Th + Pf—10 g/plant	46 14 (42.79)	56.00 (48.45)	17.812	45.11 (42.19)
Neem cake + Ridomil MZ—50 g + 0.2 %	53.85 (47.21)	64.01 (53.14)	21.614	76.09 (60.73)
Ridomil MZ—0.2 %	38.44 (38.32)	52.00 (46.15)	17.829	45.25 (42.27)
Check—plain water	0.05 (1.28)	0.05 (1.28)	12.274	0.05 (1.28)
CD at 5 %	*6.042*	*6.128*	–	*5.914*

The figures given in parentheses are angular transformed values

7.3.2.1 Symptoms

Tuber rot could be a serious problem in elephant foot yam if the planting material used for raising the crop is infected. Injuries during harvest and transport further aggravate the problem of tuber rot. Infestation of the roots and tubers by the root-knot nematode *Meloidogyne incognita* acts as a predisposing factor for infection by a large number of fungi.

7.3.2.2 Management

1. *Cultural Methods*: The tubers need to be stored properly during the off-season for using them as seed material for next planting season.

   This disease can be very well managed by adopting the following practices:
   - Sanitation by removing and burning of infected tubers.
   - Deep plowing with chisel plow.
   - Ridge planting.
   - Proper drainage and controlled irrigation.
2. *Chemical Methods*: *Amorphophallus* seed carmel treatment with Carbendazim (0.05 %) and stored on cemented floor gave minimal carmel rotting (7.57), minimal sprouting (9.60 %), and maximum shelf life (148.65 days). The following fungicides have been recommended for the management of tuber rot (Table 7.8):
3. *Biological Methods*: The conidial mixture of 2 bioagents *Trichoderma pseudokoningii* (effective in reducing the mycelial growth) and *T. virens* (good colonizer of sclerotia) was effective in reducing the incidence of tuber rot.
4. *Integrated Methods*: Use of healthy tubers, treatment of planting material with cow dung slurry mixed with *Trichoderma viride* (Fig. 7.4); mulching with paddy straw or locally available leaves/twigs or black polythene mulch; periodic removal of mosaic-infected plants, if any from the seed crop; and one foliar spraying with Mancozeb (0.2 %) and any systemic insecticide (0.05 %) at 60 days after planting provide effective control of all the major diseases of elephant foot yam.

7.3.3 Leaf Blight/Leaf Rot, *Phytophthora colocasiae*

Leaf blight and leaf rot caused by *Phytophthora colocasiae* is responsible for heavy reduction in yield and qualitative degradation of the crop. The disease occurs in regions having moderately high temperature (22–23 °C), high relative humidity (85–100 %), and frequent rainfall.

7.3.3.1 Symptoms

Symptomatic plants ranged from chlorotic and stunted to completely blighted as the disease progressed. Necrotic root symptoms began at the tip, but progressed quickly, eventually killing the whole root (Fig. 7.5). The cortex of severely affected roots could be easily sloughed off, leaving only parts of the vascular system intact.

Table 7.8 Fungicides commonly used to control post-harvest decay of edible aroids

Fungicide	Effectiveness
Benomyl	Effective where *Botryodiplodia theobromae* is the predominant decay organism
	Ineffective against Phycomycetous fungi
	Recommended in countries where *Phytophthora colocasiae* and *Pythium splendens* cause major storage losses
Copper oxychloride	Control corm rots caused by *P. colocasiae*
Captafol, Mancozeb	Only delays *B. theobromae* rots by ~10 days
Sodium hypochlorite	Effective against all common storage decay fungi in the Pacific except *Sclerotium rolfsii*

Source: Cooke et al. (1988)

7.3.3.2 Management

1. *Chemical Methods*: Seed tuber dip treatment with 0.05 % Agrimycin for 12 h decreased the disease incidence (19.73 % as compared to 65.91 % in control) and increased the tuber yield (23.95 MT/ha as compared to 7.81 MT/ha in control).

   Three sprays of 100 ppm Agrimycin at 15 days interval after the first appearance of the disease significantly reduced the disease incidence (24.65 % as compared to 50.88 % in control) and increased the yield (36.25 t/ha as compared to 17.81 t/ha in control).

2. *Integrated Methods*
   (a) *Biological and Cultural Methods*: The mixture of *Trichoderma harzianum* + *Pseudomonas fluorescens* resulted in maximum reduction of the disease (Table 7.9). The corm treatment with antagonists has resulted in the management of *Phytophthora colocasiae*-like pathogen whose spread completely depends upon airborne zoosporangia and zoospores. The possible reason behind

Fig. 7.4 Tuber treatment with *Trichoderma* and cow dung slurry

Fig. 7.5 Root rot and leaf blight of elephant yam. (**a**) Field showing symptoms in various stages of disease progression. (**b**) Single plant showing severely blighted symptoms. (**c**) Severely affected corm with rot symptoms

Table 7.9 Effect of application organic amendment, antagonists, and fungitoxicant on leaf blight and yield

Treatment	Rate and mode of application	% Leaf blight intensity (% disease control)	Yield (t/ha) (% yield increase)
Trichoderma harzianum	100 g/10 L water (seed treatment)	54.00 (46.63)	176.66 (57.41)
T. harzianum + *Pseudomonas fluorescens*	(3:1) 75 + 25 g/10 L water (seed treatment)	43.00 (56.70)	177.95 (57.69)
T. harzianum + *P. fluorescens*	(1:1) 50 + 50 g/10 L water (seed treatment)	50.00 (49.66)	170.71 (51.27)
Ridomil MZ (Mancozeb + Ridomil)	25 g/10 lit water (seed treatment)	68.00 (31.54)	127.30 (12.81)
FYM + *T. harzianum* + *P. fluorescens*	4.50 g + 30 g/27 pit (pit application)	49.33 (50.33)	167.82 (48.72)
FYM + *T. harzianum*	4.50 g + 30 g/27 pit (pit application)	58.33 (41.27)	149.01 (32.04)
FYM alone	4.50 kg/1+ pit (pit application)	71.66 (27.85)	134.54 (19.23)
Control	Water	99.33	112.84
CD at 5 %		12.59	

reducing the severity of blight may be due to induction of systemic resistance due to *T. harzianum* or *P. fluorescens* applied through corm (Singh et al 2005).

(b) *Biological and Cultural Methods*: IDM package (selection of healthy tubers + cormel treatment with *Trichoderma viride* at 5 g/kg mixed with cow dung slurry before planting + application of *T. viride* enriched compost before planting + mulching with paddy straw after planting + spraying of CTCRI formulation at 60 and 90 DAP) was found effective against leaf blight (Table 7.10).

(c) *Cultural, Chemical and Biological Methods*: The disease can also be managed by
   - Early planting to avoid heavy monsoon rains.
   - Use of healthy planting materials.
   - Treatment of the seed tubers with biocontrol agent viz. *Trichoderma viride*.
   - Spraying with fungicides, viz., Mancozeb (0.2 %) or metalaxyl (0.05 %).

Table 7.10 Effect of integration of cultural, organic amendments, and bioagents on the incidence of leaf blight on elephant foot yam

Treatment	Leaf blight incidence (%)[a]
IDM package	28.75 (32.20)
Farmers' practice	53.10 (46.86)
Foliar spray of dimethoate at 0.05 %	38.21 (38.06)
CD (P = 0.05)	*5.59*

[a]Figures in brackets are arc sin values

Fig. 7.6 Dasheen mosaic disease symptoms on Amorphophallus

7.3.4 Dasheen Mosaic Virus

Among the diseases, the viral disease caused by the dasheen mosaic virus seems to be the most destructive, resulting in reduced tuber yield.

7.3.4.1 Distribution and Economic Importance

The disease is distributed in Himachal Pradesh, foothills of Himalayas, where ornamental plants are grown, and West Bengal. Disease incidence varies from 5 to 10 %. Yield loss up to 30–35 % has been reported from Uttar Pradesh.

7.3.4.2 Symptoms

The mosaic, mottling, leaf puckering, rolling, leaf narrowing, distortion of leaf lamina, and stunting type of symptoms are often noticed in elephant foot yam (Fig. 7.6). Corms produced by the mottled plants are much smaller than those without mottled leaves. The symptoms are transient and expression of symptoms varies from time to time depending upon weather conditions. Virus causing mosaic in elephant foot yam has been identified as *dasheen mosaic virus*.

7.3.4.3 Particle Morphology

A virus with flexuous, filamentous particles (750Ã–12 nm) was observed (Fig. 7.7) under electron microscope.

7.3.4.4 Transmission and Spread

The disease is primarily spread through infected planting materials and secondary spread is through different species of aphids.

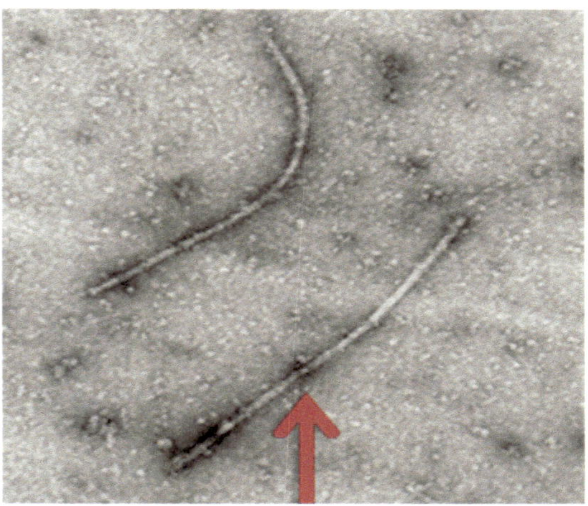

Fig. 7.7 Virus particles (flexuous rods) of dasheen mosaic virus

The virus is transmitted in a nonpersistent manner by aphid vectors *Aphis craccivora*, *A. gossypii*, *Myzus persicae*, and *Pentalonia nigronervosa*. Virus is also transmitted by mechanical inoculation and by grafting but not transmitted by contact between plants or by seed.

7.3.4.5 Host Range

The host range includes *Aglaonema* sp., *Philodendron* sp., *Colocasia esculenta*, *Zantedeschia* sp., *Datura metal*, and *D. alba*.

7.3.4.6 Management

Using planting material from mosaic-free plants is the best way to avoid the disease. Tubers free from mosaic virus should be used for planting. Spraying of systemic insecticides should be undertaken to prevent secondary spread of the virus.

Hot air treatment of mosaic-infected tubers at 55 °C for 10 min before planting followed by two sprayings of Monocrotophos (0.05 %) or any other broad-based insecticide at 60 and 90 days after planting significantly reduced mosaic disease incidence, and there was a corresponding increase in tuber yield.

Mosaic-free plants are produced using micropropagation techniques in several crops including *Amorphophallus*.

7.3.5 Leaf Spot, Collar Rot, and Mosaic

Elephant foot yam is affected by three major diseases, namely, collar rot caused by *Sclerotium rolfsii*, leaf blight caused by *Phytophthora colocasiae*, and mosaic caused by dasheen mosaic virus.

7.3.5.1 Management

1. *Integrated Methods*: IDM package recorded the lowest disease incidence (8.2, 11.5, and 4.4 % compared to 13.4, 17.1, and 10.0 % in control of leaf spot, collar rot, and mosaic diseases, respectively) and disease intensity (4.3, 6.9, and 2.1 % compared to 7.7, 11.7, and 4.5 % in control of leaf spot, collar rot, and mosaic diseases, respectively) and the highest corm yield (33.00 t/ha compared to 22.5 t/ha in control) in Bihar (Kumari et al 2013) (Table 7.11).

In Andhra Pradesh, IDM package gave significantly highest sprouting (97.68 %) which was on par with chemical treatment (97.18 %). The chemical treatment gave the lowest disease incidence of leaf blight (PDI – 15.8), mosaic disease (PDI – 11.8), and collar rot (PDI – 3.9) and the highest cormel weight (43.3 t/ha) which was on par with IDM package. However, the highest benefit/cost ratio was obtained with IDM package (1.82) followed by chemical treatment (1.53) (Table 7.12).

7.4 Nematodes

The root-knot and the lesion nematodes are important pests of *Amorphophallus*.

The lesion nematodes (*Pratylenchus* spp.) have been recorded in high numbers from Tamil Nadu, leading to crop failure in many farms. The surface of the infested tubers was black in color with cracks. The blackening extended further deep inside the tubers reducing the edible portion. The infested tubers were smaller in size.

Table 7.11 Effect of IDM package on the incidence and intensity of leaf spot, mosaic, and collar rot of elephant foot yam

	Leaf spot (%)		Mosaic (%)		Collar rot (%)		Tuber yield (t/ha)
Treatment	Incidence	Intensity	Incidence	Intensity	Incidence	Intensity	
IDM package[a]	8.2	4.3	4.4	2.1	11.75	6.9	33.0
Farmers' practice[b]	13.4	7.7	10.0	4.5	17.1	11.7	22.5
Chemical control[c]	10.9	6.4	8.2	2.9	13.5	8.8	27.4
CD (P = 0.05)	*1.91*	*1.60*	*1.45*	*1.28*	*1.45*	*2.10*	*1.50*

[a]*IDM package*: Planting of healthy seed corms treated with *Trichoderma viride* at 5 g/kg in cow dung slurry before planting, application of neem cake at 200 g/pit along with *T. viride* enriched compost at 1 kg/pit, mulching with paddy straw after planting, and spraying of CTCRI formulation (prepared by mixing cow urine, 2 L; neem cake, 1 kg; and leaves of neem, bael, peepal, eucalyptus, and custard apple, 21 leaves each in 10 L of water taken in earthen vessel, allowed to decompose for 30 days under anaerobic conditions, filtered through two layers of muslin cloth, and then filtrate diluted in 100 L of water) at 60 and 90 DAP

[b]*Farmers' practice*: Planting of healthy seed corms without treatment or use of chemical pesticides/CTCRI formulation

[c]*Chemical control*: Seed corm treatment with Mancozeb (0.2 %) and dimethoate (0.05 %) before planting and 2 sequential spraying of Mancozeb (0.2 %) and dimethoate (0.05 %) applied at 60 and 90 DAP

Table 7.12 Effect of IDM package on the incidence and intensity of leaf spot, mosaic, and collar rot of elephant foot yam

Treatment	Sprouting (%)	Leaf blight (PDI)	Mosaic (PDI)	Collar rot (PDI)	Cormel yield (t/ha)	Benefit: cost ratio
IDM package	97.68 (81.33)	18.2 (24.58)	12.04 (19.59)	3.53 (11.51)	42.8	1.82
Farmers' practice	92.97 (74.69)	25.6 (28.62)	17.66 (23.32)	6.81 (15.60)	34.5	1.42
Chemical control	97.18 (80.44)	15.8 (23.80)	11.86 (19.81)	3.97 (11.35)	43.3	1.53
CD (P = 0.05)	*1.661*	*3.717*	*2.539*	*2.304*	*6.026*	

The stunt nematode, *Tylenchorhynchus indicus*, was found to infect the roots and corms and cause severe postharvest loss in combination with fungi like *Fusarium* sp. and *Sclerotium* sp.

7.4.1 Root-Knot Nematode, *Meloidogyne incognita*

Amorphophallus spp. are highly susceptible to *M. incognita*. The nematode infestation is very high in Kerala, Andhra Pradesh, and Tamil Nadu. The root-knot nematodes have been recorded in high numbers from Tamil Nadu leading to crop failure in many farms. The nematode is causing serious loss in major *Amorphophallus*-growing areas in Kerala, Tamil Nadu, and Andhra Pradesh.

7.4.1.1 Symptoms

In *Amorphophallus* the dry rot of tubers is caused by the root-knot nematode. Infestation sets in early attacking the roots and then infests the tuber tissue. The nematode multiplies inside the tuber tissue and cause dry rot. In store the dry rot continues to be serious resulting in damage of the whole infested tuber. No resistance has so far been reported in *Amorphophallus* to the nematode.

The roots produce typical root-knot symptoms. In corms, the galls appear as irregular projections which harbor adult females and eggs. The area of infestation in tuber was discolored when the infestation was severe, the infested area dried up resembling dry rot. Infested tubers were deformed and smaller. The degree of infestation was very high in cormels compared with that in corms.

7.4.1.2 Management

Continuous cropping of a particular crop results in buildup of the nematodes in fields. Hence, crop rotation is essential. Farmers rotate *Amorphophallus* with resistant *Colocasia* in Tanjore district of Tamil Nadu. In Kerala, *Amorphophallus* is grown only once in three years in the same field. In East and West Godavari District of Andhra Pradesh, *Amorphophallus* is rotated with sugarcane, paddy, cereals, banana, etc.

Taking up cassava and sweet potato in between susceptible crops are also found to be effective in reducing root-knot nematode population. Cassava leaves and its dry leaf powder have been found to be nematicidal. Incorporation of the powder was found to be effective in reducing the root-knot nematode problem in *Amorphophallus*.

7.5 Future Thrusts

- There is a need to intensify the breeding efforts toward developing high-yielding collar rot-resistant/rot-tolerant varieties of elephant foot yam.
- The development of transgenic elephant foot yam resistant against dasheen mosaic virus with due attention toward increasing transformation efficiency and developing more effective gene constructs.
- Techniques developed in the selection and preparation of healthy planting material of *Amorphophallus* should be popularized in those areas in Andhra Pradesh, Bihar, and Gujarat where intensive cultivation of the crop is practiced.
- Popularization of IDM package developed for collar rot and other diseases of elephant foot yam by the Regional Centre of CTCRI, Bhubaneswar, and recommended by AICRP (Tuber Crops) through demonstrations and awareness activities.
- Quick diagnostic methods to be developed for dasheen mosaic virus in elephant foot yam.

- Studies initiated on the use of biocontrol agents (*Trichoderma harzianum*, *Pseudomonas fluorescens*, etc.) should be intensified so as to formulate technology for management of soil-borne diseases in *Amorphophallus*.
- Disease-free elephant foot yam mother cultures should be established for all the released cultivars and maintained for further mass multiplication.
- Refinement of IDM technologies developed for the management of collar rot needs to be undertaken.
- Safe control measures and storage techniques are to be developed to ensure longer storage life and better viability of planting materials of *Amorphophallus*. The studies are to be intensified to develop environmentally friendly control measures through use of biopesticides, attractants, and repellants.
- Emerging pests and diseases due to climate change and introduction of invasive pests and pathogens are also main challenges. Appropriate technologies to contain such biotic stresses have to focus on refined eco-friendly methods. In vitro elimination of dasheen mosaic in *Amorphophallus* has to be standardized using meristem and nodal culture combined with chemotherapy and thermotherapy. Strategy for mass multiplication of disease-free planting materials has to be improvised. Production of large quantities of polyclonal antibodies and developing cost-effective, easy, and sensitive diagnostic kits for indexing the planting materials are necessary. The important fungal disease like collar rot of *Amorphophallus* is disseminated through planting materials and soil. Efforts to provide healthy planting material by employing diagnostic techniques to identify pathogens need to be strengthened.

References

Cooke RD, Rickard JE, Thompson AK (1988) The storage of tropical root and tuber crops – cassava, yam and edible aroids. Exp Agric 24:457–470

Gogoi NK, Phookan AK, Narzary BD (2002) Management of collar rot of Elephant's foot yam. Indian Phytopathol 55(2):238–240

Kumari R, Singh PP, Rai RC (2013) Integrated management of diseases of elephant foot yam (*Amorphophallus paeoniifolius*). In: Misra RS, Neduchezhiyan M (eds) Aroids: opportunities and challenges. Allied Publishers Pvt Ltd, New Delhi, pp 240–245

Mukherjee PK, Raghu K (1997) Effect of temperature on antagonistic and biocontrol potential of *Trichoderma* sp. on *Sclerotium rolfsii*. Mycopathologia 139(3):151–155

Palaniswami MS (1994) Pests of edible aroids, yams and Chinese potato. In: Chadha KL, Nayar GG (eds) Advances in horticulture, vol. 8 – Tuber crops. Malhotra Publishing House, New Delhi, pp 490–491

Rajamma P, Jayaprakas CA, Palaniswami MS (2002) Bio-ecology of storage pests and their natural enemies in aroids and yams. In: Annual report 2001–2002. Central Tuber Crops Research Institute, Thiruvananthapuram, Kerala, p 54

Rajamma P, Jayaprakas CA, Palaniswami MS (2007) Bio-ecology of storage pests and their natural enemies in aroids and yams. In: Annual report 2006–2007. Central Tuber Crops Research Institute, Thiruvananthapuram, Kerala, 69 pp

Singh R, Yadav RS, Singh V, Singh PP (2005) Integrated management of leaf blight of *Amorphophallus paeoniifolius* Blume. Veg Sci 32(2):169–17

Singh R, Singh PP, Singh V (2006) Integrated management of collar rot of *Amorphophallus paeoniifolius* Blume caused by *Sclerotium rolfsii* Saccardo. Veg Sci 33(1):45–49

8 Yam Bean, *Pachyrhizus erosus*

Abstract

Distribution, economic importance, symptoms/damage, host range, biology, survival and spread, and management of insect and mite pests and fungal, bacterial, viral, and nematode diseases of yam bean (*Pachyrhizus erosus*) using regulatory, physical, cultural, chemical, biological, host resistance, and integrated pest management methods are discussed.

Keywords

Yam bean • *Pachyrhizus erosus* • Pests • Diseases • Nematodes

8.1 Introduction

Yam bean, *Pachyrhizus erosus* (family: Fabaceae), also called "jícama," is a high-yielding leguminous root crop with good nutritional value. The genus includes five species, of which two are wild and three are cultivated types. *P. erosus* is a native of tropical America, and it is widely cultivated in India, Mexico, China, Singapore, the Philippines, Hawaii, and Indonesia. In India, it is grown in parts of West Bengal, Tripura, Bihar, Orissa, and Assam.

8.1.1 Description

Yam bean is a climbing bean that has hairy stems. It can grow up to 6 m tall. The stems are woody at the base. It has a white-fleshed tuber with a rough, sandy-colored skin. The leaves are alternate and made up of three leaflets. These leaflets have large teeth. The flowers are violet or white (Fig. 8.1). The pod is 8–15 cm long, curved and hairy, and contains 8–11 flattened seeds. The seeds are almost black.

8.1.2 Distribution

It grows in warm places, like coastal areas in Papua New Guinea, and up to about 70 m altitude in the tropics. A light rich sandy soil (well drained) is suitable. It cannot tolerate frost. Plants need 11–13 h of daylight for tubers to form. It suits hardiness zones 10–11. It is a minor food in the Solomon Islands.

P.P. Reddy, *Plant Protection in Tropical Root and Tuber Crops*,
DOI 10.1007/978-81-322-2389-4_8

Fig. 8.1 Yam bean plants and tubers

8.1.3 Cultivation

It is grown from seeds and also grows wild. Seed should be presoaked for 12 h in warm water to encourage rapid germination. Seeds germinate within 2 weeks. Plants can be grown by dividing the root clump and then growing plants from the thickened roots. Cuttings will grow. A spacing of 50 cm between plants is suitable. Topping the plant by picking out the growing point and removing the flowers is said to help tuber formation.

8.1.4 Production

Tubers are ready by about 6 months after sowing. Individual tubers can be up to 20 kg in weight.

8.1.5 Uses

The young tuber is eaten either raw or cooked. It can also be pickled. The young pods can be eaten, provided they are well cooked. Old pods and mature seeds can be poisonous.

8.2 Insect Pests

A number of insect pests are reported to cause leaf, tuber, and seed damage in *P. erosus*. The crop is highly susceptible to the leafhopper, *Empoasca dolichi* (Duke et al. 1978).

The insect damage observed elsewhere also seems to be a problem in Manabí, where "the whitefly complex" (*Aleurotrachelus* sp., *Bemisia* sp., *Aleurothrixus* sp.) (Homoptera: Aleyrodidae) has been observed in yam bean plants.

In Mexico and Central America, flower buds and the young pods may be damaged by *Thecla jebus*. In Central America, the yam bean is reported to be subject to attacks from the larvae of *Thecla jebus* and *Ferrisia virgata*.

The mealybug (*Ferrisia virgata*) is a severe pest of yam beans grown under greenhouse conditions (Duke 1981). It effectively destroys inflorescences of the yam bean (Sorenson 1990).

The American thrips, *Frankliniella occidentalis*, causes the premature abscission of flower buds (Sorenson 1990).

Phyllophaga (n.v. "gallina ciega") may cause severe tuber damage (Heredia 1985).

Plants in a recent field trial (1994) at Maracay, Venezuela, were heavily attacked by insect pests previously not considered to cause serious problems in *P. erosus*. Termites (Termitidae) hollowed the stems of young plants, destroying 15–20 % of the plants.

Anarector spp. (Gelerucidae) and *Nezara viridula* (Pentatomidae) caused severe leaf damage.

In Tonga, seed damage caused by the coffee bean weevil (*Araecerus fasciculatus*) has been recorded in seed belonging to all species during storage.

Species belonging to the genus *Diabrotica* are serious pests in many humid areas, and the leaf damage is often quite extensive, although differences in susceptibility between accessions have been recorded, i.e., more hairy genotypes tend to suffer less from attacks. In very humid conditions such as those in the province of Esmeraldas, Ecuador, leaf-eating insects such as the species belonging to the genus *Diabrotica* and *Disonycha glabrata* may cause extensive damage.

In Tonga the rose beetle (*Adoretus versutus*) has been identified as the main cause of leaf damage (Grum et al. 1991).

8.2.1 Bruchids

The most serious pests endangering cultivation of *P. tuberosus* in small local communities are without a doubt the various bruchids, which may destroy as much as 80 % of the locally stored seeds. In Mexico, the seeds are often attacked by weevils. The following species have been isolated from collected seeds: *Caryedes icamae* in the Department of Loreto, Peru, and in the province of Esmeraldas, Ecuador, and *Acanthoscelides obtectus* in the Department of La Paz, Bolivia. The bruchids, *Acanthoscelides sanfordi* and *A. taboga*, have been isolated from seeds from landraces collected in the Mexican state of Yucatan in 1985.

The bruchid, *Caryedes icamae*, was identified in five seed samples from different localities in Bolivia in 1994. Bruchids are known to damage the seeds of several legume genera containing toxic constituents.

Even though the mature seeds have the highest amount of endogenous rotenone in any part of the plant, it is the seed which suffers the most serious attack by an insect pest, e.g., the bean weevil (Bruchidae).

8.2.2 Spotted Pod Borer, *Maruca vitrata*

8.2.2.1 Damage

The *Maruca* pod borer, *Maruca vitrata* (syn. *M. testulalis*) (Lepidoptera: Pyralidae), is a serious pest of vegetable legumes in Asia and Africa. *Maruca* larvae attack flowers and young pods, and 20–80 % yield losses are reported (Beaudoin 1992).

8.2.2.2 Management

In order to get effective pest control, larvae must be killed within a very brief period soon after hatching from eggs laid on leaves and before they start boring inside flowers or pods, because once the larvae enter into plant parts, they are inaccessible for most ordinary control measures.

1. *Cultural Methods*: In Dholi, Bihar, the larval population of pod borer at 3 days after the first spraying was least in dimethoate at 0.05 % (2.1/flower shoot) compared to 5.1/flower shoot in control. Yam bean seed extract (YBSE) (5 %), tobacco decoction (3 %), and neem oil (3 %) also proved effective in minimizing pod borer population (2.7–3.8/flower shoot) which were on par with each other. Ten days after first spraying, the larval population varied significantly from 2.4 to 4.5/flower shoot with minimum and maximum being in dimethoate at 0.05 % and cassava tuber rind extract (3 %), respectively, as against 5.5 larvae/flower shoot in untreated control. Among plant products, the lowest pod damage (12.1 %) was recorded with YBSE (5 %) which was on par with 3 % tobacco decoction (12.5 %). The highest seed yield was recorded with dimethoate (1.23 t/ha) followed by YBSE (1.17 t/ha) and tobacco decoction (1.01 t/ha) (Table 8.1).

Table 8.1 Effect of plant products and chemicals for the management of pod borer

Treatment	Pod borer larvae/flowering shoot		Pod damage (%)	Seed yield (t/ha)
	10 days after spraying	% reduction over initial population		
Cassava tuber rind extracts (3 %)	4.9	3.9	19.7	0.73
Neem oil (3 %)	4.2	8.7	15.9	0.83
Tobacco decoction (3 %)	2.7	18.2	12.5	1.01
Yam bean seed extract (5 %)	2.8	17.6	12.1	1.17
Dimethoate (0.05 %)	1.7	34.6	7.5	1.23
Untreated control	7.9	(+)33.9	27.2	0.65
CD (P=0.05)	*1.3*	–	*2.2*	*0.16*

Table 8.2 Effect of plant products and chemicals for the management of pod borer

Treatment	Larval population/plant	% reduction over check	No. of seeds/pod
Cassava seed extract 3 %	1.96	24.90	5.20
Neem oil 3 %	1.05	59.77	7.20
Tobacco decoction 3 %	1.25	52.10	7.00
Yam bean seed extract 5 % + soap soln.	1.43	45.21	7.64
Dimethoate 0.05 %	0.96	63.20	7.90
Control	2.61	–	4.30
CD (P=0.05)	*0.93*	–	*1.42*

At Kalyani, West Bengal, three sprayings of neem oil 3 % significantly reduced the larval population of pod borer (1.05/plant and 59.77 % reduction over check) followed by tobacco decoction 3 % (1.25/plant and 52.10 % reduction over check) and yam bean seed extract 5 % with soap solution (1.43/plant and 45.21 % reduction over check). Foliar spraying of yam bean seed extract gave significantly higher seed numbers/pod (7.64) followed by neem oil (7.2) and tobacco decoction (7.0). However, dimethoate spray at 0.05 % gave least larval population (0.96/plant), highest larval reduction over check (63.20 %), and highest number of seeds/pod (7.9) (Table 8.2).

2. *Biological Methods*: A new entomopathogenic virus [nuclear polyhedrosis virus (NPV) belonging to the family Baculoviridae] was recently found to be infecting *M. vitrata* in the field. Early larval stages of *M. vitrata* are highly susceptible than the latter stages. An infected larva becomes sluggish, pinkish in color, loses appetite, becomes fragile, and ruptures to release polyhedra. Dead larvae are often found hanging from the top of plants, a phenomenon known as treetop disease. The infected insect will die within a few days to a week, depending upon the larval stage, dose, and other environmental factors.

8.2.3 Rose Beetle, *Adoretus versutus*

8.2.3.1 Distribution

A. versutus is native to the Indian region (Lever 1945). The pest is reported in Sri Lanka, India, Indonesia, Pakistan, Madagascar, Mauritius, Réunion, St. Helena, Seychelles, Fiji, Samoa, Tonga, Wallis Islands, and Cook Islands. It is not currently recorded in Australia and New Zealand.

8.2.3.2 Damage

A. versutus can cause heavy damage in nurseries and young yam bean plantations. Seedlings are very vulnerable to attack because they can be defoliated rapidly. Rose beetles attack 3- to 4-month-old seedlings in nurseries and young plants after planting.

Fig. 8.2 Leaf damage caused by insect *Adoretus* sp.

A. versutus adults feed by perforating the leaflets, starting from the middle and without destroying the ribs. The leaflets are eaten away in very small but numerous patches, giving a skeletal appearance to the leaflet. The attacks are more numerous at the apex of the leaflets than at the base. Besides this characteristic feeding behavior, the adult rose beetles make depressions in the border of the areas eaten, which is typical of *Adoretus* spp. and distinguishes them from the damage caused by other foliage pests (Fig. 8.2). *A. versutus* feed in the early hours of the night.

Although *A. versutus* attacks sometimes occur on yam bean plantations which are more than 3 years old, the damage caused to the host plants after this age is negligible. The effect of adult feeding on mature yam bean is not important, but plants up to 18 months old may be severely damaged or even killed.

8.2.3.3 Host Range

A. versutus is a polyphagous insect that has many major (18 species) and minor (22 species) hosts. Taro is one of the major hosts and 18 taro varieties get affected by this pest. Apple, pears, orange, lemon, and papaya are among the many minor hosts of *A. versutus*. Besides fruits and cultivated crops, the pest also attacks many ornamental plants. *A. versutus* is also a pest of *Eucalyptus tereticornis* and on fresh papaya fruit in Fiji.

8.2.3.4 Biology

Adult *A. versutus* are nocturnal and feed mainly in the early hours of the night. After feeding, they hide 5–10 cm deep in the ground and disappear completely during daylight hours. If disturbed during feeding, the beetles fall to the ground. The generations are continuous. The fecundity of *A. versutus* females is not known, but comparison with related beetles of similar size suggests the possibility that each female may produce about 40 eggs (Waterhouse and Norris 1987). Eggs lay in the soil where the life cycle completes by about 3 months (Waterhouse and Norris 1987). The larvae feed on roots and decaying vegetation and sometimes on branches. The dark brown body of the adult insect (12.8 mm long, 6.8 mm wide) is covered with dense grayish white scales dotted with brown-red hairs surrounding small blackish brown alveolus on the wing cases. The strongly developed forelegs and hind legs are used for burrowing. Males have a smaller last sternite than females.

8.2.3.5 Development and Spread

The infested plant materials (especially leaf) carried out by people are the most common pathway to disperse *A. versutus* in new areas. The larvae in infested nursery could also be dispersed early in new areas through soil movement.

8.2.3.6 Management

1. *Cultural Methods*: *A. versutus* is difficult to control because a considerable part of its life cycle occurs underground.

Rose beetle damage can be prevented by putting up structural barriers (such as coconut fronds or bamboo fences) around each plant, provided that those are at least as tall as the foliage. In Malaysia, Samoa, and Fiji, this method has been employed successfully to preserve cocoa from damage during the first year after planting out. It then becomes impracticable as the plants grow taller (Entwistle 1972).

Handpicking of adult beetles by lantern light has also been advocated (Lever 1945).

No practical measures are known for attacking the underground stages of *A. versutus*, but the larva's habit of emerging at night and traversing the soil surface to attack the host plant could be exploited by handpicking.

8.3 Diseases

P. tuberosus is reported to be infected by the following fungi: *Colletotrichum pachyrrhizi* (although only the species *C. truncatum* has been confirmed to attack *Pachyrhizus*), *Phakopsora pachyrhizi*, and *Pythium aphanidermatum*.

Several fungi have been reported to cause severe damage in *P. erosus*. A high mortality rate has been recorded in young plants as a result of "root attacks" by *Pythium* spp., *Corticium* spp., and *Macrophomina* spp. in multilocational field trials in Senegal.

In China, *Pythium aphanidermatum* is the cause of root rot in *P. erosus* (Yu et al. 1945).

Mohanty and Behera (1961) reported a severe leaf spot disease observed at Bhubaneswar, India, and succeeded in identifying the fungus as *Cercospora canescens*.

In Mexico, the bacterial disease called bean halo blight, caused by *Pseudomonas syringae* pv. *phaseolicola*, has been observed on several occasions on *P. erosus* (Diaz 1979). The same disease was reported in Hawaii by Birch et al. (1981). However, the disease does not appear to reduce yields significantly (Diaz 1979).

In the Philippines, a mosaic disease has been noted, which is caused by a systemic virus, transmitted through the seed or root. A slight mottling or chlorosis and blistering of the leaves, together with the production of very small tubers, are characteristic symptoms.

Diseases associated with *Pachyrhizus* species are presented in Table 8.3.

8.3.1 Rust, *Phakopsora pachyrhizi*

8.3.1.1 Distribution

Asia: Cambodia, China, India, Indonesia, Japan, Korea, Malaysia, Nepal, the Philippines, Russia (Far East), Taiwan, Thailand, and Vietnam

Africa: Ghana, Mozambique, Nigeria, Rwanda, Sierra Leone, South Africa, Uganda, Zambia, and Zimbabwe

North America: Mexico (San Luis Potosi, Tamaulipas, Veracruz) and the United States (Alabama, Arkansas, Florida, Georgia, Illinois, Indiana, Iowa, Kansas, Kentucky, Hawaii, Louisiana, Mississippi, Missouri, Nebraska, North Carolina, Oklahoma, South Carolina, Tennessee, Texas, Virginia)

South America: Argentina, Bolivia, Brazil (Goias, Maranhão, Mato Grosso, Mato Grosso do Sul, Minas Gerais, Paraná, Rio Grande do Sul, São Paulo), Paraguay, and Uruguay

Oceania: Australia and Papua New Guinea

8.3.1.2 Symptoms

The first symptoms of rust caused by *Phakopsora pachyrhizi* begin as very small brown or brick-red spots on leaves. In the field, these spots usually begin in the lower canopy at or after flowering, although seedlings can be infected under certain circumstances. Often the first lesions appear toward the base of the leaflet near the petiole and leaf veins. This part of the leaflet probably retains dew longer, making conditions more favorable for infection. Lesions remain small (2–5 mm in diameter) but increase in number as the disease progresses. Pustules (Fig. 8.3), called uredinia, form

Table 8.3 Diseases associated with *Pachyrhizus* species

Pest name (Order: Family)	Distribution[a]	Plant part affected	References
Alternaria dauci (mitosporic fungi)	ES, HO, NI, US	Leaves	ARS (2000) and CABI (2000)
Chaetoseptoria wellmanii (mitosporic fungi)	ES, US	Leaves	ARS (2000), CMI (1986), and Wellman (1977)
Colletotrichum sp. (probably *C. gloeosporioides*) (Ascomycota: Phyllachorales)	ES, US	Leaves, pod, and fruit	ARS (2000) and Wellman (1977)
Erysiphe communis (Ascomycota: Erysiphales)	ES, US	Leaves, stems	ARS (2000) and CABI (2000)
Erysiphe polygoni (Ascomycota: Erysiphales)	ES, US	Leaves, stems	CABI (2000) and Crandall et al. (1951)
Leveillula taurica (Ascomycota: Erysiphales)	HO, NI, US	Leaves, stems	CABI (2000) and CMI (1978)
Erysiphe sp. (*Oidium* sp.) (mitosporic fungi)]	ES, US	Leaves	ARS (2000) and Wellman (1977)
Phaeoisariopsis griseola (*Isariopsis griseola*) (mitosporic fungi)	ES, GU, NI, US	Leaves, stems	ARS (2000), CABI (2000), CMI (1986), and Crandall et al. (1951)
Phytophthora sp. (Oomycota: Pythiales)	ES, US	Root	ARS (2000) and Wellman (1977)
Rhizopus (*niger*) (Zygomycetes: Mucorales)	ES	Root	Wellman (1977)

[a]*ES* E Salvador, *HO* Honduras, *NI* Nicaragua, *US* the United States

Fig. 8.3 Yam bean rust urediniospores on leaf

in these lesions, mostly on the lower leaf surface, and they can produce many urediniospores (Fig. 8.3). The raised pustules can be seen with the unaided eye, especially when sporulating. Even though the lesions are small, each lesion often has several pustules (uredinia). Lesions can be completely covered in urediniospores when the pustules are active. Rust urediniospores are pale yellow-brown to colorless, with an echinulate (short spines) surface ornamentation. This coloration is different from many other rust pathogens whose spores are often reddish brown (rust colored). Germination of *P. pachyrhizi* urediniospores occurs through an equatorial (central) pore, producing a germ tube that ends in an appressorium, which the fungus uses to penetrate the host directly or through a stoma.

As more and more lesions form on a leaflet, the affected area begins to yellow, and eventually the leaflet falls from the plant. While rust usually begins in the lower canopy, it quickly progresses up the plant until all of the leaves have some level of disease. Severely diseased plants may become completely defoliated. The loss of effective leaf tissue results in yield reductions from both fewer and smaller seeds. Yield losses as high as 30–80 % have been reported, but the amount of loss depends on when the disease begins and how rapidly it progresses. Besides leaves, rust can also appear on petioles, stems, and even cotyledons, but most rust lesions occur on leaves.

Lesions may be either tan or red-brown. Tan lesions have many pustules that produce numerous urediniospores. Red-brown lesions, thought to be a moderate resistance reaction, have only a few pustules that produce only a few urediniospores.

As pustules age, they may turn black. This is caused by the formation of a layer of teliospores in the pustules, turning pustules from uredinia into telia. Teliospores have two functions: survival of the fungus in the absence of a living host (over seasoning) and sexual reproduction. The thick walls of the teliospores protect the fungus from the environment and attack by other organisms. In rusts, the teliospores germinate forming a basidium and four basidiospores during which sexual recombination occurs. Germination of *P. pachyrhizi* teliospores has been observed only in the laboratory and does not seem to make a significant contribution to the perpetuation of this disease in the field.

8.3.1.3 Biology

Rust pathogen produces only two types of spores: urediniospores and teliospores. The uredinial stage is the repeating stage. This means that urediniospores can infect the same host on which they were produced during the same season. Epidemics can develop quickly from only a few pustules because spore-producing pustules are produced in as little as 7–10 days after infection, and each pustule can produce hundreds of urediniospores (Fig. 8.4).

8.3.1.4 Epidemiology

Rust epidemics begin with the arrival of airborne inoculum (urediniospores). This pathogen is unique among rusts because it has many alternative hosts, which may serve as sources of inoculum. Alternative hosts are other plants that can become infected with the same pathogen but are not required to complete the pathogen's life cycle. Alternative hosts are not to be confused with alternate host, which is a plant other than the principal host that is needed for a pathogen to complete its life cycle. In frost-free areas, such as South America, Central America, the Caribbean basin, southern Texas, and Florida, the inoculum source could be nearby on volunteer soybean plants, kudzu, or some other alternative host. In

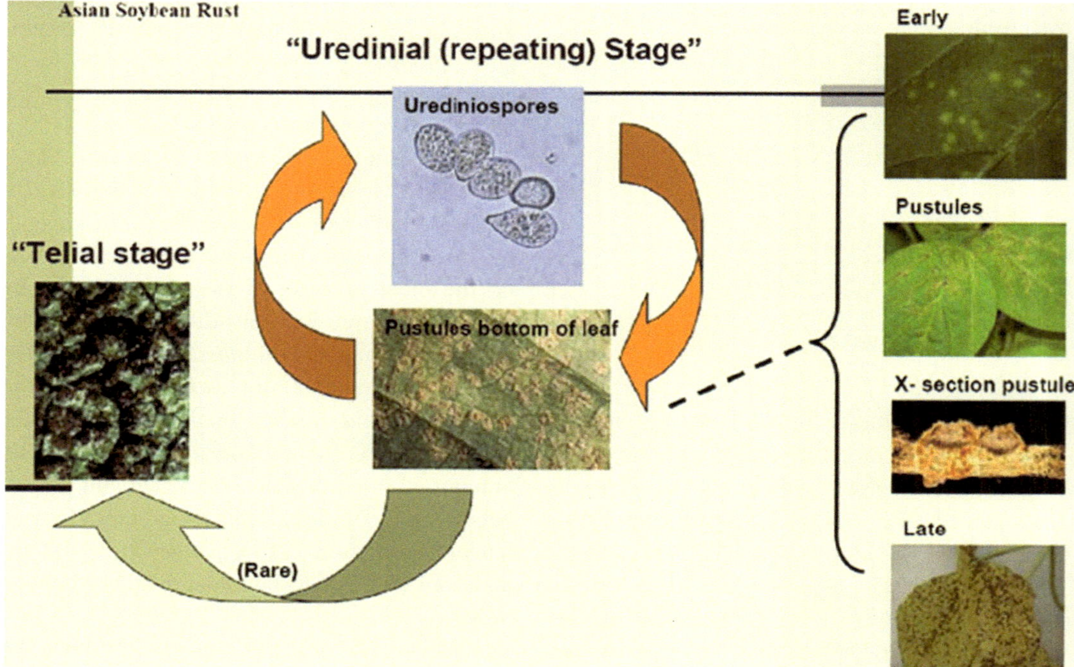

Fig. 8.4 Rust life cycle

areas that experience frost, such as the Midwestern United States, inoculum must be blown in from overwintering sources that may be hundreds of miles away. Because spores of *P. pachyrhizi* are sensitive to ultraviolet radiation, long-distance movement of these rust spores probably occurs in storm systems where clouds protect the spores from the sun.

Once viable spores have landed on the leaf surface of a suitable host, infection and subsequent epidemic development are dependent on environmental conditions. Generally, infection occurs when leaves are wet and temperatures are between 8 and 28 °C, with an optimum of 16–28 °C. At 25 °C, some infection occurs in as little as 6 h of leaf wetness, but 12 h is optimal. After infection, lesions and pustules with urediniospores can appear within 7 or 8 days, and the next infection cycle is set to begin. This short life cycle means that, under the right conditions, rust epidemics can quickly build up from almost undetectable levels to very high levels. Rust epidemics can progress from below detectable levels to defoliation within a month. Epidemics may seem to progress even faster than that, because early infections occur in the lower canopy and are hard to find. Besides the environment, plant age affects rust epidemics. Usually, rust lesions are not found on yam bean until flowering, unless there are high inoculum levels early in the season. This may be due to greater susceptibility of plants to rust as the host enters the reproductive stages, it may be because in lower parts of the canopy spores are more protected from UV radiation, or it may be because conditions in the canopy become more humid as the canopy closes. In any event, lesions can form at any growth stage, but major increases in the disease do not occur until after flowering.

8.3.1.5 Host Range

The rust pathogen has a wide host range (Table 8.4).

8.3.1.6 Spread

Over long distances, *P. pachyrhizi* is mainly spread by wind-borne spores (e.g., in the United States, it is considered that Hurricane Ivan transported it from South America to Southern United States). Trade of host plants cannot be excluded as a pathway (e.g., leafy vegetables, ornamentals, pods). Planting material of ornamental cut foliage and vegetables of host plants may ensure dissemination of the pathogen.

Table 8.4 Known hosts of rust caused by *Phakopsora pachyrhizi*

Common name	Scientific name
Bean, common, dry (field, kidney, navy, pinto)[a]	*Phaseolus vulgaris* var. *vulgaris*
Bean, common, succulent (garden, green, snap, and wax)[a]	*Phaseolus vulgaris* var. *vulgaris*
Bean, fava, or broad bean	*Vicia faba*
Bean, lablab, or hyacinth[a]	*Lablab purpureus*
Bean, lima[a]	*Phaseolus lunatus* var. *lunatus*
Bean, mung[a]	*Vigna radiata*
Bean, scarlet runner[a]	*Phaseolus coccineus*
Bean, yam[a]	*Pachyrhizus ahipa, P. erosus*
Black-eyed pea, cowpea, or yard-long bean[a]	*Vigna unguiculata*
Crotalaria[a]	*Crotalaria anagyroides, C. spectabilis*
Crown vetch	*Securigera varia*
Florida beggarweed[a]	*Desmodium tortuosum*
Kudzu[a]	*Pueraria montana* var. *lobata*
Pigeon pea[a]	*Cajanus cajan*
Siratro[a]	*Macroptilium atropurpureum*
Soybean (including edamame)[a]	*Glycine max*
Sword bean[a]	*Canavalia gladiata*
Udid or black gram[a]	*Vigna mungo*
Wild soybean[a]	*Neonotonia wightii*

[a]Includes field observations of infection, in addition to infection resulting from artificial inoculation

8.3.1.7 Management

There are three basic management tactics that can play a role in reducing rust epidemics: fungicides, genetic resistance, and cultural practices. At present, fungicides are the only highly effective tactic, but long-term management will probably depend more on resistance, in combination with fungicides and changes in cultural practices.

1. *Chemical Methods*: At present, the most effective means of managing rust is the use of fungicides. However, to be effective, selecting the right fungicide and applying it at the right time are crucial. Fungicides for rust control can be classified into three groups: Chloronitriles, Strobilurins, and Triazoles. Chlorothalonil is the one Chloronitrile fungicide with protectant mode of action which affects many biochemical pathways in the pathogen, but it is not taken up by the plant, not even by the cuticle. As a result, it is more subject to weathering than the Strobilurins or the Triazoles, and complete coverage of the leaf surface is critical. To be effective, Chlorothalonil may need to be reapplied several times if new growth or weathering occurs.

Strobilurin fungicides are modeled after a natural antifungal compound, Strobilurin, produced by certain mushrooms. Strobilurins (also known as QoI fungicides) inhibit mitochondrial respiration in the pathogen. Strobilurins are typically absorbed by the cuticle and act as protectant fungicides. A protectant fungicide prevents infections from taking place, but it has little effect on disease development once infection has occurred. Therefore, to be effective, protectants like the Strobilurins must be applied before infection occurs. Depending on the rate applied, Strobilurins are effective for up to 2 weeks after an application, but they will not protect newly developing leaves. Strobilurins control a broad range of pathogens.

Triazoles inhibit sterol production, which disrupts cell membrane function in the pathogen. Triazoles are absorbed and translocated upward in the plant. While they generally do not prevent infection, the Triazoles can kill the fungus in the plant and prevent pustules and spores from forming. The extent to which these chemicals are translocated depends on the Triazole, but all of them move up the plant into new growth to 1° or another. Still, systemicity of Triazoles in plants is incomplete and does not approach the level of systemicity associated with certain herbicides or insecticides. Triazoles are effective for 3 or 4 weeks after application and give some protection to new growth. While highly effective against rust, the Triazoles are not as effective as the Strobilurins against other pathogens. Some fungicide products (premixes) contain both a Triazole and a Strobilurin. The premixes provide protection against a broader range of pathogens and reduce the possibility of pathogens developing resistance to either product.

2. *Cultural Methods*: There are several cultural practices that may help manage rust. In most areas of the United States where rust must be introduced each year for an epidemic to occur, changing planting and harvest dates may avoid the disease. Planting early with an early-maturing cultivar may avoid the rust until the crop has either been harvested or is so far along that the disease will have little impact on yield. Planting dates may also be delayed so that the vulnerable reproductive period occurs during dry conditions that do not favor rust. In areas where the weather is marginal for rust development, wider row spacing along with lower plant populations may hasten canopy drying, thus reducing the dew period enough to prevent or at least slow the disease development. It may also allow better fungicide penetration into the canopy, increasing the effectiveness of chemical control. Research is needed to confirm this. However, because the more open canopy provides less weed suppression, weed problems may be more severe with this strategy, and this method is unlikely to affect rust significantly if weather conditions are very favorable for the disease. Adjusting soil fertility, particularly potassium and phosphorus levels, may help increase disease resistance, but there is little research in this area yet. While it is unlikely that cultural control measures alone will be enough to control rust, they may increase the effectiveness of host resistance or fungicide applications.

8.3.2 Sincama Mosaic Virus (SMV)

The sincama mosaic disease is doubtless the most serious disease of the yam bean; it is so far restricted to *P. erosus* cultivars. It is probably identical to the bean common mosaic virus (BCMV), although when originally describing the disease, Fajardo and Marañon (1932) claimed that this virus was not transmitted to the common bean.

8.3.2.1 Symptoms

The bean common mosaic virus (BCMV) or sincama mosaic virus (SMV) may become a serious problem locally, particularly in fields bordering on wild vegetation with frequent *P. erosus* escapes. The typical symptoms are irregular chlorosis of the leaves, young shoots becoming brittle, and the seed set being reduced as a result of atrophied pollen (the pollen fertility of infected plants is reduced from 95 to 100 % to less than 10 %). Tuber growth is also affected and yield will decrease by 20–40 %. Whether the virus may also be seed-borne is uncertain, although during the 6 years of introducing new accessions through the quarantine screen house in Tonga, neither BCMV nor SMV has been identified. It has occasionally been recorded in Tonga. As the disease leads to complete deformation of the reproductive shoots, seed contamination is not a problem. The pathogen must "survive" in wild vegetation (Fajardo and Marañon 1932).

The yam bean plantlets were found to be infected with BCMV in greenhouse during the initial fortnight following germination. In infected plants, growth of the tuberous root did not take place, and, as mentioned, pollen fertility was greatly reduced (Sørensen 1990).

A reduction of reducing sugars, total sugars, pentosans, and dry matter had taken place in infected material, but the starch percentage had apparently increased (Fajardo and Marañon 1932).

8.3.2.2 Transmission

Transmission of the disease appears to be restricted to contamination through mechanical wounding and insect attack. The aphids and possibly spider mites are capable of transmitting the disease. Transmission does not occur through physical contact between healthy and infected plants neither above nor below ground level (Fajardo and Marañon 1932).

8.3.2.3 Management

The implementation of prophylactic control will be an obvious advantage to such tests. All seed material must be germinated and kept under quarantine for a minimum of 1 month prior to transfer to final growth site. If tuber material is to be included in field trials, they should likewise be kept under quarantine until their disease-free condition can be ascertained.

8.3.3 Witches' Broom

Another disease spread by insect vectors is the witches' broom disease, first diagnosed by Thung and Hadiwidjaja (1957) and probably caused by mycoplasma-like organisms.

8.3.3.1 Symptoms

The symptoms of the disease are excessive branching, dwarfed leaves or atrophy of the leaves, and antholysis, i.e., deformation of the flowers.

8.3.3.2 Transmission

Witches' broom was found not to be transmittable through sap inoculations, but there are some indications that the disease may possibly be transmitted by insects such as whitefly, *Orosius argentatus*. The most likely vectors of these diseases are a number of species of leaf-piercing insect pests. These are aphids (*Aphis rumicis, Brevicoryne brassicae, Oregma lanigera*), mealybug (*Ferrisia virgata*), and whiteflies (*Orosius argentatus*). Witches' broom disease has never been observed to affect more than a few plants in any one field.

8.4 Nematodes

8.4.1 Root-Knot Nematode, *Meloidogyne arenaria*

8.4.1.1 Symptoms

Nematodes may be a problem locally, and when evaluating a Bolivian accession in Esmeraldas, Ecuador, the nematode *Meloidogyne* sp. completely destroyed all tubers in the test plants. In Bolivia, the most severe tuber damage observed

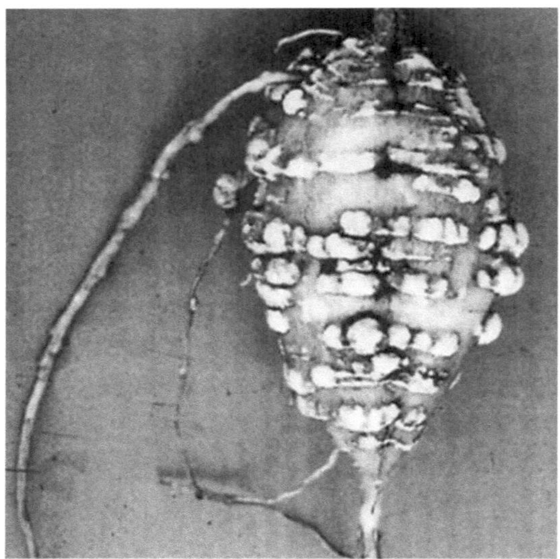

Fig. 8.5 Root-knot nematode on yam bean tuber

was rotting due to lack of irrigation management and/or nematodes. The nematode *Meloidogyne arenaria* is cited by Duke (1981) as the cause of tuber damage (Fig. 8.5).

Yam bean is also reported to be infected by the nematodes *Meloidogyne arenaria* and *Pratylenchus* sp., a cause of dramatic local yield reduction, especially on sandy soils (Noda 1979; Duke 1981). The two latter reports have been confirmed, by personal observations in the field trials at Jardín Tropical, Esmeraldas, Ecuador. Farmers also cite nematodes as one of the most important problems affecting tuber quality in the jíquima cultivar group. The symptoms, which they describe as "warts" (locally known as "verruga" or "peste"), give the affected tubers a bitter taste.

Several nematodes have been reported as being the cause of significant yield reductions in Brazil. Noda et al. (1991) observed serious damage caused by attacks by *Meloidogyne* and *Pratylenchus*. This observation was confirmed by the first field trial conducted during the rainy season in Esmeraldas, Ecuador, where one accession belonging to the jíquima cultivar group (*P. tuberosus*) did not produce any marketable tubers, owing to nematode (*Meloidogyne*) damage. Duke (1981) mentions the species *Meloidogyne arenaria* as an important cause of tuber damage.

Of the farms monitored by INIAP, 57 % reported the presence of the nematode *Meloidogyne* spp. in their soils.

8.4.1.2 Management

Although nematodes may cause tuber damage in Manabí, only crop rotation and the use of noninfested fields are used as control measures, i.e., nematicides are not used.

References

ARS (2000) Fungal data base. Systematic Botany and Mycology Laboratory, Agricultural Research Service, USDA. http://nt.ars-cxxgrin.gov/SBMLweb/Databases/DatabaseHome.htm

Beaudoin L (1992) Feasibility study into the biological control of the rose beetle Adoretus versutus Harold within South Pacific. CIRAD-IRCC/MinistFre de l'agriculture, de l'élevage, des foréts et des poches (Vanuatu). A component of Project No. 5: Biological of the taro beetles in South Pacific, September 1991–September 1992. CIRAD-CP, Montpellier

Birch RG, Alvarez AM, Patil SS (1981) A bacterial leaf spot caused in yam bean by *Pseudomonas syringae* pv. *phaseolicola*. Phytopathology 71:1289–1293

CABI (2000) Crop protection compendium, Global module, 2nd edn. CAB International, Wallingford

CMI (1986) Commonwealth mycological institute distribution maps of plant diseases, No. 328, *Phaeoisariopsis griseola* (Sacc.). Commonwealth Agricultural Bureau International, Wallingford

Crandall BS, Abrego L, Patiño B (1951) A check list of the diseases of economic plants of El Salvador, Central America. Plant Dis Report 35:545–554

Diaz A (1979) El cultivo de la Jícama en el Estado de Guanajuato. SARH/CIAB Publication 116, Guanajuato, Mexico, pp 1–4

Duke JA (1981) Handbook of legumes of world economic importance. Plenum Press, New York/London

Duke JA, Okigbo BN, Reed CF (1978) Resistance of 11 tropical grain legumes to *Empoasca dolichi* (Paoli). Trop Grain Legum Bull (11/12):3–7

Entwistle PF (1972) Pests of cocoa. Longman, London, pp 531–533

Fajardo TG, Marañon J (1932) The mosaic disease of Sincamas, *Pachyrrhizus erosus* (L) Urban. Philipp J Sci 48(2):129–140

Grum M, Stölen O, Sørensen M (1991) Yam bean (*Pachyrhizus* Rich. ex DC) variety trials in Tonga, South Pacific: fresh tuber yields, dry matter and nitrogen contents. In: Proceedings of the 26th annual meeting of the Caribbean Food Crops Society, Mayagüez, Puerto Rico, July 29–August 4, 1990.

Caribbean Food Crops Soc & UADA-ARS TARS, Mayagüez, Puerto Rico 26:407–418
Heredia ZA (1985) Guía para cultivar jicama en el Bajío. SARH Folleto para Productores 15:1–11
Lever RJAW (1945) Entomological notes. Agric J Fiji 16:98–103
Mohanty NN, Behera BC (1961) Leaf spot of *Pachyrhizus angulatus* Rich. Sci Nat 27(5):54
CMI (Commonwealth Mycological Institute) (1978) Distribution maps of plant diseases, No. 217, *Leveillula taurica* (Lév) Arnaud. Commonwealth Agricultural Bureau International, Wallingford
Noda H (1979) Potencialidade da cultura de feijâo-macuco (*Pachyrrhizus tuberosus*). In: von der Pahlen A (ed) Introdução à horticultura e fruticultura no amazonas. CNPq/INPA-SUFRAMA, Manaus, pp 58–65
Noda H, Bueno CR, Silva DF (1991) Genetic erosion threatens native Amazonian vegetable crops. Diversity 7(1/2):62–63
Sørensen M (1990) Observations on distribution, ecology and cultivation of the tuber-bearing legume genus *Pachyrhizus* Rich. ex DC (Fabaceae: Phaseoleae). Wagening Pap 90–3:1–38
Thung TH, Hadiwidjaja T (1957) De Heksenbezemziekte bij Leguminosen. T Pl Ziekten 63:58–63
Waterhouse DF, Norris KR (1987) Biological control: Pacific prospects. Inkata Press, Melbourne, 454 pp
Wellman FL (1977) Dictionary of tropical American crops and their diseases. The Scarecrow Press, Inc, Metuchen, 285 pp
Yu TF, Chiu WF, Cheng NT, Wu TT (1945) Studies on *Pythium aphanidermatum* (Edson) Fitz in China. Lignan Sci J 21(1–4):45–62

Achira, *Canna edulis* 9

Abstract
Distribution, economic importance, symptoms/damage, host range, biology, survival and spread, and management of insect and mite pests and fungal, bacterial, viral, and nematode diseases of Achira (*Canna edulis*) using regulatory, physical, cultural, chemical, biological, host resistance and integrated pest management methods are discussed.

Keywords
Achira • *Canna edulis* • Pests • Diseases

9.1 Introduction

The genus *Canna* includes about 25 species of herbs, widely distributed throughout the tropics and subtropics. It belongs to the family Cannaceae. Achira, *C. edulis*, originated in the Andean region or Peruvian coast and extended from Venezuela to Northern Chile in South America. It is commercially cultivated in Australia for the production of starch, whereas in India, it is grown for the edible, tuberous rhizome. The plant is hardy and has low incidence of pests and diseases.

9.1.1 Description

Canna is a broad-leafed plant with purple leaf sheaths, growing about 1.5–2.5 m tall. It grows as a perennial (from year to year). The stems are in clusters (Fig. 9.1). Flowers are red and produced at the top of the plant. The petals are small and red and 4–10 mm wide. The fruit is a 3-celled capsule with black seeds.

Underground, it has a much-branched root or rhizome with fattened sections covered with leaf scars. These are often light red on the outside and yellowish white inside. A clump of 15–20 suckers often develops. These starchy tubers are from 6 cm across to 15 cm long (Fig. 9.1). The shape varies.

9.1.2 Distribution

It is grown and used in some of the coastal areas of the tropics and subtropics. It will grow from sea level up to 1,600 m. In the Andes, it grows between 1,000 and 2,500 m altitude. It needs a heavy fertile soil. It cannot tolerate strong winds

P.P. Reddy, *Plant Protection in Tropical Root and Tuber Crops*,
DOI 10.1007/978-81-322-2389-4_9

Fig. 9.1 *Canna* plant and rhizomes

or waterlogging. It does best with an evenly distributed rainfall. It can tolerate some shade. It needs to be in a frost-free location. It does best where days and nights are warm.

9.1.3 Cultivation

The end section of the rhizome is planted. Well-developed tubers with one or two undamaged buds should be used. A spacing of 1 m × 1 m is suitable. Tubers are planted about 15 cm deep and need to be kept weed-free during early growth. About 2.5 tons of tubers are required to plant a hectare.

9.1.4 Production

Harvesting occurs from 6 to 19 months after planting. Parts of the underground root are harvested as needed. The tubers can be stored in cool and dry places. They need to be processed immediately for starch manufacture. High yields of tops and rhizomes are possible. Up to 38 tons/ha of rhizomes and 50 tons/ha of tops have been achieved.

9.1.5 Uses

The rhizomes are eaten after cooking. They are boiled or baked. The leaves and rhizomes are used as feed for animals. Starch can be extracted from the roots. This is achieved by rasping the tubers, then washing the starch out, and straining out the fibers. The large starch grains are very digestible. The starch is used to make transparent noodles. It is a popular root crop in some places.

9.2 Insect, Mite, and Vertebrate Pests

Cannas are largely free of pests, but in the United States, plants sometimes fall victim to the canna leaf roller moth, and the resultant leaf damage, while not fatal to the plant, can be most distressing to a keen gardener (Cook 2001).

9.2.1 Leaf Roller, *Calpodes ethlius* (Lepidoptera: Hesperiidae)

The worst pest of *Canna* is a caterpillar known as the lesser canna leaf roller, which is primarily found in the Southern United States. Two leaf-rolling caterpillars are very destructive to *Canna* in Florida, the larger canna leaf roller, *Calpodes ethlius*, whose adult form is variously called the Brazilian skipper, the canna butterfly, or the arrowroot butterfly (Moore 1928), and the lesser canna leaf roller, *Geshna cannalis* (Quaintance).

9.2.1.1 Distribution

Calpodes ethlius is native to Mexico, Central America, and the northern part of the South American Continent, being found as far south as the northern border of Argentina (Scudder 1989). It is found in the Galapagos Islands in the Pacific Ocean (Onore and Mielke 1988) and throughout the West Indies in the Caribbean Sea. In the United States, it is well established in Florida, Texas, and southern Arizona (Scudder 1989). It is an occasional visitor to other Gulf coastal states and can be found in coastal Georgia and South Carolina. To date, *C. ethlius* has been collected in at least 24 states, although many of those records are considered to be "strays" rather than indications of establishment.

Fig. 9.2 *Canna* leaf damage caused by larvae of the larger leaf roller, *Calpodes ethlius*

9.2.1.2 Damage

The *Canna* leaf roller moth lays eggs in the bud of the developing stalk. These hatching caterpillars use a sticky webbing to keep the leaf from unfurling, which protects them from predators and insecticide sprays. They feed and pupate inside the rolled-up leaf and can cause significant damage to the developing stalks (Fig. 9.2). Leaf rolling reduces the aesthetic appeal of ornamental *Canna* and voracious leaf feeding by larger larvae may be so severe that plants do not flower (Young 1982). In food crops, such as arrowroot, severely defoliated plants may produce little of the harvestable rhizome.

9.2.1.3 Biology

The adults of *C. ethlius* are brownish skipper butterflies of medium size (Fig. 9.3). The female's forewing ranges from 22 to 27 mm in length, while that of the male is somewhat smaller at 23 to 24 mm (Scudder 1989). The forewings are rich dark brown except for five vitreous (=transparent) spots and have a fringe of dirty golden yellow hairs on the posterior margin. The hind wings are a little darker than the forewings and have three vitreous spots arranged in an almost straight line. The dorsal surface of the thorax is olive colored, while the ventral surface is yellowish gray. The abdomen is dark brown dorsally and pale yellow whitish ventrally. Adult skippers are strong, fast fliers and can usually only be collected while they suck nectar at flowering plants. In southern Florida, adults are usually first noticed in May with a second generation of adults appearing in early June. It is not known how many generations per year occur in Florida, but adults have been collected in all months of the year (Kimball 1965) and all larval stages (Fig. 9.3) have been collected in January. It is thought that *C. ethlius* can only overwinter in the very southernmost region of Florida and must recolonize more in northern regions each year (Moore 1928).

Female *C. ethlius* lay eggs singly or in widely spaced groups of five to seven eggs on the upper and lower surfaces of *Canna* leaves (Scudder

Fig. 9.3 Larva and adult of the larger *Canna* leaf roller, *Calpodes ethlius*

1989). The egg measures about 1.25 mm in diameter and 0.68 mm in height. It is a pale green in color when first deposited and, within a day, becomes a distinct pink color (Moore 1928).

The first instar larvae close after 4–5 days under Florida summer conditions and are approximately 4 mm in length and yellowish in color (Scudder 1989; Cockerell 1992). Its head is very large, black, glossy, and bilobed. The thoracic shield, legs, and anal tip are also black. The larva consumes most of its egg shell (King 1980) and then moves to the margin of the leaf where it eats two narrow strips from the margin of the leaf toward the midrib producing a leaf flap (Moore 1928). The larva fastens silken strands between the outer edge of the leaf flap and the main portion of the leaf. As the silk shrinks or condenses as it dries, the leaf flap is drawn over the caterpillar, making a flattened tubular shelter with exits at either end. The larva keeps the majority of its body within the leaf roll, only extending its head to feed. The leaf roll may serve as protection from both predators and direct sun. The first instar molts to the next stage after 3 days.

The remaining four larval instars range in length from 5 to 50 mm. The second through fourth instars have a dark brown, almost black, head and a dark green body that is semitransparent, allowing the organs and tracheal system to be seen through the integument. The larvae appear dark green because of the leaf material within their guts. There is a white stripe on each side of the dorsum of the larva running the length of the body. The fifth instar has an orange head with a dark triangle on the frontal region. The larvae consume most of the leaf around their leaf roll and then move to another area of the leaf, when necessary. They roll a bigger leaf roll as they outgrow the previous one. Several large larvae may defoliate an entire *Canna* plant in this manner. The larvae have an interesting behavior of flicking away frass so that their leaf roll remains free of pellets and, instead, frass accumulates on the ground. Fecal firing is facilitated by the anatomy of the larva's anal comb and the buildup of hemostatic pressure within its terminal abdominal segment (Caveney et al. 1998). It is hypothesized that frass removal may reduce discovery by parasites and predators which often use the odor and sight of frass as clues to the potential presence of hosts or prey.

The larva feeds voraciously during the final fifth instar stage (Fig. 9.3). As it prepares for pupation, the larva ceases feeding and empties its gut of leaf material. It shortens in length and becomes bright bluish green. This prepupal stage lasts from 1 to 8 days in Florida, depending on the ambient temperature (Moore 1928). The larva pupates within its leaf roll after first having spun a thin silken mat covering the "floor" of its shelter. The chrysalis is bright green in color (Cockerell 1992) and is approximately 36 mm in length (Scudder 1989). It has a pronounced black spine at the anterior end, and the long proboscis of the adult is enclosed in a case that extends beyond the cremaster at the tip of the abdomen. The pupal stage lasts 7–8 days in the summer in Florida.

9.2.1.4 Host Range

The host plants of *C. ethlius* include members of the Cannaceae family, which contains the single genus, *Canna. Calpodes ethlius* has also been collected from West Indian arrowroot, *Maranta arundinacea*, of the closely related family Marantaceae.

9.2.1.5 Management

The key to controlling *Canna* leaf rollers is vigilance. Leaf rollers can be a problem as early as spring, so keep a close eye for the first sign of webs holding the newly emerging leaves together. Opening the leaves and removing the offending caterpillars will work on a small scale, but in larger plantings, you can simply clip off the top half of the rolled leaf.

Some *Canna* spp. are more susceptible to damage by *Canna* leaf rollers than others. *Canna glauca*, for example, is virtually untouched. In 1981, 39 *Canna* varieties were evaluated for host plant resistance to egg laying (= oviposition) and larval feeding by *C. ethlius* (Reinert et al. 1983). In general, the red-leaved varieties (especially "Wyoming," "Black Velvet," and "Ambassador") received many more eggs than the green-leaved varieties. Varieties with scarlet, red, or orange flowers, regardless of leaf color, tended to experience more oviposition and larval feeding damage than varieties with other flower colors. Although varieties have been identified that are particularly susceptible to the Brazilian skipper, none has been identified as resistant.

Little attention has been paid to biological control of *C. ethlius* in *Canna*. However, this skipper attacks arrowroot, *M. arundinacea*, which is an important food crop in the Caribbean region. A very small (0.8–1.2 mm in length) parasitic wasp, *Ooencyrtus calpodicus* (Hymenoptera: Encyrtidae), was found in Trinidad attacking eggs of *C. ethlius* and was introduced into other Caribbean islands for the control of this species. However, *Ooencyrtus calpodicus* has not been recorded from the United States. Three Hymenopteran parasites have been recorded from *C. ethlius* in Florida: *Brachymeria incerta* (Chalcididae) parasitizing pupae, *Xenufens ruskini* (Trichogrammatidae) (Krombein et al. 1979), and *Trichogramma minutum* (Trichogrammatidae) parasitizing eggs (Moore 1928). Parasitic tachinid flies, including *Eucelatoria australis*, *Eusisyropa blanda*, and *Lixophaga diatraeae*, have emerged from pupae in Florida (Arnaud 1978). Infection by pathogens has been noted which cause the larvae to turn black, liquefy, and disintegrate. There is probably considerable natural mortality of *C. ethlius* in Florida due to parasitism, predation, and diseases but obviously not enough to keep populations from heavily defoliating cannas in certain years.

Biopesticides such as Dipel (*Bacillus thuringiensis*) can be sprayed into the bottom half of the leaf so that it reaches the caterpillars. If the leaf roller population is high, sprays are required throughout the growing season, but as the moth population reduces, the need to spray lessens. Stickers will probably need to be added to the spray because the thick waxy coating on *Canna* leaves repels water. Again, the key is to monitor plants and not allow the larvae to mature, which starts the cycle over again.

9.2.2 Japanese Beetle, *Popillia japonica*

Japanese beetles would fall into the range of minor pests on *Canna*. These beetles can also ravage the leaves if left uncontrolled. Japanese beetles typically prefer the *Canna* flowers.

9.2.2.1 Damage

These insects damage plants by skeletonizing the foliage by consuming only the leaf material between the veins. It is about 15 mm long and 10 mm wide, with iridescent copper-colored elytra and green thorax and head. It is not very destructive in Japan, where it is controlled by natural predators, but in America it is a serious pest of about 200 species of plants, including rose bushes, grapes, hops, *Canna*, and others.

9.2.2.2 Biology

The life cycle of the Japanese beetle is typically 1 year in most parts of the United States, but this

Fig. 9.4 Japanese beetle adult

Fig. 9.5 Bird cherry oat aphid identified in greenhouse on *Canna* leaves

can be extended in cooler climates; for instance, in its native Japan, the beetle's life cycle is 2 years long as a result of the higher latitudes of the grasslands required for the larval stage. During the larval stage, the white grubs can be identified by their V-shaped raster pattern (Fig. 9.4).

9.2.2.3 Management

Japanese beetles can be easily picked up and killed.

During the larval stage, it is susceptible to a fatal disease called milky spore disease, caused by a bacterium called milky spore, *Paenibacillus* (formerly *Bacillus*) *popilliae*. The USDA developed this biological control and it is commercially available in powder form for application. Standard applications (low density across a broad area) take from 1 to 5 years to establish maximal protection against larval survival (depending on climate), expanding through the soil through repeated rounds of infection.

On field crop, floating row covers can be used to exclude the beetles; however this may necessitate hand pollination of flowers. Kaolin sprays can also be used as barriers.

Research performed by many US extension service branches has shown that pheromone traps attract more beetles than they catch. Traps are most effective when spread out over an entire community, and downwind and at the borders (i.e., as far away as possible, particularly upwind), of managed property containing plants being protected. Natural repellents include catnip, chives, garlic, and tansy as well as the remains of dead beetles, but these methods have limited effectiveness. Additionally, when present in small numbers, the beetles may be manually controlled using a soap-water spray mixture, shaking a plant in the morning hours and disposing of the fallen beetles, or simply picking them off attractions such as rose flowers, since the presence of beetles attracts more beetles to that plant.

Natural predators of the beetle are the anchor bug and the blue-winged wasp.

9.2.3 Bird Cherry Oat Aphids

9.2.3.1 Damage

Bird cherry oat aphids (Fig. 9.5) have been identified in stored rhizomes and in the foliage of potted plants. Although these aphids are most commonly found on cereals, they seem to have a presence in greenhouse grown cannas. They appear in the cooler months in the early spring. They do pose a threat requiring a viable management strategy.

9.2.3.2 Management

There are a number of natural enemies to aphids including parasitic wasps, *Aphidoletes*, and lady beetles that can be quite effective for controlling

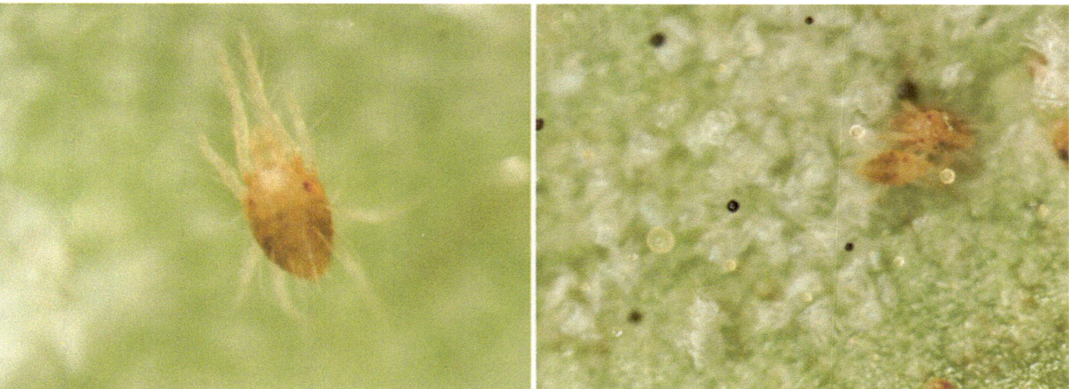

Fig. 9.6 Two spotted spider mite on *Canna* leaf. There are an abundance of white eggs and the leaf looks silvery from spider mite damage

aphid populations in the greenhouse. These can be purchased online from a number of merchants.

Combining weekly treatments of neem oil extract with organic pesticides that contain *Beauveria bassiana* can protect *Canna* foliage from damage by aphids.

Carbaryl (1-naphthyl methylcarbamate) is effective for controlling both caterpillars and aphids.

9.2.4 Whiteflies

9.2.4.1 Damage

Whiteflies will rarely attack *Canna* in the garden, but can sometimes be a problem indoors or in a greenhouse.

9.2.4.2 Management

Stress reduction goes a long way to prevent such attacks, but when appropriate, these pests can be killed with insecticides (see your extension office for recommendations). It is better to try to prevent insect infestation by removing dead foliage and providing a humid environment.

9.2.5 Two-Spotted Spider Mite

9.2.5.1 Damage

The two-spotted spider mite is a tiny moving yellow or bronze mite, which is related to spiders, and can produce webbing that covers leaves and flowers. They lay eggs on the underside of leaves. Spider mites are a major nuisance on cannas in the greenhouse and most often appear in the summer months. The nymphs and adults produce webbing, which can damage flowers and make leaves unattractive. Severe infestations can cause severe discoloration of *Canna* leaves and stunt plant growth. Leaves will look silvery and can become dried and scorched when infested (Fig. 9.6). The damage can be so severe that some might think the scorched leaves are caused by a fungus or virus.

9.2.5.2 Management

There are natural enemies, but it is very important to consider which ones to introduce into the greenhouse. For example, western flower thrips, *Frankliniella occidentalis*, is a predator that feeds on the mite eggs, but thrips are also known to transmit plant viruses, so this is not an advisable predator to introduce into the greenhouse. A better choice might be the western predatory mite, *Galendromus occidentalis*, or minute pirate bugs (*Orius* spp.), which do not transmit plant viruses. Minute pirate bugs feed on eggs of a range of insects including aphids and caterpillars. These are all commercially available online.

Combining weekly treatments of neem oil extract with organic pesticides that contain *Beauveria bassiana* can devastate spider mite colonies. Neem oil extract (100 %) is a miticide that can be used to treat *Canna* foliage. Neem oil is a natural disinfectant and routine weekly treatments in the spring and fall can control spider mite infestations.

9.2.6 Slugs and Snails

9.2.6.1 Damage

Slugs and snails are fond of cannas and can leave large holes in the leaves, preferring the tender young unfurled leaves. Slugs and snails would fall into the range of minor pests on *Canna*. They damage the plants by chewing holes in the leaves or feasting on the flowers.

9.2.6.2 Management

Problems with slugs and snails can be virtually eliminated with good organic soil preparation and by avoiding the use of chemical fertilizers, which kill off many of the natural snail and slug predators.

9.2.7 Rodents

9.2.7.1 Damage

Rodents may eat the stored rhizomes.

9.2.7.2 Management

Rodents can be treated with baits or traps.

9.3 Diseases

Cannas are remarkably free of diseases, compared to many genera. However, they may fall victim to *Canna* rust, a fungus resulting in orange spots on the plant's leaves, caused by over moist soil. They are also susceptible to certain plant viruses, some of which are *Canna* specific, which may result in spotted or streaked leaves, in a mild form, but can finally result in stunted growth and twisted and distorted blooms and foliage.

The flowers are sometimes affected by a gray, fuzzy mold called *Botrytis*. Under humid conditions, it is often found growing on the older flowers. Treatment is to simply remove the old flowers, so the mold does not spread to the new flowers.

9.3.1 Rust, *Puccinia thaliae*

In hot, humid climates, *Canna* can develop a fungal problem called *Canna* rust. The rust disease caused by *Puccinia thaliae* is conspicuous and commonly occurs in Hawaii. It afflicts *Canna* lily and several other plant genera.

9.3.1.1 Symptoms

Symptoms of *Canna* rust have been reported throughout Hawaii since the early 1970s (Gardner and Martinez 1985). The disease produces numerous small, yellow, irregularly shaped, powdery rust pustules on leaves, petioles, and flowers. These pustules are found primarily on the lower surface of leaves, with corresponding 1–2 mm-diameter chlorotic lesions on the upper surface (Kessler 2007). In advanced stages of the disease, spots on the upper leaf surface coalesce and turn dark brown (Fig. 9.7). Heavily diseased leaves eventually desiccate, collapse, and fall.

Golden urediniospores form in the uredinia. These small eruptions can occur on both sides of a leaf but mostly develop on the undersides. The erumpent uredinia, covered initially by the epidermis, later push through it to expose their powdery golden spores. The urediniospores have bristly walls, may vary from egg to pear shape, measure $28–40 \times 20–25$ μm, and range in color from golden to pale yellow orange. Sparse, minute, round reproductive cells called telia also form in the uredinia and are covered by the epidermis. The cylindrical to club-shaped teliospores (Gardner and Martinez 1985) measure $50–83 \times 14–21$ μm, have acute or truncate apices that gradually thicken, and have short pedicels (Saccardo 1902). Pycnia and aecia, two other spore stages of a rust fungus, have not been reported for *P. thaliae* (Gardner and Martinez 1985).

9.3.1.2 Host Range

The host range of *P. thaliae* includes *Canna indica* (syn. *C. edulis*) and some hybrids of *C. generalis* in the order Zingiberales, family Cannaceae, as well as *Maranta arundinacea*, *Stromanthe tonckat*, *Thalia dealbata*, *T. geniculata*, *Calathea* sp., *Ctenanthe* sp., and *Ischnosiphon* sp. in the family Marantaceae

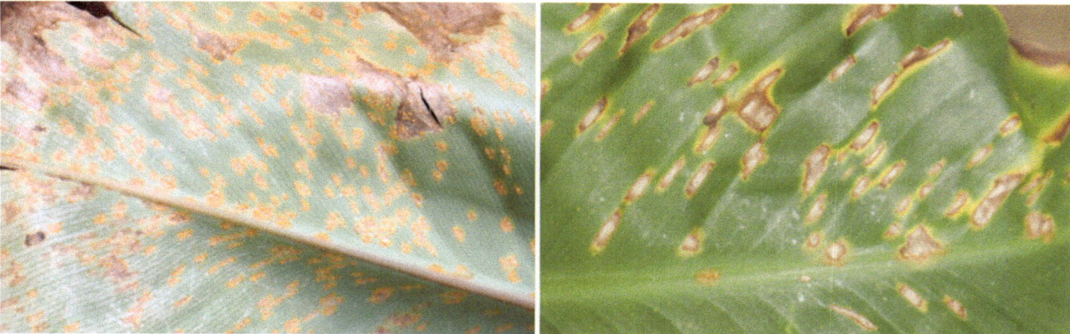

Fig. 9.7 Rust symptoms on *Canna* leaves

(Zingiberales) (Bhasabutra et al. 2011; Jeeva et al. 2004; Kaur et al. 2011; Padamsee and McKenzie 2012).

9.3.1.3 Epidemiology

Infection and disease development are favored by factors that create high relative humidity including high soil moisture, tall weeds, poor air circulation, and dense or shaded plantings of *Canna* lilies.

9.3.1.4 Management

Integrate the following practices for the best management of rust of *Canna* lily:

- Grow *Canna* cultivars resistant to this disease. Broshat et al. (1983) evaluated 38 cultivars of *Canna* for susceptibility to rust. The cultivars "Halloween" and "Yellow King Humbert" were highly susceptible, and even the most resistant cultivars "Louis Cayeux" and "La Boheme" were still moderately susceptible. However, some hybridized varieties may show a high level of resistance to the rust in Hawaii. Growers should observe the varieties growing in their area and select only those with few or no symptoms of rust.
- Plant *Canna* lilies in well-drained, fertile soils in well-ventilated areas that receive full sunlight. Apply a layer of compost or mulch several inches thick around plants. Avoid planting cannas in shady or poorly drained locations with high relative humidity.
- Apply a moderate amount of fertilizer (e.g., NPK 5-10-5) monthly to ensure rapid plant growth.
- Prune cannas periodically to reduce crowding and relative humidity in the canopy. Practice "deadheading" by cutting old or spent flowers from the plant. When the flowering is complete, remove the entire stem and leaves to ground level, because no more flowers will develop on these stems. This practice reduces the leafy appearance of the plant, increases aeration in the canopy, and permits more light and fungicide spray to reach the plant.
- Avoid sprinkler irrigation in order to prevent wetting of the plant leaves.
- Remove and destroy infected foliage. Do not compost symptomatic leaves.
- Destroy all unwanted "volunteer" *Canna* plants.
- Copper fungicides may provide some control of the rust. However, thorough coverage of both upper and lower leaf surfaces and frequent applications (every 2–3 weeks) may be required to adequately control the disease. Note that fungicide sprays for rust of *Canna* lily may be relatively costly in labor, equipment, and fungicide product. Although other more effective or systemic fungicides having modes of action with higher specificity may be available, we do not recommend their use due to the higher probability of the development of fungicide resistance in populations of *P. thaliae*. Control *Canna* rust with fungicides applied in the first weeks following planting of rhizomes in fresh soil. For commercial greenhouses, liquid copper-based sprays are recommended to control *Canna* rust.

9.3.2 *Canna* viruses

Canna viruses are the most serious cultivation problem. They are easily passed from plant to plant by sucking insects such as aphids and then spread by unsanitary division techniques. Virus can cause spotted or streaked leaves, stunted growth, and distorted blooms. Low levels will not kill the plant, but they may reduce its vigor. In most cases, low levels of virus are undetectable and unnoticeable, except during cool weather. High virus loads, on the other hand, can render *Canna* so unattractive that they must be discarded. Because cannas are often carelessly divided, viruses can easily spread and multiply. Cannas are also one of the few plants in which viruses can also be transmitted by seed. While many of the seed strains are fairly clean of virus, this is not a guarantee of a virus-free plant. There are several viruses that can infect *Canna* including bean yellow mosaic virus, *Hippeastrum* mosaic virus, tomato aspermy virus, cucumber mosaic virus, *Canna* yellow streak virus, and, most seriously, *Canna* yellow mottle virus. Most *Canna* cultivars tolerate a certain amount of viral load and will grow and thrive despite being infected.

9.3.2.1 Symptoms

- It manifests itself in rust-colored streaks or mottled markings on the leaves and in color breaks on the flowers (Fig. 9.8).
- Sometimes leaves are slightly distorted and puckered.
- Like many plants under stress, affected cannas will flower very early in the season and before the plant is of full height.
- Over the years the *Canna* will lose vigor and become increasingly unsightly.
- Some are spread by aphids and others by sap-sucking insects.
- At one time it was thought that cannas may have the ability to outgrow the virus, but that is not the case.

9.3.2.2 Management

As late as 2005, there were virtually no virus-free cannas grown in cultivation, but the savior came in the form of tissue culture. There had been efforts trying to persuade labs to tackle the issue of cleaning up the *Canna* virus in the laboratory, and finally, AgriStarts of Florida stepped forward. During the tissue culturing process, the *Canna*'s sterile tissue is subjected to high heat which causes the developing plant to stretch. The stretched part of the new tissue is then re-cultured before the virus has a chance to reinfect it. Each new culture is then checked to make sure the virus is not still present, a process called virus indexing. This is a time-consuming and expensive process since it often takes several tries to make sure the tissue is finally clean. AgriStarts is continuing their work and it is our hope that other labs will join them in their goal to bring less virus-infected stock to the market.

Despite selling clean stock, there is no guarantee that the plants will stay clean once

Fig. 9.8 *Left* – *Canna* yellow mottle badnavirus (CYMV) infected foliage. *Center* – bean yellow mosaic virus (BYMV) infected foliage. *Right* – virus causes color breaks in flowers

Table 9.1 Management of pests and diseases of Achira

Pest and disease	Natural enemy	Chemical control
Spotted spider mites	Western predatory mite, minute pirate bugs, *Beauveria bassiana*	Neem oil
Bird cherry oat aphids	Lady bugs, minute pirate bugs, *Aphidoletes* parasitic wasp	Carbaryl (1-naphthyl methylcarbamate)
Caterpillars	Lady bugs, lacewings	Carbaryl (1-naphthyl methylcarbamate)
	Minute pirate bugs	
Canna rust		Copper fungicides
Canna viruses		Virkon, bleach

they are exposed to the environment, but at least we have got a much better quality plant than we had available in the past. There is no cure for virus infections in *Canna* other than the procedure mentioned above or to destroy the infected plant.

Pests and diseases of *Canna* and their management are presented in Table 9.1.

References

Arnaud PH Jr (1978) A host-parasite catalog of North American Tachinidae (Diptera). Misc Publ No. 1319. USDA, Washington DC

Bhasabutra T, Choobamroong W, Beasley DR, Shivas RG (2011) Canna rust (*Puccinia thaliae*). Updated on 2/1/2011. PaDIL – http://www.padil.gov.au. http://www.padil.gov.au/ thai-bio/Pest/host%20type-thai-bio/140433. Accessed 24 June 2012

Broshat TK, Reinert JA, Donselman HM (1983) Resistance of *Canna* cultivars to canna rust and *Hippeastrum* mosaic. Hortic Sci 18:451–452

Caveney S, McLean H, Surry D (1998) Faecal firing in a skipper caterpillar is pressure-driven. J Exp Biol 201:121–133

Cockerell TDA (1992) Notes on the life-history of *Calpodes ethlius*. Entomol News 3:78–80

Cook I (2001) The gardener's guide to growing canna. Timber Press, Portland

Gardner DE, Martinez AP (1985) Occurrence of canna rust (*Puccinia thaliae*) in Hawaii. Plant Dis 69:1101

Jeeva ML, Hegde V, Makeshkumar T, Sriram S, Nair RR, Edison S, Mathur N (2004) Rust of Queensland arrowroot (*Canna edulis*) caused by *Puccinia thaliae*: a new record for India. Plant Pathol 53:261

Kaur R, Rush TA, Ferrin DM, Aime MC (2011) First report of *Puccinia thaliae* rust on canna lily in Louisiana. Plant Dis 95:353

Kessler JR (2007) Canna lilies for Alabama gardens. Alabama Cooperative Extension System ANR-1315. http://www.aces.edu/pubs/docs/A/ANR-1315/ANR-1315.pdf. Accessed 24 June 2012

Kimball CP (1965) The Lepidoptera of Florida: an annotated checklist. Division of Plant Industry, State of Florida Department of Agriculture, Gainesville

King HS (1980) Internal organization of *Hesperia ethlius* Cram as observed in the living animal. Psyche 3:322–342

Krombein KV, Hurd PD Jr, Smith DR, Burks BD (1979) Catalog of Hymenoptera in America north of Mexico. Vol. 1, Symphyta and Apocrita (Parasitica). Smithsonian Institution Press, Washington, DC

Moore MB (1928) A study of the life history and habits under Florida conditions of the canna butterfly (Brazilian skipper), *Calpodes ethlius*(Cramer), an insect pest of the canna. MS thesis, University of Florida, Gainesville, 95 pp

Onore G, Mielke O (1988) *Calpodes ethlius* (Cramer) (Lepidoptera, Hesperiidae) em Gal pagos. Rev Bras Zool 5:619–620

Padamsee P, McKenzie HC (2012) First report of *Puccinia thaliae* on canna lily in New Zealand and in some Pacific Island countries. Aust Plant Dis Note 7:139–141

Reinert JA, Broschat TK, Donselman HM (1983) Resistance of *Canna* spp to the skipper butterfly, *Calpodes ethlius* (Lepidoptera: Hesperiidae). Environ Entomol 12:1829–1832

Saccardo PA (1902) Sylloge fungorum omnium hucusque cognitorum, vol 16. Sumptibus Auctoris, Pattavii, pp 306–307

Scudder SH (1989) *Calpodes ethlius* – the Brazilian skipper. In: The butterflies of the Eastern United States and Canada with special reference to New England. Vol II. Lycaenidae, Papilionidae, Hesperidae. SH Scudder Publication, Cambridge, pp 1750–1757

Young AM (1982) Notes on the interaction of the skipper butterfly *Calpodes ethlius* (Lepidoptera: Hesperiidae) with its larval host plant *Canna edulis* (Cannaceae) in Mazatlan, state of Sinaloa, Mexico. N Y Entomol Soc 90:99–114

10 Winged Bean, *Psophocarpus tetragonolobus*

Abstract

Distribution, economic importance, symptoms/damage, host range, biology, survival and spread, and management of insect and mite pests and fungal, bacterial, viral, and nematode diseases of winged bean (*Psophocarpus tetragonolobus*) using regulatory, physical, cultural, chemical, biological, host resistance, and integrated pest management methods are discussed.

Keywords

Winged bean • *Psophocarpus tetragonolobus* • Pests • Diseases • Nematodes

10.1 Introduction

The genus *Psophocarpus* includes five species, and they are found throughout the tropics. *Psophocarpus tetragonolobus* is a perennial climber with tuberous roots cultivated in India, Myanmar, tropical Africa, Asia, and Madagascar for its edible pods and tuberous roots (Fig. 10.1). The origin of winged bean is uncertain, although it is believed to be a native of either India or Mauritius or African region where all the species of this genus are found in the wild. In India, it is cultivated in Maharashtra, Karnataka, Bihar, Orissa, Tripura, West Bengal, Assam, Tamil Nadu, and Kerala.

The winged bean is well recognized by farmers and consumers in the Asian region for its variety of uses and disease tolerance. It is exceptional among all food plants in that practically all parts of the plant are edible and a usable product is provided at every stage of its life cycle, like leaves, flowers, stem, pods, roots, etc. Winged bean is nutrient rich, and all parts of the plant are edible. Leaves can be eaten like spinach, flowers can be used in salads, tubers can be eaten raw or cooked, and seeds can be used in similar ways as the soybean. Winged beans have exceptionally high protein content and have been suggested as a potential food source for the tropics. It is unique among leguminous crops in that several parts of the plant—leaves, pods, seeds, and tubers—are edible and rich in protein. Young seedpods cooked for a few minutes taste like French beans (*Phaseolus vulgaris*). Mature seeds are eaten cooked. They are very nutritious, being rich in oil (up to 17 %), protein, vitamin E, and calcium. The seed can also be roasted and eaten like peanuts or fermented and used as tempeh. An edible

P.P. Reddy, *Plant Protection in Tropical Root and Tuber Crops*,
DOI 10.1007/978-81-322-2389-4_10

Fig. 10.1 Winged bean plant and tubers

oil is obtained from the seed. Roots are cooked like potatoes. Slightly sweet, it contains 20 % or more protein, which is about 12 times that of potatoes. Leaves and young shoots are cooked. Flowers and flower buds— raw or cooked—have a flavor similar to mushrooms when sautéed. The light blue flowers are also used for coloring in foods. The roasted seed is a coffee substitute.

When grown in mixed garden cultivation or shifting agriculture, the winged bean is unusually free of pests and diseases. Nevertheless, the plant is known to be susceptible to a number of pests and diseases, and the importance of these may become greater with larger stands of single varieties. Until disease- and insect-resistant varieties are identified, the danger of pests and diseases in winged beans grown in monoculture can be limited by planting more than one variety, as in the more extensive acreage now planted with winged bean in Thailand.

10.2 Insect Pests

It has been reported that a wide range of insects belonging to Hemiptera, Thysanoptera, Diptera, Coleoptera, Lepidoptera, Orthoptera, and Acarina have been identified as pests on winged bean, mainly in the Southeast Asian countries. The most important ones include *Maruca testulalis* and *Helicoverpa armigera* which cause pod and flower damage; *Leucoptera psophocarpella* has been found to cause extensive leaf damage; *Mylabris afzelii* and *M. pustulata* cause extensive flower damage (Khan 1982). Caterpillars of *Podalia* spp. have been reported from Brazil as serious, and also the mites *Polyphagotarsonemus latus* and *Tetranychus urticae*. The black bean aphid *Aphis craccivora* is commonly found on shoots of young plants in Papua New Guinea, but it causes more severe damage in Guam where 2–20 % damage may occur. The ladybird *Henosepilachna signatipennis* also causes extensive leaf damage through the feeding activities of both adult and larvae.

10.2.1 Bean Pod Borer, *Maruca testulalis*

10.2.1.1 Distribution

The bean pod borer *Maruca testulalis* is the most widespread flower pest in Papua New Guinea and Thailand.

10.2.1.2 Damage

It attacks the stem end of the pods (Fig. 10.2).

Fig. 10.2 Larva and moth of bean pod borer, *Maruca testulalis*

10.2.1.3 Biology

Eggs are normally deposited on floral buds and flowers, although oviposition on leaves, leaf axils, terminal shoots, and pods has also been recorded. A female may lay up to 400 eggs. Eggs are usually deposited in batches of 2–16. There are five larval instars (Odebiyi 1981). Larval development is completed in 8–16.3 days, and prepupal period lasts for 1–2 days. Pupation occurs in the soil in a pupal cell and lasts for 6.4–11 days. Life cycle is completed in 18–35 days.

10.2.1.4 Host Range

The most frequent host plants are *Cajanus cajan*, *Vigna unguiculata*, *Phaseolus lunatus*, and *Pueraria phaseoloides*.

10.2.1.5 Management

Intercropping maize with bean (one third bean-two thirds maize) was considered useful as a cultural method for controlling pod borers in winged bean.

Four high-volume sprays of Cypermethrin at 0.008 % (1st spray at initiation of flowering, 2nd spray at 50 % flowering, 3rd spray at 100 % flowering, and 4th spray at 100 % pod setting) were effective for protecting the crop against *Maruca*.

Bacillus thuringiensis (Bt) is effective in controlling pod borers. Neem seed powder and neem kernel extract were also effective against the pod borer.

10.2.2 Bean Fly, *Ophiomyia phaseoli*

10.2.2.1 Damage

The bean fly is sometimes a serious seedling pest. Larvae bore through the stems and mine the leaves. Young plants are attacked by the bean fly (Fig. 10.3).

10.2.2.2 Biology

Pupation takes place inside the stem. The barrel-shaped pupae are black and about 3 mm long. The total life cycle takes 2–3 weeks. As many as seven generations of this pest have been reported during the active season of the pest infestation.

10.2.2.3 Management

Apply neem cake 250 kg/ha immediately after germination. Do not delay, particularly during Kharif period.

As soon as a few adults are noticed hovering over the crop, spray Endosulfan 35 EC at 2 ml/l or acephate 75 WP at 0.75 g/l or pulverized NSPE 4 % or neem soap 1 % or neem formulation with 10,000 ppm azadirachtin at 2 ml/l. However, the botanicals get washed away by rain, and they will be ineffective if it rains within 1–2 days of spray.

10.2.3 Pea Blue Butterfly, *Lampides boeticus*

10.2.3.1 Symptoms

The pea butterfly (Fig. 10.4) larvae not only feed within developing flowers but also on seeds

Fig. 10.3 Bean fly damage and the adult

Fig. 10.4 Pea blue butterfly, *Lampides boeticus*

within the seed pods of the host. Frass is deposited at one end of the pod, where it may cause decay. This often shows up externally as dark discoloration.

10.2.3.2 Management

1. *Chemical Methods*: Sprays of deltamethrin and thiosulfuric acid S, S′-[2-(dimethylamino)-1, 3-propanediyl] ester disodium salt when eggs of the third and sixth generations were hatching gave 87.2–88.8 % and 80–81.3 % control, respectively, in China.

   By day three posttreatment, carbaryl gave complete control of larvae of *L. boeticus* in Egypt.

   In Thailand, cyhalothrin has been used to effectively control *L. boeticus* (Sepswasdi et al. 1991).

2. *Cultural Methods*: In field experiments in Dharwad, Karnataka, India, the lowest incidence of *L. boeticus* was recorded when the crop was sown in the first week of October. The percentage of seed and pod damage was also lowest in an early October sowing (Jagginavar et al. 1990).

10.2.4 Southern Green Stink Bug, *Nezara viridula*

10.2.4.1 Distribution

This pest is also known as the green vegetable bug in some countries. It occurs throughout the Americas and in various Pacific Islands, Japan, Australia, New Zealand, and Africa. First seen in Hawaii in 1961 on Oahu, it has spread to all major islands.

10.2.4.2 Damage

The bugs feed by piercing plant tissue with needlelike stylets. The actual feeding puncture is not immediately visible. Adults and nearly all nymphal stages (2nd to 5th nymphal stages) feed on a variety of plant tissue. Succulent parts of the plant and the developing flowers or fruit are preferred. Feeding injury becomes visible sometime after actual feeding. Nymphal feeding on flower buds results in premature abscission. Feeding injury on leguminous pods results in seed damage and ultimately distorted development of the pods. Soft and spongy tissue under otherwise intact epidermis results when the stink bug feeds on vegetable tissue.

10.2.4.3 Biology

The life history and habits of *N. viridula* have been extensively studied. Life cycle studies conducted at different temperatures, on several hosts, was summarized by Harris and Todd (1980). Eggs hatch in 4–5 days, and the nymphal development is completed in 25–43 days (Harris and Todd 1980). In Hawaii the life cycle (eggs–adult) is completed in 35–45 days.

10.2.4.4 Management

1. *Cultural Methods*: Growers who continually experience damage should monitor weed hosts surrounding the crop field. It may be possible to utilize trap-borders of preferred hosts such as Crotalaria (rattle pod) to attract and hold stink bug populations. Stink bugs will usually remain on the plants where parasites can readily find them. It is important that these borders should not be allowed to dry before control occurs.
2. *Biological Methods*: The stink bug is controlled by natural enemies that were deliberately introduced. Several species of egg parasites were introduced.
   (a) *Parasitoids*: In many places biological control by introduced parasites is so effective that chemical control is seldom necessary (Waterhouse and Norris 1987). In Hawaii, the introduced parasites *Trissolcus basalis*, *Trichopoda pilipes*, *and T. pennipes* have become established. The first two mentioned above are plentiful and are generally effective in controlling Southern green stink bug populations in Hawaii.

   Trichopoda pennipes and *T. pilipes* are flies that parasitize adult stink bugs. Both parasites are attracted to male stink bugs but parasitize both sexes (Mitchell and Mau 1971). They lay oval white eggs on the adult. Maggots that hatch from the eggs penetrate into the stink bugs and feed within. The maggots emerge when mature and the stink bug host dies.

   Trissolcus basalis is a wasp that parasitizes eggs. Female wasps deposit their eggs into the stink bug eggs and the larvae complete their development within. The wasp is very efficient and usually parasitizes all of the stink bug eggs in the egg cluster. The average rate of parasitization is 95 %.

   (b) *Predators*: Predation by the big headed ant, *Pheidole megacephala*, has been reported. They prey upon eggs and young gregarious nymphs and carry them back to their nest. Ant predation can be high during periods of low rainfall (Nishida 1966). Spiders are also known to feed upon the young nymphs.
3. *Chemical Methods*: Insecticidal applications are usually not required; however, sprays may be needed if stink bug populations are high. This pest can be chemically controlled by the use of carbamate and organophosphate compounds. However, because most of these compounds persist on the treated plant for a relatively short period, the crop is vulnerable to reinfestation from nearby areas. Insecticidal control may be improved by using ground equipment to get maximum coverage and by taking advantage of times when the pest is active and not hidden deep within the foliage so that insecticides will make contact.

10.3 Diseases

Probably the most serious disease, which occurs in Papua New Guinea, Java, the Philippines, and Malaysia, is false rust, caused by *Synchytrium psophocarpi*, which may be controlled by copper fungicides. In plantations *Cercospora* spp. causes leaf spot, though not to the extent of producing defoliation. Leaf spot caused by *Pseudocercospora psophocarpi*, powdery mildew caused by *Erysiphe cichoracearum* (Price 1980), and collar rot caused by *Macrophomina phaseolina*, *Fusarium semitectum*, *F. equiseti*, *F. moniliforme*, and *Rhizoctonia solani* (leading to high seedling mortality) have been identified as the major fungal diseases of the winged bean. Viral symptoms have been observed in Nigeria and the Ivory

Coast. Necrotic mosaic virus, ring spot mosaic virus, and leaf curl diseases have been recorded (Khan 1982). In addition, the witches' broom, a mycoplasma disease, has also been identified on this crop (Khan 1982). It is probable that as plantation-type plantings increase, disease problems will become more severe.

Thanatephorus cucumeris causes leaf blight in Papua New Guinea, Sarawak, Sabah, and West Malaysia. *Mycosphaerella* sp. has been associated with a concentric ring spot in Papua New Guinea. *Choanephora cucurbitarum* is associated with flower blight in Papua New Guinea and Sarawak and *Myrothecium roridum* with a leaf spot in Malaysia.

Colletotrichum lindemuthianum occurs in Papua New Guinea and *C. gloeosporioides* (*Glomerella cingulata*) in the Ivory Coast.

Viruslike symptoms have been found in winged beans growing in Papua New Guinea, Indonesia, Ghana, Ivory Coast, Trinidad, and the Philippines. Knowledge of virus diseases and their control is still incomplete, but severe damage has occurred in Indonesia, Ivory Coast, and Ghana. Three virus diseases were observed on winged bean. These include Psophocarpus necrotic mosaic virus, Psophocarpus ring spot mosaic virus, and leaf curl and little leaf of unknown etiology. *Aphis craccivora* can transmit Psophocarpus ring spot mosaic virus and therefore has a dual role as pest and vector.

10.3.1 False Rust, *Synchytrium psophocarpi*

False rust is one of the important diseases infecting winged bean. *S. psophocarpi* has only been found on the genus *Psophocarpus*. However, almost every plant of *P. grandiflorus* observed in the North Kivu area of Zaire, at Lake Bunyonyi in South Western Uganda, and in the Kefa province in Ethiopia was infected by the false rust with the appearance of *S. psophocarpi*.

10.3.1.1 Distribution

Synchytrium psophocarpi has been recorded on winged bean in Papua New Guinea, Indonesia, Philippines, Malaysia, and to a minor extent in Bangladesh (Drinkall 1978). It has been suspected but not confirmed in West Africa. However, there is no record of it in Myanmar or India.

10.3.1.2 Symptoms

"False rust" or "orange gall" symptoms include the appearance of bright-orange pustules along the veins of young leaves and on stems, pods, and sepals of flowers (Fig. 10.5). Infection leads to hyperplasia and galling, with abnormal branching at the nodes. The fungus attacks only *P. tetragonolobus*. *P. scandens* is immune. The disease affects pod production and possibly seed yield. The disease is a serious problem during the rainy season, when it can assume epidemic proportions.

S. psophocarpi attacks plants at all growth stages. In *P. tetragonolobus* plants, galls are produced on both surfaces of the leaf lamina and are grouped close to and along both sides of the main vein and veinlets (Fig. 10.5). Heavily infected leaflets turn inwards and become distorted, eventually drying and falling off. As they mature, the galls change color from green to yellow, finally becoming orange to orange-brown. Leaf petioles become twisted and covered with erumpent orange galls, but the infection is limited to the sepals of floral buds, although flowers often abort. Pods are produced only sparingly, and those that do develop are malformed and spotted with erumpent galls. Seedlings are also attacked and frequently killed, the surviving plants becoming gnarled, twisted, and stunted with deformed leaf primordia and twisted lateral roots that are bunched at the nodes (Vera-Chaston 1977).

Synchytrium psophocarpi has produced only sporangia in leaf tissues and is therefore a microcyclic species. The dimensions and shape of the sporangia agree well with the description of *S. psophocarpi* (Karling 1964), which is believed to be restricted to *Psophocarpus* spp.

10.3.1.3 Management

All Papua New Guinea varieties are susceptible, but some disease-resistant varieties have been found in Indonesia. Resistant plants have been

Fig. 10.5 Rust pustules on leaf and stem of winged bean plant

found in Thailand (Price et al. 1982) and Indonesia (Thompson and Haryono 1979) and appear to result from a hypersensitivity response (Parman and Thompson 1981). *Psophocarpus scandens* is resistant to both false rust and leaf spot, caused by *Pseudocercospora psophocarpi* (Price et al. 1982), and may provide a source of resistance for breeders.

10.3.2 Dark Leaf Spot, *Pseudocercospora psophocarpi*

10.3.2.1 Distribution

Dark leaf spot is common throughout Southeast Asia and in Papua New Guinea, the Philippines, Malaysia, and Indonesia, which is particularly serious in hot and humid areas.

10.3.2.2 Symptoms

Pseudocercospora psophocarpi attacks the leaves; the first symptoms are yellow spots on the upper surface. The undersurfaces have a whitish bloom, which becomes gray and finally black as the leaves mature. This is followed by necrosis of the entire leaf. It is especially destructive during the rainy season. All Papua New Guinea lines are susceptible. The winged bean is the only known host at present. Here, too, *Psophocarpus scandens* is immune. A similar disease is also caused by *Cercospora psophocarpicola* in Singapore and by *Cercospora canescens* in Bangladesh.

10.3.2.3 Management

Staked plants performed better than unstaked plants in terms of number of pods per plant, length of dry and fresh pods, dry pod, and grain yield. Staking, however, did not have any effect on the severity of leaf spot disease complex of the winged bean. It is, therefore, advised that winged bean plants be staked for a better performance.

Benomyl spray at fortnightly intervals has been found to give good control.

Psophocarpus scandens is resistant to dark leaf spot.

10.3.3 Powdery Mildew, *Oidium* sp. (*Erysiphe cichoracearum*)

Powdery mildew may be potentially important in relatively cooler areas with dry season but high humidity.

10.3.3.1 Distribution

The disease occurs in the Highlands of Papua New Guinea and in Indonesia, but there is little information on its economic importance.

10.3.3.2 Symptoms

The symptoms of powdery mildew are powdery white patches on the leaves. Only the imperfect *Oidium* stage of the fungus has been recorded.

10.3.3.3 Management

Improve air circulation by thinning and pruning.

10.3.4 Collar Rot, *Macrophomina phaseolina*, *Fusarium semitectum*, *F. equiseti*, *F. moniliforme*, and *Rhizoctonia solani*

10.3.4.1 Symptoms

Collar rot affects seedlings of 3–4 weeks old. The symptoms are wilting of the leaves followed by the death of the plant. The hypocotyl region of affected plants is usually constricted, with black necrotic lesions at soil level. This disease caused severe field losses in Papua New Guinea. The disease is influenced by soil type, depth of sowing, and inoculum density.

10.3.4.2 Management

Shallow planting in well-drained soils has been recommended for control.

10.3.5 *Choanephora* Blight, *Choanephora cucurbitarum*

10.3.5.1 Symptoms

Flowers were susceptible to infection at all stages of development, and wounding substantially increased infection, especially of young flower buds.

10.3.5.2 Management

Triadimenol, Vinclozolin, and Bitertanol were most effective in inhibiting spore germination and mycelial growth. Monthly applications of these fungicides in a field trial reduced blight. Triadimenol was more effective in controlling blight and increasing the pod yield than Vinclozolin and Bitertanol. The highest pod yield in SLS 7 accession was obtained with Triadimenol followed by Vinclozolin and Bitertanol treatments (Table 10.1) (Gunasekera and Shanthichandra 1990).

Table 10.1 Effect of different fungicides on pod yield of winged bean

Treatment	Pod yield ($\times 10^3$ kg ha^{-1})
Triadimenol	8.2
Vinclozolin	7.2
Bitertanol	6.9
CD (P=0.01)	**0.3**

10.3.6 Necrotic Mosaic Virus

10.3.6.1 Symptoms

It affected 9 % of the young plants distributed throughout the entire field. Leaves of diseased plants were necrotic and distorted, with an occasional yellow mosaic pattern. The leaf surfaces were noticeably reduced, and the plants were underdeveloped. Severity of the disease appeared to be correlated with the time of infection. On older plants, the number of flowers was reduced, and consequently the yield was greatly reduced.

10.3.6.2 Transmission

The disease was found to be mechanically transmissible from *Psophocarpus* to *Psophocarpus*, as well as to a limited number of other plants belonging to the families Leguminosae and Solanaceae. Apart from *Psophocarpus*, the only host plant showing symptoms was *Canavalia ensiformis*; other infected plants remained symptomless.

There was no transmission of the disease by seeds collected from infected *Psophocarpus*. *Aphis gossypii* and *A. craccivora* failed to transmit the disease in either a nonpersistent or a persistent manner.

10.3.7 Ring Spot Mosaic Virus

10.3.7.1 Symptoms

Every plant in both fields of winged bean in Kpouebo village was infected. Contamination apparently occurred early, because symptoms could be observed on the first leaves. Leaves of diseased plants had light-green ring spots. The spots often coalesced to form a yellow mosaic.

The exact reduction in yield caused by the disease is not known but is estimated at about 10–20 %. All plants were infected and the total loss was rather severe.

10.3.7.2 Transmission

The virus was easily transmitted mechanically from *Psophocarpus* to *Psophocarpus* by using a phosphate buffer 0.1 NI at pH 7.1. Every inoculated seedling became diseased. The host range of this virus was found to be rather wide because it was able to infect plants within the families Leguminosae, Solanaceae, and Chenopodiaceae. Seeds harvested from naturally infected *Psophocarpus* had a virus. Transmission rate was about 1 %. *Aphis craccivora* transmitted the disease in a nonpersistent manner, producing the same symptoms.

10.3.8 Leaf-Curl Disease

10.3.8.1 Symptoms

This disease appeared only at Kpouebo. All plants were infected, but not before the beginning of flowering. Leaves became dark green, thickened, and coiled downward. The apex stopped growing and many flowers fell off. Green-pod and seed production appeared to be severely affected by the disease. Young leaves of the aftergrowth, however, were free of symptoms; symptoms reappeared later. This demonstrates that the development of the causal agent was rather slow.

10.3.8.2 Transmission

All of the usual methods of mechanical transmission, grafting, and purification failed to prove the presence of a virus. Preliminary transmission by seed or insects was also unsuccessful, and no virus was observed on electron microscope preparations. Consequently, it cannot yet be concluded that there is a viral agent for this disease. Similarities with okra leaf curl, however, which is whitefly transmitted, support the virus hypothesis. In addition, it is improbable that this disease is an artificially induced physiological disorder because no chemical fertilizer or pesticide was used.

10.3.8.3 Management of Virus Diseases

- Virus management aims at prevention through integrated management practice that involves controlling the virus source, aphid populations, and virus transmission into pulse crops.
- Rotate legume crops with cereals to reduce virus and vector sources and where possible avoid close proximity to perennial pastures (e.g., Lucerne) or other crops that host viruses and aphid vectors.
- Eliminate summer weeds and self-sown pulses "green bridge" that are hosts for viruses and a refugee for aphids.
- Aphid activity is influenced by seasonal conditions and will require early monitoring in nearby crops and pastures and possible use of chemical or cultural controls to reduce numbers.
- Since bare soil is more attractive to some aphid species, sow winged bean seeds directly into cereal standing stubbles and encourage rapid canopy cover through early high planting density.
- Obtain virus-tested seed since PSbMV, CMV, BYMV, and AMV depend largely on seed transmissions for survival.
- The chemical Gaucho 350SD is now registered and when applied as seed treatment will help protect seedlings from early season aphid attack and reduce virus spread.

10.3.9 Witches' Broom, Phytoplasma

10.3.9.1 Symptoms

Witches' broom disease of winged bean was not transmitted by sap; however, it could easily be transmitted by grafting. The disease could infect other leguminous hosts like *Phaseolus vulgaris*, *P. lunatus*, *P. angularis*, *Vigna sinensis*, and *Vicia faba*. Electron microscopy of infected plant parts, viz., leaf mid rib, petiole, floral parts, stem, and roots, revealed the presence of a large number of pleomorphic bodies resembling typical mycoplasma-like organisms (MLOs). The MLOs were restricted to sieve elements only. The size of the MLOs varied from 60 to 950 nm in diameter (Singh 1991).

10.3.9.2 Management

Tetracycline treatment of infected plants at 250, 500, and 700 ppm showed a very conspicuous response in terms of suppression of disease symptoms.

10.4 Nematodes

10.4.1 Root-Knot Nematodes, *Meloidogyne* spp.

Meloidogyne incognita, *M. javanica*, and *M. arenaria* have been identified to be responsible for nematode disorders in the winged bean.

Root-knot nematodes have been reported to cause up to 70 % losses in tuberous roots: *M. incognita* has been recorded as more aggressive than *M. javanica*.

10.4.1.1 Symptoms

The root-knot nematodes (*M. incognita and M. javanica*) can cause stunting of plants, yellowing of leaves, and severe galling of roots; this not only damages the roots but also reduces tuber production and may affect pod and seed yield. The nematodes are cosmopolitan and have a wide host range. Damage to winged bean has been reported in Papua New Guinea, the Philippines, Ivory Coast, Indonesia, and Mauritius, but the problem arises wherever the plant is grown and is especially troublesome in sandy soils. Most of the Papua New Guinea lines are susceptible, and a screening for root-knot resistance of the winged bean world germplasm collection is needed. No assessment of reduction in yield of pods, seeds, or leaves was made.

A survey in January 1978 showed that the field was very heavily infested. Many plants had been killed and tubers were reduced in number (an estimated 50 % of the tubers of the Papua cultivar and 60–70 % of the Ghana cultivar failed to develop). At harvest time, in November, heady galled roots were found in the field plot. On some roots, the attack was severe enough to render the tubers unfit for consumption.

Root-knot nematode damage on tubers was evident, even in a field that had a very low level of infestation at sowing time. Reduction in yield of the other edible parts of the plant (pods, beans, and so forth) probably occurred.

10.4.1.2 Management

It appeared imperative to control the nematodes either by chemicals, or by using resistant cultivars, or by rotation with nonhost plants.

At Adiopodoume, the plot was treated with 5 g of 3 % carbofuran (2, 3-dihydro-2, 2-dimethyl-7-benzofuranyl methylcarbamate) per plant at planting and showed no infestation during the growing season. However, it is not known whether this was due to the nematicide or to a low level of infestation in the soil.

Flooding for 30–40 days has been recommended as control method in the Philippines.

References

Drinkall MJ (1978) False rust disease of the winged bean. Pest Abstr News Summ 24:160–166

Gunasekera SA, Shanthichandra WKN (1990) Pod yield of winged bean after chemical treatment for *Choanephora* blight disease. J Nat Sci Counc Sri Lanka 18(1):27–30

Harris VE, Todd JW (1980) Duration of immature stages of the southern green stink bug, *Nezara viridula* (L), with a comparative review of previous studies. J Georgia Entomol Soc 15(2):114–124

Jagginavar SB, Kulkarni KA, Lingappa S (1990) A note on seasonal abundance of cowpea pod borer complex. Karnataka J Agric Sci 4:56–58

Karling JS (1964) Synchytrium. Academic, New York

Khan T (1982) Winged bean production in the tropics. Food and Agriculture Organization of the United Nations, Rome, p 1

Mitchell WC, Mau RFL (1971) Response of the female southern green stink bug and its parasite, *Trichopoda pennipes*, to male stink bug pheromones. J Econ Entomol 64(4):856–859

Nishida T (1966) Behavior and mortality of the southern stink bug *Nezara viridula* in Hawaii. Res Popul Ecol 8:78–88

Odebiyi JB (1981) Studies on the biology of the cowpea pod-borer, *Maruca testulalis* in Kenya – 1. Determination of the larval instars. Insect Sci Appl 1:339–341

Parman, Thompson AE (1981) Resistance to false rust in winged beans. Hortic Sci 16:63–65

Price TV (1980) Diseases of the winged bean. In: The winged bean. Papers presented in the first international symposium on developing the potentials of the winged bean January 1978, Manila. Sponsored by the Philippine Council for Agriculture and Resources Research, Los Banos, Laguna, Philippines

Price TV, Drinkall MJ, Munro PE (1982) Studies on the incidence, severity and distribution of false rust (*Synchytrium psophocarpi*), leaf spot (*Pseudocercospora psophocarpi*) and powdery mildew (*Oidium* sp.) on winged bean (*Psophocarpus tetragonolobus*) in Papua New Guinea. Ann Appl Biol 101:473–482

Sepswasdi P, Pitaksa S, Chareonrak T, Phapoom V, Heuef-Rolf B (1991) Crop loss assessment for major mungbean pests in rice based cropping system. In: Proceedings of the mungbean meeting. Tropical Agriculture Research Center, Bangkok, pp 259–267

Singh SJ (1991) Winged bean witches' broom: a new mycoplasma disease in India. Int J Trop Plant Dis 9(2):201–206

Thompson AE, Haryono SK (1979) Sources of resistance to two important diseases of winged bean, *Psophocarpus tetragonolobus* (L) DC. Hortic Sci 14(4):532–533

Vera-Chaston HP (1977) *Synchytrium psophocarpi* (Rac.) its development and anatomy of infection on *Psophocarpus tetragonolobus* (L) DC Kalikasan. Philipp J Biol 6(3):183–198

Waterhouse DF, Norris KR (1987) *Nezara viridula* (Linnaeus), Hemiptera: Pentatomidae, green vegetable bug (Australia, New Zealand), southern green stink bug (USA). In: Waterhouse DF, Norris KR (eds) Biological control: Pacific prospects. Inkata Press, Melbourne, pp 81–89

11 Tannia, *Xanthosoma sagittifolium*

Abstract

The distribution, economic importance, symptoms/damage, host range, biology, survival and spread, and management of insect and mite pests and the fungal, bacterial, viral, and nematode diseases of tannia (*Xanthosoma sagittifolium*) using regulatory, physical, cultural, chemical, biological, host resistance, and integrated pest management methods are discussed.

Keywords

Tannia • *Xanthosoma sagittifolium* • Pests • Diseases

11.1 Introduction

Tannia (*Xanthosoma sagittifolium*) is an edible aroid grown commercially in the United States. It is a major crop which originated in northern South America. It is known by several common names and in the United States is marketed as malanga (its Cuban name). Yautia and tannia are common names for the crop in the Caribbean. Tannia is known for its love of a humid environment and its ability to flourish in shaded conditions. It is among the most shade tolerant of terrestrial food crops.

Tannia is an herb which grows up to 2 m tall. It has a short stem and produces large leaves at the top of the stem. A corm is produced at the base of the plant. It produces about ten cormels on the underground corm. These are about 15–25 cm long and flask shaped. They get wider toward the tip. Leaves are large and the stalk joins to the edge of the leaf. The leaves stand erect on stout petioles, or leaf stalks. There is a vein around the edge of the leaf. The leaf stalks can be 1 m long. The leaf blade is oval and 50–75 cm long. The leaf has triangular lobes at the bottom (Fig. 11.1).

The flower is produced below the leaves. The large bract around the flower is pale green and about 20 cm long. The bases of this bract overlap. The closely arranged spike of flowers is about 15 cm long. The smaller female section is at the bottom of the spike and the larger male section toward the top.

In the world, 0.362 million tons of tannia tubers are produced mainly in Cuba, Venezuela, El Salvador, Peru, and Dominican Republic (Table 11.1).

P.P. Reddy, *Plant Protection in Tropical Root and Tuber Crops*,
DOI 10.1007/978-81-322-2389-4_11

Fig. 11.1 Tannia plant, cormels, and flower

Table 11.1 Top tannia producers (2011)

Country	Productions (t)
Cuba	132,100
Venezuela	90,336
El Salvador	42,961
Peru	34,771
Dominican Republic	32,918
World total	**362,415**

Source: FAOSTAT Database (2012)

11.2 Insect Pests

Insect pests which sometimes attack tannias include *Euetheola bidentata* in Surinam; *Graphocephala propior*, *Quinta cannae*, and *Cacographis ortholatis* in Venezuela; *Aspidiotus destructor* in the Antilles and Polynesia; and *Tetraleurodes ursorum* and *Corythucha gossypii* in the Antilles.

11.2.1 Dynastid Beetle, *Ligyrus ebenus*

The Dynastid beetle has been for many years an important pest of tannias (*Xanthosoma* spp.) in the northwest of British Guiana, where this crop is very valuable.

11.2.1.1 Distribution

In Trinidad, Venezuela, and Guyana, the dynastid beetle has been reported to attack the tannia crop.

11.2.1.2 Damage

They attack the crop either by boring into the growing plants and shoot from above or below or by feeding directly on the sets below ground and corms and inflict serious damage in both cases. They eat into the tannia corms shortly before the crop is ready for reaping and may do serious injury. If early planting is carried out, the sets may be injured by this same group of larvae prior to pupation.

11.2.1.3 Biology

The adult beetles are strongly attracted to humus and decaying vegetable matter, the females deposit their eggs in such substances, and the larvae feed on them and the rootlets of succulent plants. The adults emerge from April to July, being usually most abundant in May and June, but after emergence, they may remain in the ground until the beginning of the wet season, when they appear in the fields. By the end of July, a large number of beetles die, while others burrow into the ground and continue to live below the surface for several months. Oviposition begins from October to November. The entire life cycle of the insect occupies about a year, and almost the whole of its life is spent below ground.

11.2.1.4 Management

1. *Cultural Methods*: The control measures recommended are late planting (the planting, which should normally take place in May, is usually delayed until about July), treatment of the cut surfaces of the sets with hot tar prior to planting, and the removal of all rotting tannias and other decaying vegetable matter from the fields. A short flooding of about 10 h given to the fields after the tannia crop has been removed will completely destroy both larvae and pupae.

   The application of 5 % aqueous yam bean seed extract for effective control of yam bean pod borer is recommended.
2. *Chemical Methods*: It can be effectively controlled by spraying with Malathion.
3. *Host Resistance*: The yellow variety of tannia appears to be attacked in preference to the white.
4. *Biological Methods*: The fungus, *Metarhizium anisopliae*, may cause a high mortality among the larvae, pupae, and adults.

11.2.2 Aphid, *Aphis gossypii* and *Pentalonia nigronervosa*

11.2.2.1 Damage

Aphis gossypii attack tannias in the Antilles and Surinam and *Pentalonia nigronervosa* in the Antilles. They are present on the underside of the leaves (Fig. 11.2). The number of corms per plant and the size/weight of the corms are reduced because of weakening of plants due to downward wilting and curling of leaves.

11.2.2.2 Management

The following management practices may be followed to keep the aphids under control:

- Provide hedges to limit movements of aphids from one crop to another and to encourage natural enemies, e.g., beetles and hoverflies, *Aphidius*.
- Uproot of host plants from neighboring fields.
- Sprinkler irrigation or sustained rain can reduce infestation.

Fig. 11.2 Aphids on tannia leaf

Fig. 11.3 Whiteflies on tannia leaf

- Control ants in the field, as these will disrupt natural enemy activities.
- Treatment with selective insecticides: soap solutions and neem products.
- Spraying of yam bean seed extract at 5 and 2 % gave effective control of mustard aphid and increased cormel yield.
- Destruction of infested leaves after harvesting.

11.2.3 Whitefly, *Bemisia tabaci* and *B. argentifolii*

11.2.3.1 Damage

Whiteflies are present on the underside of the leaves (Fig. 11.3). The number of corms per plant and the size/weight of the corms are reduced because of weakening of plants due to downward wilting and curling of leaves.

11.2.3.2 Management

The following management practices may be followed to keep whiteflies under control:

- Provide hedges to limit movements of whiteflies from one crop to another and to encourage natural enemies.
- Sprinkler irrigation or strong rain will limit the spread of the pest.
- Systemic insecticides to control the pupae (puparium): soap and oils.
- Selective insecticides (to limit the negative impact on natural enemies) used alternately (to limit the risks of resistance) to control adults (neem, molting disruptors).
- Encourage natural enemies: parasitic wasps (*Encarsia formosa*, *Eretmocerus* sp.), predatory mites (*Amblyseius* spp.), ladybird beetles, and lacewings.
- Destruction of harvest residues to prevent buildup and spreading.

Fig. 11.4 Giant African slug feeding damage on tannia

11.2.4 Slugs

11.2.4.1 Damage

The slugs wound corms (Fig. 11.4) and provide entry points for spoilage microorganisms. Losses can reach up to 60 % of corms.

11.2.4.2 Management

Prevention is by the use of disease-free planting material, weeding and hilling, and treatment with copper-based pesticides (Opara 2000).

11.3 Diseases

Important diseases include root and corm rot; dasheen mosaic potyvirus (DMV); southern blight, *Athelia rolfsii*; and leaf spot and blight, *Xanthomonas axonopodis* pv. *dieffenbachiae*.

Several root rots can affect tannias, with varying degrees of severity. During the 1930s, root rot was particularly serious in Ghana, and investigations suggested that the causal agent could have been a fungal or bacterial infection or a combination of both and that the nutritional status of the crop was also a factor influencing the severity of attack. Tannias are susceptible to various *Pythium* rots; in Puerto Rico, a soft rot known as "curru-taca" due to *Pythium ultimum* affects the crops, while in New Caledonia, *P. irregulare* is reported occasionally to cause serious losses. In Venezuela, the crop is sometimes infected by *Cercospora chevalier*, *C. verruculosa*, and *Punctellina solteroi*. Viruses have been reported but do not appear to be serious.

11.3.1 Root Rot, *Pythium myriotylum*

The root rot disease (RRD) is the most devastating disease in cocoyam at present (Tambong et al. 1998) and can cause a total loss of yield (Saborío et al. 2004a). The pathogen that seems to be the main causal agent is *Pythium myriotylum* (Tambong et al. 1999).

11.3.1.1 Economic Importance

Root rot of cocoyam caused by *Pythium myriotylum* is the most devastating disease of this important tropical tuber crop with yield reductions of up to 90 %.

11.3.1.2 Symptoms

The most serious tannia disease is called "leaf-burning disease" or "root rot disease." The aboveground symptoms of the disease are stunting of plants with rotting of roots and corms (Fig. 11.5). Hence, the descriptions for this disease indicate that the outer leaves of the plant gradually turn yellow from margin to midrib and finally the leaf dies. The roots of the plant also die. Production of marketable cormels is severely affected. Corms become soft with malodorous flesh and do not preserve well in storage.

This fungus attacks the roots, especially in hydromorphic soil. On the parts above ground, withering and stunting of leaves, shortening of leafstalks, and chlorosis of the blade (greenish-yellow color) may be noticed. When an affected corm is cut, the diseased parts are faded and soft with a marked separation from healthy areas, which remain white or colored, depending on the cultivar. The healthy roots are cream or rose colored and turgescent, whereas the diseased roots are dark and flaccid before complete necrosis.

11.3.1.3 Spread and Survival

The main sources of the inoculum are infected soil and infected planting material. The pathogen is soilborne and spread by hydromorphy. The development of the disease is helped by poorly drained soils. The pathogen can persist in the soil for many years (Nzietchueng 1984).

11.3.1.4 Management

1. *Cultural Methods*: Wide spacing, high mounds, regulation of the time of planting (Onwueme and Charles 1994), drainage improvement, use of disease-free planting material through meristem tip culture (Fig. 11.6) (Saborío et al. 2004b), planting on ridges, crop rotation (Giacometti and León 1994), and use of organic manures (Torres-Portuguez 1996) have been suggested in order to control the disease.

   The following cultural management practices may be followed to keep corm rot under control:
   - Avoid infected fields.
   - Planting on ridges (balks).
   - Use of healthy (disease-free) cuttings.
   - Former banana plantations offer favorable conditions to control the disease.
   - High soil fertility.
   - Planting in well-drained soils.
   - A prompt harvest.

Fig. 11.5 Damages on collar and leaf stalks and wilting due to corm rot

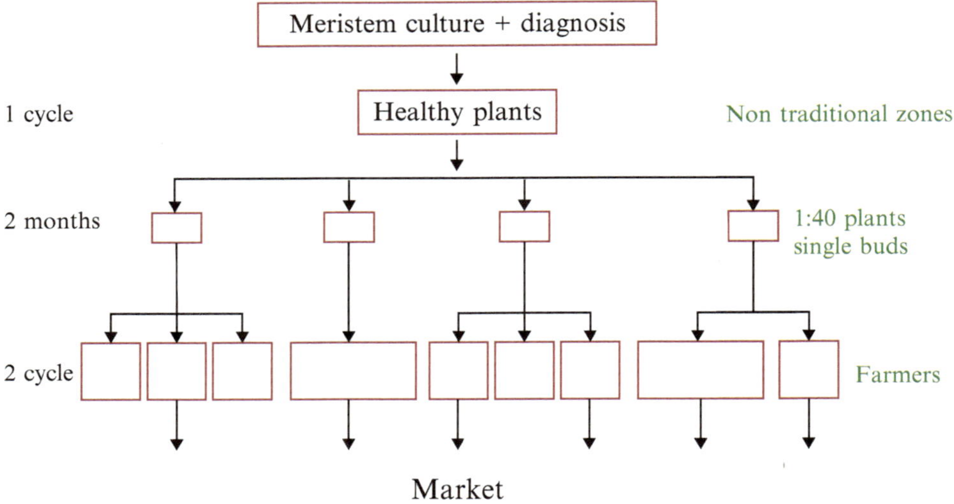

Fig. 11.6 Scheme for the use of disease-free plants obtained from meristem culture

2. *Chemical Methods*: Disinfection of the corms before packing through soaking and fungicide application (Acrobat) as soil drench gives effective control of the disease.
3. *Biological Methods*: *Pseudomonas aeruginosa* PNA1 strongly reduced root rot disease of tissue-culture-derived cocoyam plantlets. The biocontrol activity of PNA1 against the pathogen involves phenazines. The efficacy of PNA1 to control root rot on cocoyam was improved when the strain and the pathogen were allowed to interact for 24 h prior to transplanting cocoyam plantlets (Tambong and Hofte 2001).

11.3.2 Cormel Rot, *Corticium rolfsii*

11.3.2.1 Economic Importance

The 11 % drop in Nigeria's national production figures between 2000 and 2004 was due to microbial attack by *Sclerotium rolfsii* in storage (FAO 2001, 2004).

11.3.2.2 Symptoms

In storage, serious losses due to rotting of the corms and cormels are a major factor affecting adversely the quantity and quality of cormels for consumption and planting.

11.3.2.3 Management

The leaf powders of *Cassia alata* (candle bush) and *Dennettia tripetala* (pepper fruit) were found to be more effective in reducing rot development in cocoyam cormels (Nwachukwu and Osuji 2008).

Cormel rot on tannia was prevented by 200 ppm of mancozeb, benomyl, and iprodione (Ohazurike 1996).

11.3.3 Dasheen Mosaic Virus (DMV)

Dasheen mosaic virus (DMV) is the most important virus that infects cocoyam.

11.3.3.1 Economic Importance

The yield per virus-free plant was significantly higher than yield obtained in a virus-infected plant. The estimated yield ha^{-1} for virus-free plants was 18.2 and 13.4 t for virus-infected plants. The virus-free plants produced 25 % higher yield compared to the virus-infected plants. The estimated yield obtained from virus-free plants (18.2 t ha^{-1}) was 2.5 times higher than

Fig. 11.7 Symptoms of DMV in cocoyam. "Feathering" along the veins, leaf distortion, and stunted plant

the current national yield average (7.2 t ha^{-1}) and very close to the former national yield average (19–22 t ha^{-1}). Yield losses of up to 90 % sometimes occur.

11.3.3.2 Symptoms

Plants infected with DMV show a wide variety of mosaic types: small, irregular, and sparse, with spots ranging from green to yellow, grey or white, in the shape of a feather on either side of the main veins. Generally the plants overcome these attacks and regain a normal appearance, but some severe attacks from which plants have not recovered have been reported. The whole plant is invaded after transmission by the aphid vector. The number of corms per plant and the size/weight of the corms are reduced due to stunting of plants.

The visible symptoms on the plants includes leaf distortion, vein chlorosis, mosaic feathering along the veins (Zettler et al. 1989), and, in the case of a severe attack, stunted plants (Fig. 11.7). DMV is transmitted to a noninfected plant in a nonpersistent manner exclusively by aphids (Brunt et al. 1996), and the virus transmission in the field can be very rapid, probably in less than 1 h (Pernezny et al. 1993). Although DMV is not lethal, it retards the plant growth and reduces the yield (Zettler et al. 1989).

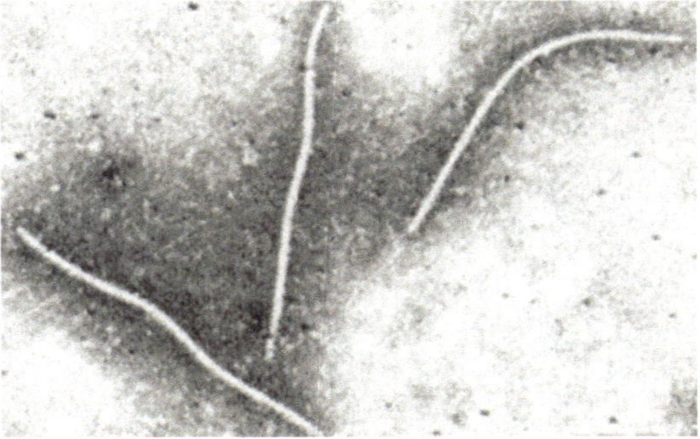

Fig. 11.8 Flexuous rods of dasheen mosaic virus

11.3.3.3 Particle Morphology

The virus particles are flexuous filaments (Fig. 11.8) of about 750 nm long (Zettler et al. 1970).

11.3.3.4 Management

The following management practices may be adopted to keep the disease under check:

- Use virus-free seed stock.
- Isolate and protect plots (clearing).
- Destroy diseased plants by burning or burying.
- Use cuttings from healthy mother plants.
- Virus-free plants were produced through meristem tip culture, and protocols for the regeneration of plants via callus and protocorm-like structures were established. Although a

genotypic response to tissue culture was observed, all the tested genotypes produced shoots from meristem and from meristem-derived calli.

References

Brunt AA, Crabtree K, Dallwitz MJ, Gibbs AJ, Watson L, Zurcher EJ (eds) (1996 onwards) Plant viruses online: descriptions and lists from the VIDE database. Version: 20th August 1996. http:// biology.anu.edu.au/ Groups/MES/vide/

FAO (2001) Food and Agriculture Organization of the United Nations, Rome, Italy. FAO Bull Stat 2(2):336

FAO (2004) Food and Agriculture Organization of the United Nations Agricultural Development in Nigeria. FAO, Rome

FAOSTAT Database (2012) UN Food & Agriculture Organization, Rome

Giacometti DC, León J (1994) Tannia Yautia (*Xanthosoma sagittifolium*). In: Hernaldo JE, León J (eds) Neglected crops: 1492 from a different perspective, Plant production and protection series no. 26. FAO, Rome, pp 253–258

Nwachukwu EO, Osuji JO (2008) Evaluation of plant extracts for antifungal activity against *Sclerotium rolfsii* causing cocoyam cormel rot in storage. Res J Agric Biol Sci 4(6):784–787

Nzietchueng S (1984) Root rot of *Xanthosoma sagittifolium* caused by *Pythium myriotylum* in Cameroon. In: Terry ER, Doku EV, Arene OB, Mahungu NM (eds) Tropical root crops: production and uses in Africa. Proceedings of the second triennial symposium of the international society for tropical root crops, Douala, Cameroon. International Development Research Centre, Ottawa, pp 185–188

Ohazurike NC (1996) Effect of some fungicides on extracellular enzymes of *Sclerotium rolfsii* sacc. http:// www.ncbi.nlm.nih.gov/entrez/query.fcgi?cmd=Retrieve&db=pubmed&dopt=Abstract&list_uids=0

Onwueme IC, Charles WB (1994) Cultivation of cocoyam. In: Tropical root and tuber crops. Production, perspectives and future prospects, FAO plant production and protection paper 126. FAO, Rome, pp 139–161

Opara LU (2000) Edible aroids – post operation. FAO, Rome

Pernezny K, Lamberts M, Ramos L (1993) Tropical vegetable diseases: 1. Series of the Plant Pathology Department. Institute of Food and Agricultural Sciences, University of Florida, Gainesville

Saborío F, Umaña G, Solano W, Ureña G, Muñoz G, Hidalgo N, Brenes A (2004a) Mejoramiento genético del tiquisque (*Xanthosoma sagittifolium*) contra el Mal Seco. Memoria REDBIO (2004) (Talleres.www.redbio.org. 21 Sept 2005)

Saborío F, Umaña G, Solano W, Amador P, Muñoz G, Valerin A, Torres A, Valverde R (2004b) Induction of genetic variation in *Xanthosoma* spp. In: International Atomic Energy Agency (ed) Genetic improvement of under-utilized and neglected crops in low income food deficit countries through irradiation and related techniques. International Atomic Energy Agency, Vienna, pp 143–154

Tambong JT, Hofte M (2001) Phenazines are involved in biocontrol of *Pythium myriotylum* on cocoyam by *Pseudomonas aeruginosa* PNA1. Eur J Plant Pathol 107:511–521

Tambong JT, Poppe J, Höfte M (1999) Pathogenecity, electrophoretic characterization and in planta detection of the cocoyam root rot disease pathogen, *Pythium myriotylum*. Eur J Plant Pathol 105:597–607

Tambong JT, Sapra VT, Garton S (1998) *In vitro* induction of tetraploids in colchicine-treated cocoyam plantlets. Euphytica 104:191–197

Torres-Portuguez SE (1996) Estudio preliminar de la utilización del complejo orgánico sobre el "mal seco" en cultivo de tiquisque blanco (*Xanthosoma sagittifolium*). In: Congreso Nacional Agronómico y de Recursos Naturales, San José, Costa Rica, Memoria 2:111

Zettler FW, Jackson GVH, Frison EA (1989) Technical guidelines for the safe movement of edible aroid germplasm. FAO/IBPGR (ed) Food and Agriculture Organization of the United Nations, Rome/ International Board for Plant Genetic Resources, Rome

Zettler FW, Foxe MJ, Hartman RD, Edwardson RD, Christie RG (1970) Filamentous viruses infecting dasheen and other araceous plants. Phytopathology 60:983–987

12 Enset, *Ensete ventricosum*

Abstract

The distribution, economic importance, symptoms/damage, host range, biology, survival and spread, and management of insect and mite pests and fungal, bacterial, viral, and nematode diseases of enset (*Ensete ventricosum*) using regulatory, physical, cultural, chemical, biological, host resistance and integrated pest management methods are discussed.

Keywords

Enset • *Ensete ventricosum* • Pests • Diseases

12.1 Introduction

Enset, also called as "false banana," is a crop of major economic and sociocultural importance for a wide range of smallholder households in Ethiopia. It is a staple food for over 12 million people in the southern highland areas of Ethiopia. It grows best at altitudes ranging from 2,000 to 2,700 m (Brandt et al. 1997). Enset-based farming system is an indigenous and sustainable agricultural system in Ethiopia. About 224,000 ha of land is assumed to be covered with enset cultivation in the country (CSA 1994). It supports more than 20–25 % of Ethiopia's population as staple and co-staple food and is also used as a traditional medicine (Tabogie 1997). In addition to this, enset constitutes one of the cheap sources of carbohydrate in Ethiopia, providing more calcium and iron than most cereals, tubers, and root crops (Kefale and Stephen 1991). It has been known to play a role of a barrier food deficit for human and animal feed during the dry spell and recurrent drought due to its resistance to fluctuating rainfall patterns after establishment. Currently, enset starch is being used for textile, paper, and adhesive industries in saving the foreign exchange rate which could have been spent to import the raw material (Edwards 1991). The edible starch is extracted from the corm and kocho, and this is an important food for millions of people in Ethiopia (Argent 1976). A tree takes 4–5 years to mature, at which time the root can weigh around 80 lb. The fresh corm is cooked like potatoes before eating.

Enset is large non-woody plant, a gigantic monocarpic evergreen perennial herb (not a tree) which grows up to 6 m (20 ft) tall. It has a stout pseudo stem of tightly overlapping leaf bases and large banana-like leaf blades of up to 5 m tall by 1 m

P.P. Reddy, *Plant Protection in Tropical Root and Tuber Crops*,
DOI 10.1007/978-81-322-2389-4_12

Fig. 12.1 Enset plant

wide, with a salmon-pink midrib (Fig. 12.1). The flowers, which only occur once from the center of the plant at the end of that plant's life, are in massive pendant thyrses covered by large pink bracts. The fruits are similar to those of the domestic banana and are edible but insipid, with hard, black, rounded seeds. After flowering, the plant dies back.

Wild enset plants are produced from seeds, while most domesticated plants are propagated from suckers. Up to 400 suckers can be produced from just one mother plant. In 1994, 3,000 km^2 of enset were grown in Ethiopia, with a harvest estimated to be almost 10 t per hectare. Enset is often intercropped with sorghum, although the practice among the Gedeo is to intercrop it with coffee.

12.2 Insect and Vertebrate Pests

Insect pests such as spider mites, leaf hoppers, aphids, and root mealybugs are a serious problem in enset cultivation. Porcupines, mole rats, and wild pigs will eat enset corms and pseudo stems

12.2.1 Root Mealybug, *Cataenococcus ensete*

The root mealybug is a major pest in the enset-growing regions of southern Ethiopia. According to Addis (2005), more than 30 % of the sampled enset farms were infested with root mealybugs.

12.2.1.1 Damage

Enset root mealybugs were found exclusively on the roots and corm of enset (Fig. 12.2). Damage by enset root mealybugs was nonspecific, including common symptoms of slow plant growth, lack of vigor, and subsequent death, especially under moisture stress. Infested plants displayed retarded growth where most lateral leaves were desiccated, but with a green central shoot. All roots were found to be vulnerable to mealybugs attack. It was observed that enset plants attacked by root mealybugs have a significantly lower number of roots as compared to healthy plants. In addition, mealybug-damaged enset plants were more easily uprooted.

The pest attacks enset plants at any age, with infestations being most serious on 2–4-year-old plants (Anon 2001). Mealybug-infested enset plants exhibit retarded growth, loss of vigor, and dried lateral leaves but green central shoot and eventually plant death (Addis 2005).

12.2.1.2 Biology

Different development stages of enset root mealybugs are recognized:

- Bright-orange- to yellow-orange-colored "crawlers" or rapidly moving first instars that are creamy white but barely visible, being 0.5–2.7 mm long. They greatly resemble the adults but are significantly smaller. Average duration of the first, second, and third instar nymphs is 16.2 ± 0.5, 18.2 ± 0.7, and 19.8 ± 0.4 days, respectively.
- Second and third instar mealybugs begin to develop distinct lateral and caudal spines, increase in body size, and start to produce large amount of honeydew excretion (Addis 2005; Addis et al. 2008).

Fig. 12.2 Root mealybug damage to enset plants and roots

- Adult female mealybugs have an elongate-oval body covered with wax secretions on the dorsal a nd lateral sides. Adult females show pronounced crosswise grooves running down their body and give birth to live young ones on a shallow pile of waxy secretions. Each female mealybug produces 156–383 nymphs in its lifetime. The adult life of a mealybug is 50.0±0.5 days. The total life span of mealybugs is 94–113 days. The body size of adult mealybugs ranges from 2.8 to 4.0 mm in length and 2.85–3.70 mm in width. There is generation overlap.

12.2.1.3 Spread

Crawling first instar nymphs can play a key role in dispersal over the host plant and can be responsible for spreading the mealybug population to new host plants. Occasionally, these crawlers are observed on the lower part of the pseudo stem just above the soil. Mealybugs can also be dispersed by water, when flooding occurs.

The main dispersal route is via the distribution of enset suckers from infested corms, especially in nursery sites established below 2,000 MSL where there is an environment conducive to mealybug reproduction. Planting material is commonly exchanged between farmers without discrimination of infested suckers, and subsequent transplanting of infested plants leads to further dispersal.

Ants' activities have also been implicated in mealybug dispersal. The literature documents many examples of the symbiotic relationship between ants and mealybugs, in which the ants protect the mealybugs to harvest their honeydew. Malsch et al. (2001) mentioned that when the mealybugs are disturbed at the time of cultivation, weeding, transplanting, and harvesting, their attendant ants carry and take them to new plants or root parts. Even though dispersal of mealybugs by ants is a short distance, ants help them to find hidden places in the roots and corm, which are very difficult to reach even with insecticides.

Finally, mealybugs can also be transported by machinery, tools, equipment, and soil movement during cultivation and repeated transplanting operations conducted at different times.

12.2.1.4 Management

1. *Regulatory Methods*: Quarantine measures could be designed to prevent further spread of the insect to different parts of the region. If practical, a clean sucker certification scheme

could be developed, in which farmers who produce planting materials would be monitored.

2. *Physical Methods*: Enset plants can be disinfested from root mealybugs by using hot water. Immersion of the seedlings in boiling water for 10–30 s prior to planting will most easily be adopted by farmers (Lemawork 2008).
3. *Cultural Methods*: The first resort for mealybug control is the use/production of clean planting material.

Repeated plowing and sanitation of enset fields has also been reported as a control option for reducing root mealybug population (Tadesse et al. 2003).

Crop rotation (during one or two cropping seasons) and/or removal of grasses and weeds in enset fields will also help to control this pest.

Removal of ant vectors has also proven successful in reducing mealybug infestation levels in enset.

The application of farm yard manure (20 kg/plant/year) could enhance enset plant growth and make the plant more resistant to root mealybug attack. In addition, the manure could directly inhibit mealybug development. Farm yard manure can also improve the soil and root health conditions.

Adult mealybugs are unable to survive for more than 3 weeks in the absence of any plant material (Addis 2005; Addis et al. 2008). While transplanting, planting pits should be left open for about a month so that any mealybug present in the vicinity of the planting hole will die of starvation. Repeated plowing and removal of weeds and grasses in enset field is believed to eradicate the root mealybug (Tadesse et al. 2003).

Drenching a seed-water suspension of 10 % *Millettia ferruginea* (botanical) into the root zone of infested enset plants in the field was found to be effective against the root mealybugs.

4. *Biological Methods*: Under field conditions, the entomopathogenic white muscardine fungus (*Beauveria bassiana* PPRC-56) caused the highest levels of mortality. However, the maximum mortality rate was only 54 %. Entomopathogens have the potential to be used as a component in the integrated pest management of enset root mealybugs.
5. *Chemical Methods*: Among the insecticides tested, the application of chlorpyrifos 48 % EC and diazinon 60 % EC at a rate of 1.7 L of solution (after dilution of the insecticide in 1:5 L of water) per field-grown enset plant, poured on the root collar area, resulted in high levels of root mealybug mortality (Tadesse 2006). More than 90 % of the adult mealybugs were killed within 2 weeks of application. For both chemicals the mortality rate reached about 98 % at 45 days after application of the chemicals.
6. *Integrated Methods*: Enset nurseries should be established, perhaps at appropriate altitudes (plant material from below 1,400 and above 2,200 MSL), for the production of clean seedlings/planting materials. Better hygiene measures, particularly at the nursery level, could be adopted, perhaps in conjunction with some form of quarantine and monitoring system. Boiling-water treatment should especially be included as a major disinfestation technique. Appropriate nutrition, perhaps using manure, will also help in producing healthier, more robust plants.

For already infested enset plants, insecticidal plant/seed extracts of *Millettia ferruginea* could be used (instead of synthetic insecticides) and entomopathogenic fungus (*Beauveria bassiana*) may also become a realistic integrated option.

12.2.2 Rodents

12.2.2.1 Damage

The mole rats, the size of large rats, tunnel and move from plant to plant eating the roots and corm, inciting variable damage including immediate decline and death of the plant. Cases were observed where serious infestation has caused 10–20 plants/week to be salvaged for what little pulp was possible to extract.

12.2.2.2 Management

Control consists of flooding the tunnel and killing the emerging mole rats. A large fish-hook type snare trap with string trigger through the tunnel is being introduced by the Ministry of Agriculture.

12.2.3 Vertebrate Pests

12.2.3.1 Damage

Porcupines and wild pigs will eat enset corms and pseudo stems.

12.2.3.2 Management

Farmers use woven fences and ditches as protection around enset fields and use traps to catch the animals.

12.3 Diseases

The main biotic stresses are bacterial wilt, fungi, and nematodes.

12.3.1 Leaf Spot Diseases

12.3.1.1 Symptoms

Leaf spot diseases caused by the fungi *Phyllosticta* sp., *Pyricularia* sp., and *Drechslera* sp. commonly affect suckers, seedlings, and young plants.

12.3.1.2 Management

Control measures include thinning overcrowded suckers, regular weeding, and avoiding intercropping suckers with tall plants.

12.3.2 Root Rot, *Sclerotium rolfsii*

12.3.2.1 Symptoms

The fungus *Sclerotium rolfsii* may cause diseases of root, corm, and pseudo stem.

12.3.2.2 Management

Fungicidal dips would be an appropriate method for controlling fungal root pathogens.

12.3.3 Bacterial Wilt, *Xanthomonas campestris* pv *musacearum*

The most serious disease of enset is bacterial wilt which is present in virtually all enset-growing areas. It is very destructive and kills enset plants at all stages of growth. Bacterial wilt was first reported in Ethiopia by Yirgou and Bradbury (1968) and is currently found in all the enset-growing regions and on wild enset plants, although it has not been reported on enset in other countries.

12.3.3.1 Symptoms

The inner folds of the drooping heartleaf are covered with grayish-brown patches. Eventually the leaf breaks at the petiole and yellowish bacterial slime oozes from the vascular bundles. All the leaves wilt, bend over, and wither, causing the death of the plant.

12.3.3.2 Spread

It is mainly spread through infected farm tools, infected planting material, repeated transplanting that damages the corm and roots, animals feeding on infected plants, and possibly insects feeding on the foliage. Since cultivated enset is harvested for its starchy pseudo stem and corm, it is not normally allowed to flower. As a result, the question of insects infecting flowers does not normally arise.

Possible sources of *Xanthomonas* spread to new plants are as follows:

- Splashing rain
- Birds, bees/wasps, and fruit flies
- Wind-blown leaves of infected plants
- Lizards, snakes, etc.
- Soil-inhabiting organisms (mole rats, insects, nematodes)
- Rainfall runoff to newly planted corms or plants

12.3.3.3 Management

Current control measures include uprooting and discarding infected plants; planting healthy, disease-free plants from less susceptible landraces; cleaning equipment that has come in contact with diseased plant material; crop rotation; avoiding overflow of water from infested to uninfested fields; removing *Canna indica* around enset as it acts as an alternate host and controlling leaf hoppers, aphids, and mole rats that may transmit the pathogen.

Early removal of the male bud is not practiced. Given that the cultivation of enset is a growing activity, focus should be put on controlling the disease and preventing its spread through timely debudding, the removal of infected plants, and the use of clean farm tools and clean planting materials.

12.4 Nematodes

Root lesion nematodes (*Pratylenchus goodeyi*) and root knot nematodes (*Meloidogyne* sp.) are commonly found and are widely distributed.

12.4.1 Root Lesion Nematode, *Pratylenchus goodeyi*

12.4.1.1 Symptoms

Pratylenchus goodeyi is often found associated in large numbers with bacterial wilt-affected plants and is therefore suspected to play a role in the development and severity of the latter disease.

12.4.1.2 Management

Root lesion nematode control includes crop rotation.

References

Addis T (2005) Biology of Enset root mealybug *Cataenococcus ensete* and its geographical distribution in southern Ethiopia. M.Sc. thesis. Alemaya University of Agriculture, School of Graduate studies, Alemaya, Ethiopia, 81 pp

Addis T, Azerefegne F, Blomme G, Kanaujia K (2008) Biology of the Enset root mealybug, *Cataenococcus ensete* and its geographical distribution in southern Ethiopia. J Appl Biosci 8(1):251–260

Anon (2001) Awassa Agricultural Research Center progress report of Entomology Research Section for the year 2000, Awassa, Ethiopia, 7 pp

Argent GCG (1976) The wild bananas of Papua New Guinea. Note R Bot Gard Edinb 35:77–114

Brandt SA, Spring A, Hiebsch C, McCabe JT, Tabogie E, Wolde-Michael G, Yntiso G, Shigeta M, Tesfaye S (1997) The tree against hunger: enset-based agricultural systems in Ethiopia. American Association for the Advancement of Science, Washington, DC, 56 pp

CSA (Central Statistical Authority) (1994) Statistical abstracts for the period 1993/94 CSA, Addis Ababa, Ethiopia

Edwards SB (1991) Crops with wild relatives found in Ethiopia. In: Engles JMM, Hawkes JG, Worede M (eds) Plant genetic resources of Ethiopia. Cambridge University Press, Cambridge, pp 42–74

Kefale A, Stephen S (1991) Enset in North Omo. Technique pamphlet no. 1. Farm Africa, Addis Abeba, Ethiopia, 49 pp

Lemawork S (2008) Evaluation of entomopathogenic fungi and hot water treatment against enset root mealybug, *Cataenococcus ensete*, Williams and Matile-Ferrero (Homoptera: Pseudococcidae) on Enset. M.Sc. thesis, Department of plant sciences, Awassa College of Agriculture, School of Graduate Studies Hawassa University, Awassa, Ethiopia, 87 pp

Malsch AKF, Kaufman E, Heckroth HP, Williams DJ, Maryati M, Maschwitz U (2001) Continuous transfer of subterranean mealybugs (Hemiptera: Pseudococcidae) by *Pseudolasius* spp. (Hymenoptera: Formicidae) during colony fission. Insect Soc 48:333–341

Tabogie E (1997) Morphological characterization of enset [*Ensete ventricosum* (Welw) Cheesman] clones and the association of yield with different traits. M.Sc. thesis presented to School of Graduate Studies, Alemaya University, Alemaya

Tadesse E (2006) Evaluation of some synthetic and botanical insecticides against the enset root mealybug (*Cataenococcus ensete*) (Homoptera: Pseudococcidae) William and Matile-Fererro in Southern Ethiopia. M.Sc. thesis, Department of plant sciences, Awassa College of Agriculture, School of Graduate Studies, Hawassa University Awassa, Ethiopia, 67 pp

Tadesse M, Anito E, Geta E (2003) Enset based farming systems of Masha Wereda, Shaka zone. EARO research report no. 51, Addis Ababa, Ethiopia, 25 pp

Yirgou D, Bradbury JF (1968) Bacterial wilt of Enset (*Ensete ventricosum*) incited by *Xanthomonas musacearum* sp. Phytopathology 58:111–112

13 Arrowroot, *Maranta arundinacea*

Abstract

The distribution, economic importance, symptoms/damage, host range, biology, survival and spread, and management of insect and mite pests and the fungal, bacterial, viral, and nematode diseases of arrowroot (*Maranta arundinacea*) using regulatory, physical, cultural, chemical, biological, host resistance, and integrated pest management methods are discussed.

Keywords

Arrowroot • *Maranta arundinacea* • Pests • Diseases

13.1 Introduction

13.1.1 Description

Arrowroot is a perennial plant belonging to Marantaceae family which grows up to 2 m high with large, fleshy underground rhizomes (Fig. 13.1). The stem is erect. There are 4–8 leaves near the base and 1–8 leaves on the stem. It has large leaves on long stalks near the base. These stalks can be 3.5–20.0 cm long. The leaf blade is 12 cm long and the leaf stalk clasps the stem. Flowers are small and white. There can be several on each leafy shoot. They occur on the ends of branches. There are 1–2 bracts under each 2–3 flowers. The rhizomes can be 20 cm long and about 3 cm wide. They are covered with fleshy scales. The fruits are capsules which are green with a red-brown tinge. They are oval and 7–8 mm long by 4–5 mm wide. The seeds are brown.

13.1.2 Distribution

It is grown in many tropical countries, including the Solomon Islands. It is a tropical plant that grows well in hot, humid climates. A temperature of 20–30 °C is best. Plants grow from the coast up to 900 m altitude in the tropics. It needs an annual rainfall of 1,500–2,000 mm. It is cultivated in China for starch. It needs deep, well-drained, and slightly acid soil.

13.1.3 Cultivation

Plants are grown from pieces of rhizome or occasionally sucker. A spacing of 1 m×0.5 m or closer is suitable. Planting material of 3,000–3,500 kg is required for 1 hectare. The pieces are planted 6–8 cm deep. It requires a fertile soil. Normally, flowers are removed to allow as much storage as possible in the rhizomes.

P.P. Reddy, *Plant Protection in Tropical Root and Tuber Crops*,
DOI 10.1007/978-81-322-2389-4_13

Fig. 13.1 Arrowroot plants and rhizomes

13.1.4 Production

Rhizomes are ready for harvest after about 11 months. Successive crops are grown in the same site for 5–7 years. The average rhizome yield is 12.5 t/ha. The rhizomes contain 8–16 % starch.

13.1.5 Uses

The rhizomes are used in soups or sauces. They can be just scraped and boiled. It is important to peel off the skin scales for they give a bitter taste to the starch. They can be used for making flour. It has easily digestible starchy flour. It is a good thickener in sauces. Arrowroot flour has a very high commercial value in the international market because it is the best ingredient in making high-quality biscuits and cookies.

13.2 Insect and Vertebrate Pests

Arrowroot is not normally subject to serious attacks by pests. In St. Vincent the only pest of any importance is the leaf roller, *Calpodes ethlius*, which has proved resistant to many common insecticides but has been controlled by deltamethrin. In Brazil and Venezuela, the crop is attacked occasionally by *Ascia monuste orseis*, *Neocurtilla hexadactyla*, and *Scapteriscus vicinus*.

Two caterpillars, in particular, are very destructive to *Canna* in Florida, the larger *Canna* leaf roller, *Calpodes ethlius* , whose adult form is variously called the Brazilian skipper, the *Canna* butterfly, or the arrowroot butterfly (Moore 1928), and the lesser *Canna* leaf roller, *Geshna cannalis* (Quaintance).

The green coconut bug *(Amblypelta cocophaga)*, bean pod borer *(Maruca vitrata)*, rice armyworm *(Mythimna loreyi)*, green vegetable bug *(Nezara viridula)*, Pacific mealybug *(Planococcus pacificus)*, and cacao armyworm *(Tiracola plagiata)* also infest arrowroot.

13.2.1 Leaf Roller, *Calpodes ethlius*

Leaf roller larvae may continue to be a problem over mild winters when plants continue to produce new foliage.

13.2.1.1 Distribution

Calpodes ethlius is native to Mexico, Central America, and the northern part of the South American continent, being found as far south as the northern border of Argentina (Scudder 1889). It is found on the Galapagos Islands in the Pacific Ocean (Onore and Mielke 1988) and throughout the West Indies in the Caribbean Sea. In the United States, it is well established in Florida, Texas, and southern Arizona (Scudder 1889).

13.2.1.2 Damage

Leaf rollers are lime green in color which feed inside rolled leaves. Plant leaves look like rolled tubes (Fig. 13.2). Leaf rollers feed on the insides of the leaf as they pupate and can decimate plants. Voracious leaf feeding by larger larvae may be so severe that plants do not flower. In arrowroot, severely defoliated plants may produce little of the harvestable rhizome. Adults feed on nectar from a variety of flowers.

13.2.1.3 Biology

Leaf roller completes three or more generations in a single season and overwinters as larvae.

13.2.1.4 Management

1. *Cultural Methods*: Maintain proper sanitation including removal of all leaf debris from the garden after frost has killed back foliage.
2. *Biological Methods*: The bacterial biopesticide *Bacillus thuringiensis* (commonly known as Bt) is effective when it is used as a spray against young larvae. Thorough coverage of all foliage is needed.
3. *Chemical Methods*: Several insecticides provide effective control of leaf rollers. Thorough coverage of all foliage is needed. Use systemic insecticides as a foliar spray (Orthene). Carbaryl (Sevin) can be applied as sprays or dusts.

13.2.2 Rodents

13.2.2.1 Damage

Rodents eat tubers left in the ground.

13.2.2.2 Management

Poison baits using zinc phosphide is recommended for the control of rats.

13.3 Diseases

There are several diseases found to be associated with arrowroot, but only one of them is reported to occur in the Philippines. In parts of the Caribbean, particularly in wet districts, arrowroot sometimes suffers from a rot caused by *Rosellinia bunodes*. Two leaf blights, caused by *Rhizoctonia solani* and *Pellicularia filamentosa*, are reported to infect arrowroot in India.

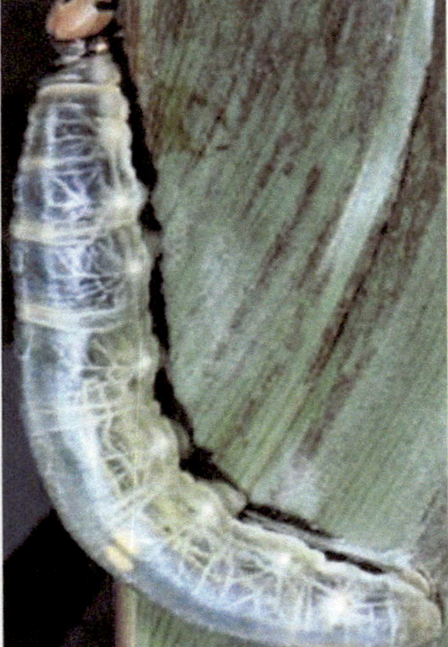

Fig. 13.2 Leaf roller damage and fifth instar larva

13.3.1 Banded Leaf Blight, *Thanatephorus cucumeris*

It was observed in North Malabar (India) in 1946 and described 2 years later after growers experienced considerable losses. The disease can be of some consequence under favorable conditions but has been reported only in India.

13.3.1.1 Symptoms

This disease, caused by *Thanatephorus cucumeris* (*Corticium solani*), showing as a chlorotic banding of the leaves with ultimate browning and rotting, spreads rapidly in close association with heavy rain.

Evidence of infection is small, water-soaked spots that may enlarge rapidly by the formation of successive but sometimes discontinuous, eccentric zones. Aerial mycelium quickly covers both leaf surfaces and basidia and basidiosphores are produced in the scant, whitish superficial growth. Affected leaves become completely necrotic and are rotten by a secondary organism. The fungus also infected leaves of several plants that include rice, corn, sugarcane, and even citrus. It is reported troublesome during periods of high rainfall.

13.3.1.2 Management

Foliar spraying (both leaf surfaces) prior to the onset of high-rainfall season with zineb, maneb, or copper fungicide using the rate recommended by the manufacturers is effective.

Satisfactory control was obtained by preventive spraying with 1 % Bordeaux mixture in July. Spraying should be carried out every year before the monsoon, using special care to obtain complete coverage of both leaf surfaces (Ramakrishnan and Ramakrishnan 1948).

13.3.2 Bacterial Wilt, *Xanthomonas marantae*

13.3.2.1 Symptoms

It is a vascular wilt disease of arrowroot reported to occur in Brazil in 1962.

13.3.2.2 Management

Use healthy propagating materials. Planting should not be located in areas where disease has occurred. The infected plants should be rouged and burned, including roots.

13.3.3 Mosaic

It is reported to occur in the Philippines but of minor importance.

13.3.3.1 Symptoms

Symptoms are similar to that of abaca mosaic, presence of whitish small dots and narrow streaks. It is transmitted either mechanically or through an insect vector – aphids in a nonpersistent manner.

13.3.3.2 Management

Control methods include rouging, selecting healthy planting materials, controlling aphids, avoiding interplanting with susceptible host, and controlling weed.

References

Moore MB (1928) A study of the life history and habits under Florida conditions of the canna butterfly (Brazilian skipper), *Calpodes ethlius* (Cramer), an insect pest of the canna. MS thesis, University of Florida, Gainesville, 95 pp

Onore G, Mielke O (1988) *Calpodes ethlius* (Cramer) (Lepidoptera, Hesperiidae) em Gal pagos. Rev Bras Zool 5:619–620

Ramakrishnan K, Ramakrishnan TS (1948) Banded leaf blight of arrowroot, *Maranta arundinacea*. Indian Phytopathol 1:129–136

Scudder SH (1889) *Calpodes ethlius* – the Brazilian skipper. In: The butterflies of the eastern United States and Canada with special reference to New England, vol II Lycaenidae, Papilionidae, Hesperidae. SH Scudder Publication, Cambridge, pp 1750–1757

Crêpe Ginger, *Cheilocostus speciosus*

14

Abstract

The distribution, economic importance, symptoms/damage, host range, biology, survival and spread, and management of insect and mite pests and fungal, bacterial, viral, and nematode diseases of crepe ginger (*Cheilocostus speciosus*) using regulatory, physical, cultural, chemical, biological, host resistance, and integrated pest management methods are discussed.

Keywords

Crepe ginger • Cheilocostus speciosus • Pests • Diseases

14.1 Introduction

Cheilocostus speciosus (syn. *Costus speciosus*) or crêpe ginger is possibly the best known cultivated species of the genus *Cheilocostus* (family: Costaceae). This plant is native to Southeast Asia and surrounding regions, from India to China to Queensland. It is especially common on the Greater Sunda Islands in Indonesia. It is also reportedly naturalized in Mauritius, Réunion, Fiji, Hawaii, Costa Rica, Belize, Melanesia, the West Indies, and Micronesia. The species reproduces vegetatively by rhizome. It is cultivated in India for its medicinal uses and elsewhere as an ornamental.

The plant is often grown in tropical and subtropical countries, because of its beautiful inflorescences and its lush vegetation. In some areas of origin, the rhizomes are also used in traditional medicine. It needs a hot and humid climate in partial shade in a drained, rich soil, which must be kept moist during the growing season.

The showy flowers emerge from dark red to reddish purple bracts on pinecone-shaped, terminal inflorescences. The individual flowers are trumpet shaped, up to 10 cm across, and have three fused, crinkled, crepe paper-like, white petals and a smaller, yellow, curved, petallike stamen. The flowers are followed by round, 1.5 cm in diameter, red seed capsules containing black seeds and white flesh. The leaves are green, alternate, and spirally arranged and have leaf blades that are narrowly elliptic to oblanceolate in shape. The cane-like stems emerge from underground rhizomes and are upright, unbranched, and green to dark red in color (Fig. 14.1).

P.P. Reddy, *Plant Protection in Tropical Root and Tuber Crops*,
DOI 10.1007/978-81-322-2389-4_14

Fig. 14.1 *Cheilocostus speciosus* plants and inflorescence

14.2 Pests

Variegated crepe ginger (*Costus speciosus* "variegatus") is susceptible to pests like mites and nematodes. Mites and nematodes can be a problem, especially on light, sandy soil.

14.3 Diseases

The following are some diseases which are observed to affect the crop.

14.3.1 *Phytophthora* Rhizome Rot, *Phytophthora solani*

14.3.1.1 Symptoms

The disease develops very quickly from July to August in the rhizomes kept in storage after the harvest. Initially, the symptoms start from the tip of the injured portion of the rhizomes. Subsequently, the rhizome becomes light brown and gives off an offensive odor.

14.3.1.2 Management

Properly harvested rhizomes without injuries can be stored for a long time. No chemical control has been worked out.

14.3.2 *Pythium* Rhizome Rot, *Pythium spinosum*

14.3.2.1 Symptoms

Infection starts from the injured portion of the rhizome, and in later stages the rhizomes turn dirty brown, and the leaves become yellowish brown and finally dry off.

14.3.2.2 Management

The prudent selection of rhizomes from the healthy crop and dipping the rhizomes in fungicidal solutions like benomyl, carbendazim, or zineb have been recommended to control the disease.

14.3.3 Leaf Blight, *Curvularia paradissi*

14.3.3.1 Symptoms

It is very severe from July to September. The symptoms develop on the leaf lamina as small, spherical to irregular spots, light brown in color, and in the advanced stages, they become dark brown. The upward cupping of young leaves is very common. In the advanced stages, all the leaves fall off, leaving bare stems in the field.

14.3.3.2 Management

This pathogen can be effectively controlled by spraying 0.3 % maneb at fortnightly intervals.

The Way Forward

15

Abstract

Low productivity due to biotic stresses is the major constraint which is still insufficiently addressed. Host plant resistance and biological control are the cornerstones of crop protection measures against biotic stresses on root and tuber crops. Transgenic improvement has been focused on cassava mosaic disease. Sustainable plant health management that considers the crop plant as a component of an agro-ecosystem holds promise for management of pests and diseases of root and tuber crops.

Keywords

Host plant resistance • Biological control • Sustainable plant health management

15.1 Introduction

The tropical root and tuber crops (cassava, sweet potato, yams, and aroids) are of utmost importance for the world food security. They are major sources of energy in developing countries with fast population growth and high urbanization rates. They are the staple food for hundreds of millions of poor people. These crops are expected to contribute significantly to the increased income generation and nutritional well-being of people in the tropics in the next decades.

Root and tuber crops are the second group of cultivated species, after cereals, in tropical countries. They are produced with very low inputs, are consumed by the poorest, contribute significantly to food security, and are also used for animal feed or as raw material for processing industries. These species belong to different botanical families but, for the purpose of scientific research, are grouped together because they are vegetatively propagated, produce underground food, and are bulky and perishable.

Roots and tubers contribute to the energy and nutrition requirements of more than two billion people; constitute an important source of income in rural and marginal areas; have multiple uses, most notably as food security crops, regular food crops, and cash crops; and are increasingly used as livestock feed and raw material for industrial purposes. They have long served as the principal source of food and nutrition for many of the world's poorest and undernourished households and are generally valued for their stable yields under conditions in which other crops may fail (Scott et al. 2000).

P.P. Reddy, *Plant Protection in Tropical Root and Tuber Crops*,
DOI 10.1007/978-81-322-2389-4_15

However, low productivity due to biotic stresses, limited added value, and poor access to markets due to their perishable nature are major constraints which are still insufficiently addressed.

Emerging pests and diseases due to climate change and introduction of invasive pests and pathogens are main challenges. Pre- and postharvest losses of these crops are very high and, depending on the species cultivated and the storage environment, may be of the order of 30–60 %. Some of the biotic stresses on tuber crops include cassava mosaic disease, sweet potato weevil, taro leaf blight, anthracnose disease and nematodes in yams, collar rot, and viral diseases in elephant foot yam. They are disseminated through planting materials and soil. Efforts to provide healthy planting material by employing diagnostic techniques to identify pathogens are in progress, and this needs to be strengthened. Ready-to-use diagnostic kits have to be developed for important fungal diseases, viz., cassava tuber rot, Amorphophallus collar rot, taro leaf blight, and yam anthracnose. Appropriate technologies to contain such biotic stresses have to focus on refined eco-friendly methods. Strategy for mass multiplication of disease-free planting materials has to be improvised. Production of large quantities of polyclonal antibodies and developing cost-effective, easy, and sensitive diagnostic kits for indexing the planting materials are necessary.

Pests and diseases of tropical tuber crops in farmers' fields are being monitored continuously to observe the occurrence of new biotic threats. This has to be continued to generate data for their effective management and prediction. The quarantine measures could be strengthened to avoid invasive pests. Significant pests and diseases are to be identified, and by developing the prediction model, decision support system could be made available to farmers.

The tropical tuber crops are highly vulnerable to viral diseases, and nonavailability of disease-free planting material to the farmers is of great concern in realizing the full potential yield of these crops. Availability of healthy planting material, particularly cassava free from mosaic disease, is a challenge to meet the emerging demands. Hence special emphasis has to be given for identification and characterization of pathogens, development of diagnostic kits/techniques, and production of virus-free mother plants of all tuber crops for mass propagation. Ready-to-use diagnostic kits have to be developed for important viral diseases with special emphasis on cassava mosaic disease, sweet potato feathery mottle virus, sweet potato leaf curl, mosaic in elephant foot yam, and other edible aroids and yam viruses. New molecular detection techniques like microarray LAMP, utilization of nanotechnology for quick diagnosis, etc. should also be explored. Production of polyclonal antibody for detecting viral and fungal diseases in large quantities and development of ELISA kit are the prime target in the future. Pathogen-specific primers are to be designed for PCR-based detection of these diseases. In vitro elimination of cassava mosaic virus, sweet potato feathery mottle virus, yam mild mosaic virus, and dasheen mosaic in elephant foot yam has to be standardized using meristem and nodal culture combined with chemotherapy and thermotherapy.

Biological control has been emerging as a promising alternative to chemical pesticides to control many pre- and postharvest diseases of these crops. The underlying mechanisms include antibiosis, competition for nutrients and space, and systemic resistance. The various strategies for use of antagonists for field uses (i.e., seed treatment, soil application, foliar spray, antagonistic mixture) have been worked out. In addition, certain mechanisms to improve performance of antagonists, such as addition of nutrients, use of antagonist mixture, and formulation of antagonist, need to be undertaken.

Studies have been carried out for biointensive integrated management of important fungal diseases of tropical tuber crops and cassava mosaic disease. Nevertheless, large-scale multiplication and delivery of biocontrol agents have to be improved. Further, studies on the effect of endophytes, vermicompost, and other organic bio-products for the development of IDM for the management of important diseases should be addressed. Parasitoids and pathogens in combination with sex pheromones and semiochemicals should be utilized for sweet potato weevil management. The potential of natural enemies for biocontrol, especially in view of environmental pollution and health hazards to people, has to be exploited to control sweet potato weevil, cassava whitefly, cassava spiraling whitefly, and storage pests. The formulation of the biopesticides extracted from cassava with other plant products, compatibility study of the biopesticides with synthetic insecticides, exploration of the utility of the biopesticides against other field and stored-product insect pests, etc. need further thrust. The role of endosymbionts in insect vectors has to be delineated to have a clear understanding of the mechanism of disease transmission.

Cassava mosaic disease being the most important disease problem in cassava, efforts on the development of transgenic cassava with resistance against cassava mosaic virus will be continued through different strategies and confirmation of resistance in transgenic cassava plants through challenge inoculation and clearance as per the biosafety rules for further field trials. Apart from this, transgenic approach needs to be explored for developing resistance against dasheen mosaic virus on elephant foot yam, sweet potato weevil, taro leaf blight, and viral diseases of taro and sweet potato. RNAi technology should also be utilized for pest and disease management. Resistant gene mapping also needs to be done to facilitate marker-assisted selection of breeding materials against major pests and diseases.

15.2 Future Thrust Areas

- Survey and surveillance of pests and diseases should be intensified for forewarning outbreak of pest and disease epidemics.
- Collection of germplasm including wild species must cater to the breeding objectives for the next 10 years especially locating sources of resistance to some of the pests and diseases.
- Allelopathic effect of different cropping systems involving tuber crops in the incidence of pests and diseases may be looked into.
- *Phytophthora* is a major devastating disease in aroids. The source of resistance in the germplasm including the wild relatives has to be studied in order to breed resistant varieties.
- Mosaic disease is an important problem in different tuber crops. As it is transmissible through vegetative propagation, utmost care has to be given to produce disease-free planting material.
- Epidemiological aspects of virus diseases to be studied by taking climatic aspects into consideration.
- Yam bean seed is a potential botanical insecticide. The effect of this biopesticide on nationally important pests needs to be explored.
- The biopesticides developed by CTCRI should be tested in large-scale field trials and popularized at national level through AICRP Centers on tuber crops. Immediate steps should be taken for the registration and commercialization of CTCRI-developed biopesticides from cassava.
- Production of disease-free quality planting materials of released and popular varieties of tuber crops which are in high demand needs to be intensified.
- Development of IPM practices in field and storage using pheromones and kairomones and biocontrol agents like fungi, bacteria, and entomopathogenic nematodes (EPN) is the need of the hour.
- Developing serological methods for detection of viruses and their use in certification of disease-free planting material should be undertaken.
- Development of biopesticides from tuber crops (cassava, yam bean, etc.), their characterization, and identification of active ingredients need to be taken up.
- Development of low-cost disease indexing in tuber crops is the need of the hour.
- Identification and utilization of bacterial metabolites from EPN for pest control in tuber crops should be undertaken.
- Studies on biotypes/races of major pests, nematodes, viruses, and fungi to help in designing proper IPM and IDM technologies have to be intensified.

- Studies on epidemiology/population dynamics of major diseases and pests should be carried out for casting diseases and developing disease management programs.
- The studies conducted on whitefly *Bemisia tabaci* and its biotypes have to be intensified so as to gather information on the spread of CMD and other vector transmitted diseases and its possible management through control of vector population.
- More studies are to be conducted on spiraling whitefly (*Aleurodicus dispersus*) and its control especially in the cassava growing areas of Tamil Nadu plains (Salem, Dharmapuri, and Namakkal districts).
- Biopesticides developed from cassava seed, leaf, rind, and yam been seed as well as starch factory affluents have to be tested under large-scale field trials. Product development has to be hastened so that the deliverables are made available to the farming community for their effective use.
- Attempts should be made to develop low-cost crop protection methods.
- Efficacy of Kairomones in combination with pheromones may be studied in the control of sweet potato weevil. Cost-effective formulation developed with kairomones is expected to provide environmentally friendly technology for the sweet potato farmer.
- The effect of climate change on geographical distribution, increased overwintering, changes in population growth rates, increases in the number of generations, extension of the development season, changes in crop-pest synchrony, changes in interspecific interactions, pest biotypes, activity and abundance of natural enemies, species extinction, increased risk of invasion by migrant pests, and efficacy of crop protection technologies needs to be investigated. Global warming will also reduce the effectiveness of host plant resistance, transgenic plants, natural enemies, biopesticides, and synthetic chemicals for pest management. Therefore, there is a need to generate information on the likely effects of climate change on pests to develop robust adaptation and mitigation technologies that will be effective in the future under global warming and climate change.
- Pests and diseases of tropical tuber crops in farmers' fields are being monitored continuously to observe the occurrence of new biotic threats. This has to be continued to generate data for their effective management and prediction. Significant pests and diseases are to be identified, and by developing the prediction model, decision support system could be made available to farmers.
- Studies have been carried out for biointensive management of important fungal diseases of tropical tuber crops and cassava mosaic disease. Nevertheless, large-scale multiplication and delivery of biocontrol agents have to be improved. Further, studies on the effect of endophytes, vermicompost, and other organic bioproducts for the development of IDM for the management of important diseases should be addressed. The potential of natural enemies for biocontrol, especially in view of environmental pollution and health hazards to people, has to be exploited to control sweet potato weevil, cassava whitefly, cassava spiraling whitefly, and storage pests. The formulation of the biopesticide extracted from cassava and yam bean with other plant products, compatibility study of the biopesticides with synthetic insecticides, exploration of the utility of the biopesticides against other field and stored-product insect pests, etc. need further thrust during XII Plan. The role of endosymbionts in insect vectors has to be delineated to have a clear understanding of the mechanism of disease transmission.
- Root-knot, lesion, and yam nematodes are gaining importance in tuber crops (sweet potato, cassava, yam, Amorphophallus, and taro). There is a need to develop biointensive integrated nematode management strategies using biocontrol agents, organic amendments, and other eco-friendly components. The role of nematodes in the development of disease complexes along with soilborne bacteria and fungi may be explored.

15.3 Conclusion

One of the persisting problems with root and tuber crops is its unrealized yield potential that could only be attained through yet-to-be-developed technologies. In the case of root and tuber crops, the potential for yield is considerably higher than the actual yield. Biophysical constraints, i.e., diseases, insect pests, and nematodes, are adversely affecting production of root and tuber crops.

Unfortunately the majority of the research on root and tuber crops remains limited to a few institutions. Of major concern are the scarcity of germplasm collections and the even greater lack of breeding programs in developing pest-resistant/pest-tolerant varieties.

Although there is room for considerable amount of breeding and selection in the root and tuber crops, progress may be slow in some instances. Genetic engineering holds promise for incorporating genes for virus immunity into most of these crops. Once accomplished, yields should increase dramatically. However, there is one note of caution. Immunities will most likely be limited to single genes. Therefore, concern for the breakdown of such immunities should be taken into consideration.

Reference

Scott G, Rosegrant MW, Ringler C (2000) Roots and tubers from the 21st century: trends, projections and policy options. Food, Agriculture and the Environment, discussion paper 31, International Food Policy Institute (IFPRI) and International Potato Institute (CIP), Washington DC, 64 pp

Index

P.P. Reddy, *Plant Protection in Tropical Root and Tuber Crops*,
DOI 10.1007/978-81-322-2389-4

Z